Genome Mapping and Molecular Breeding in Plants
Volume 7

Series Editor: Chittaranjan Kole

Volumes of the Series Genome Mapping and Molecular Breeding in Plants

Volume 1
Cereals and Millets

Volume 2
Oilseeds

Volume 3
Pulses, Sugar and Tuber Crops

Volume 4
Fruits and Nuts

Volume 5
Vegetables

Volume 6
Technical Crops

Volume 7
Forest Trees

Chittaranjan Kole (Ed.)

Forest Trees

With 18 Illustrations, 4 in Color

Springer

Chittaranjan Kole
Department of Horticulture
316 Tyson Building
The Pennsylvania State University
University Park, PA 16802
USA

e-mail: cuk10@psu.edu

ISBN 978-3-642-07092-1 e-ISBN 978-3-540-34541-1

Springer is a part of Springer Science+Business Media
springer.com

Editor: Dr. Sabine Schreck, Heidelberg, Germany
Desk Editor: Dr. Jutta Lindenborn, Heidelberg, Germany
Cover design: WMX Design GmbH, Heidelberg, Germany

Preface to the Series

Genome science has emerged unequivocally as the leading discipline of this new millennium. Progress in molecular biology during the last century has provided critical inputs for building a solid foundation for this discipline. However, it has gained fast momentum particularly in the last two decades with the advent of genetic linkage mapping with RFLP markers in humans in 1980. Since then it has been flourishing at a stupendous pace with the development of newly emerging tools and techniques. All these events are due to the concerted global efforts directed at the delineation of genomes and their improvement.

Genetic linkage maps based on molecular markers are now available for almost all plants of significant academic and economic interest, and the list of plants is growing regularly. A large number of economic genes have been mapped, tagged, cloned, sequenced, or characterized for expression and are being used for genetic tailoring of plants through molecular breeding. An array of markers in the arsenal from RFLP to SNP; tools such as BAC, YAC, ESTs, and microarrays; local physical maps of target genomic regions; and the employment of bioinformatics contributing to all the "-omics" disciplines are making the journey more and more enriching. Most naturally, the plants we commonly grow on our farms, forests, orchards, plantations, and labs have attracted emphatic attention, and deservedly so. The two-way shuttling from phenotype to genotype (or gene) and genotypte (gene) to phenotype has made the canvas much vaster. One could have easily compiled the vital information on genome mapping in economic plants within some 50 pages in the 1980s or within 500 pages in the 1990s. In the middle of the first decade of this century, even 5,000 pages would not suffice! Clearly genome mapping is no longer a mere "promising" branch of the life science; it has emerged as a full-fledged subject in its own right with promising branches of its own. Sequencing of the *Arabidopsis* genome was complete in 2000. The early 21st century witnessed the complete genome sequence of rice. Many more plant genomes are waiting in the wings of the national and international genome initiatives on individual plants or families.

The huge volume of information generated on genome analysis and improvement is dispersed mainly throughout the pages of periodicals in the form of review papers or scientific articles. There is a need for a ready reference for students and scientists alike that could provide more than just a glimpse of the present status of genome analysis and its use for genetic improvement. I personally felt the gap sorely when I failed to suggest any reference works to students and colleagues interested in the subject. This is the primary reason I conceived of a series on genome mapping and molecular breeding in plants.

There is not a single organism on earth that has no economic worth or concern for humanity. Information on genomes of lower organisms is abundant and highly useful from academic and applied points of view. Information on higher animals including humans is vast and useful. However, we first thought to concentrate only on the plants relevant to our daily lives, the agronomic, horticultural and technical crops, and forest trees, in the present series. We will come up soon with commentaries on food and fiber animals, wildlife and companion animals, laboratory animals, fishes and aquatic animals, beneficial and harmful insects, plant- and animal-associated microbes, and primates including humans in our next "genome series" dedicated to animals and microbes. In this series, 82 chapters devoted to plants or their groups have been included. We tried to include most of the plants in which significant progress has been made. We have also included preliminary works on some so-called minor and orphan crops in this series. We would be happy to include reviews on more such crops that deserve immediate national and international attention and support. The extent of coverage in terms of the number of pages, however, has nothing to do with the relative importance of a plant or plant group. Nor does the sequence of the chapters have any correlation to the importance of the plants discussed in the volumes. A simple rule of convenience has been followed.

I feel myself fortunate to have received highly positive responses from nearly 300 scientists of some 30-plus countries who contributed the chapters for this series. Scientists actively involved in analyzing and improving particular genomes contributed each and every chapter. I thank them all profoundly. I made a conscientious effort to assemble the best possible team of authors for certain chapters devoted to the important plants. In general, the lead authors of most chapters organized their teams. I extend my gratitude to them all.

The number of plants of economic relevance is enormous. They are classified from various angles. I have presented them using the most conventional approach. The volumes thus include cereals and millets (Volume I), oilseeds (Volume II), pulse, sugar and tuber crops (Volume III), fruits and nuts (Volume IV), vegetables (Volume V), technical crops including fiber and forage crops, ornamentals, plantation crops, and medicinal and aromatic plants (Volume VI), and forest trees (Volume VII).

A significant amount of information might be duplicated across the closely related species or genera, particularly where results of comparative mapping have been discussed. However, some readers would have liked to have had a chapter on a particular plant or plant group complete in itself. I ask all the readers to bear with me for such redundancy.

Obviously the contents and coverage of different chapters will vary depending on the effort expended and progress achieved. Some plants have received more attention for advanced works. We have included only introductory reviews on fundamental aspects on them since reviews in these areas are available elsewhere. On other plants, including the "orphan" crop plants, a substantial amount of information has been included on the basic aspects. This approach will be reflected in the illustrations as well.

It is mainly my research students and professional colleagues who sparked my interest in conceptualizing and pursuing this series. If this series serves its purpose, then the major credit goes to them. I would never have ventured to take up this huge task of editing without their constant support. Working and interacting with many people, particularly at the Laboratory of Molecular Biology and Biotechnology of the Orissa University of Agriculture and Technology, Bhubaneswar, India as its founder principal investigator; the Indo-Russian Center for Biotechnology, Allahabad, India as its first project coordinator; the then-USSR Academy of Sciences in Moscow; the University of Wisconsin at Madison; and The Pennsylvania State University, among institutions, and at EMBO, EUCARPIA, and Plant and Animal Genome meetings among the scientific gatherings have also inspired me and instilled confidence in my ability to accomplish this job.

I feel very fortunate for the inspiration and encouragement I have received from many dignified scientists from around the world, particularly Prof. Arthur

Kornberg, Prof. Franklin W. Stahl, Dr. Norman E. Borlaug, Dr. David V. Goeddel, Prof. Phillip A. Sharp, Prof. Gunter Blobel, and Prof. Lee Hartwell, who kindly opined on the utility of the series for students, academicians, and industry scientists of this and later generations. I express my deep regards and gratitude to them all for providing inspiration and extending generous comments.

I have been especially blessed by God with an affectionate student community and very cordial research students throughout my teaching career. I am thankful to all of them for their regards and feelings for me. I am grateful to all my teachers and colleagues for the blessings, assistance, and affection they showered on me throughout my career at various levels and places. I am equally indebted to the few critics who helped me to become professionally sounder and morally stronger.

My wife Phullara and our two children Sourav and Devleena have been of great help to me, as always, while I was engaged in editing this series. Phullara has taken pains ("pleasure" she would say) all along to assume most of my domestic responsibilities and to allow me to devote maximum possible time to my professional activities, including editing this series. Sourav and Devleena have always shown maturity and patience in allowing me to remain glued to my PC or "printed papers" ("P3" as they would say). For this series, they assisted me with Internet searches, maintenance of all hard and soft copies, and various timely inputs.

Some figures included by the authors in their chapters were published elsewhere previously. The authors have obtained permission from the concerned publishers or authors to use them again for their chapters and expressed due acknowledgement. However, as an editor I record my acknowledgements to all such publishers and authors for their generosity and good will.

I look forward to your valuable criticisms and feedback for further improvement of the series.

Publishing a book series like this requires diligence, patience, and understanding on the part of the publisher, and I am grateful to the people at Springer for having all these qualities in abundance and for their dedication to seeing this series through to completion. Their professionalism and attention to detail throughout the entire process of bringing this series to the reader made them a genuine pleasure to work with. Any enjoyment the reader may derive from this books is due in no small measure to their efforts.

Pennsylvania,
10 January 2006

Chittaranjan Kole

Preface to the Volume

Forest trees cover 30% of the land surface of the globe. They provide structural and functional habitats for about two thirds of the terrestrial species on earth. They also contain more than 90% of all terrestrial biomass. Benefits from forest tree species include renewable biomass resources, carbon sequestration, watershed protection, purification of air, and habitats for recreation. Forest trees constitute the dominant life form in many ecosystems. They also play a pivotal role in the world economy. The contribution of forest trees to international commerce is about $400 billion per annum. The importance of forest trees can be exemplified with statistical data from the USA and Canada. As far as biologically derived materials are concerned, wood-based products are second only to corn in their contribution to the US economy. Canada possesses 10% of the world's forests, and these forests form the basis of an industry annually worth about $74 billion. Therefore, systematic conservation, sustainable improvement, and pragmatic utilization of forest tree species are global priorities.

Despite the high importance of forest trees in the economy, ecology, and the environment, genetic analysis and breeding of trees have lagged behind those of field crops. In fact, it is the inherent characteristics of forest tree species that have limited endeavors to improve forest trees. Late sexual maturity, time lag required to evaluate field performance, and the cost of phenotyping traits, particularly those related to wood quality, still hamper tree improvement. Their outbred nature and high genetic load are also the limiting factors for the development of inbred lines, which in field crop species have been instrumental for the mapping of gene clusters controlling quantitative traits. Most of the important traits in forestry including biomass production, wood quality, and stress resistance are quantitative in nature and exhibit low to moderate heritability. Genome analysis and improvement in forest trees have therefore required unique concepts and strategies. Some of the familiar terms including plus tree, single family pedigree, pseudotestcross, F_2 inbred model, etc. are particular to forest trees and some fruit trees. The unique problems associated with genome analysis and improvement in tree species and the strategies developed to circumvent these problems are thoroughly discussed in the chapters of this volume on forest trees.

Genetic linkage studies in forest trees started with the application of allozymes and total protein from megagametophytes in the late 1970s and late 1980s, respectively. Genetic linkage maps using DNA markers (RFLP) could be constructed only during early 1990s, much later than those developed in field crops. In the early 1990s, genome research in forest trees were accelerated thanks to the advent of PCR-based markers. The development of SSR, AFLP, SNP, and EST marker technology created an arsenal of molecular tools for analysis and improvement of the forest tree genomes. Recently, various national and international genome initiatives have generated enormous databases. The formation of the International *Populus* Genome Consortium has led to the sequencing of the first woody plant genome, following the release of the genomes of *Arabidopsis* and rice. Today, forest trees can boast of elegant progress in the areas of structural and functional genomics on par with what has been achieved in field, orchard, and plantation crops. We expect the completion of whole-genome sequences of many more tree species and look forward to them for several byproducts, such as the modification of trees to produce biofuel that would reduce the reliance of countries on fossil fuels.

The eight chapters in this volume present thorough reviews of past, present, and future research in the leading forest tree species. The chapters include *Populus* trees, pines, spruce, eucalypts, Fagaceae trees, black walnut, Douglas fir, and *Cryptomeria japonica*. We are proud to have the contributions of 68 scientists from 17 countries included in these comprehensive reviews. I am grateful to all of them for their contributions and cooperation.

I must confess that I have never worked on tree species and was very apprehensive with respect to my ability to edit the works of the eminent scientists with lifetime involvement in and original contributions to forest tree genetics, genomics, and breeding. I feel fortunate that the authors themselves made my life easy and comfortable. Through constant communication, particularly with the corresponding authors, a cordial mental bridge developed in the course of time that paved the way for a pleasant travel through the forests! Editing of the chapters made me aware of what happened, what is happening, and what should or could happen in the vast arena of forest tree genome sciences. This helped me to remain as a student of plant genome sciences, and I do feel blessed for that unique opportunity.

This is the last volume in the series on "Genome Mapping and Molecular Breeding in Plants." It was a pleasant experience for me to work with the professional staff of the publishers and printers. I must mention here the names of Dr. Sabine Schreck and Dr. Jutta Lindenborn, Life-Science Editors of Springer, Cornelia Gründer and Martin Weissgerber, Production Editors, Le-TeX, copy editors Glenn Corey and Jardi Mullinax for their affectionate dealings and cooperative attitudes from inception to completion of this project.

We have happily noticed the appreciation of the readers for the first six volumes of this series and expect the same for this volume as well. In that case, all credit must go to the authors, and I take the responsibility for all the limitations and flaws I should and could have taken care of.

We will be looking forward to all suggestions from all corners for improvement of this volume in future editions.

Pennsylvania,
12 November 2006

Chittaranjan Kole

Contents

Contributors

H. A. Agrama
Rice Research Ext. Center,
University of Arkansas & DB National
Rice Research Center, USDA
2890 Hwy 130 E., P.O. Box 1090
Stuttgart, AR 72160, USA

Kathiravetpillai Arumuganathan
Flow Cytometry and Imaging
Core Laboratory, Benaroya Research
Institute at Virginia Mason
1201 Ninth Avenue, Seattle
WA 98101, USA

Erica Bakker
Department of Ecology and Evolution
University of Chicago
5801 South Ellis Avenue, Chicago
IL 60637, USA

Teresa Barreneche
Unité de Recherche sur les Espèces
Fruitières et la Vigne, INRA
71 Avenue Edouard Bourlaux
33883 Villenave d'Ornon, France

Maria Evangelista Bertocchi
UMR Biodiversité Gènes
& Communautés, INRA
69 Route d'Arcachon
33612 Cestas, France

Catherine Bodénès
UMR Biodiversité Gènes
& Communautés, INRA
69 Route d'Arcachon
33612 Cestas, France

Jean Bousquet
Département des Sciences du bois
et de la forêt
Université Laval
Québec G1K 7P4, Canada

Joukje Buiteveld
Alterra Wageningen UR, Centre
for Ecosystem Studies
P.O. Box 47
6700 AA Wageningen
The Netherlands

R. D. Burdon
Ensis-Genetics, Ensis
Private Bag 3020
Rotorua, New Zealand

J. E. Carlson
The School of Forest Resources
and Huck Institutes for Life Sciences
Pennsylvania State University
323 Forest Resources Building
University Park
PA 16802, USA
jec16@psu.edu

Manuela Casasoli
Dipartimento di Biologia Vegetale
Università "La Sapienza"
Piazza A. Moro 5
00185 Rome, Italy

M. T. Cervera
CIFOR - INIA
Genética y Ecofisiología Forestal
& Unidad Mixta de Genómica
y Ecofisiología Forestal INIA-UPM
Departamento de Sistemas
y Recursos Forestales
Carretera de La Coruña km 7
28040 Madrid, Spain

D. Chagné
Hort Research
Private Bag 11030
Palmerston North, New Zealand

P. Chaumeil
INRA, UMR BIOGECO
69 route d'Arcachon
33610 Cestas, France

Catherine Clark
Department of Forestry
North Carolina State University
Box 8008, Raleigh
NC 27695-8008, USA

Joan Cottrell
Forest Research
UK Forestry Commission
Northern Research Station
Roslin, Midlothian
EH25 9SY, UK

Jeremy Derory
UMR Biodiversité Gènes
& Communautés, INRA
69 Route d'Arcachon
33612 Cestas, France

Craig S. Echt
USDA - Forest Service
Southern Research Station
Southern Institute of Forest Genetics
23332 Mississippi 67, Saucier
MS 39574, USA

Jean-Marc Gion
Programme Arbres et Plantations
CIRAD Forêt TA 10/C
Campus de Baillarguet 34398
Montpellier Cedex 5, France

Dario Grattapaglia
EMBRAPA Genetic Resources
and Biotechnology SAIN Parque
Rural CP 0272
and
Graduate Program in Genomic
Sciences and Biotechnology
Universidade Catolica
de Brasília-SGAN, 916 modulo B
70790-160 DF Brasilia, Brazil

Jacqueline Grima-Pettenatti
UMR UPS/CNRS 5546
Pôle de Biotechnologies Végétales
24 chemin de Borde Rouge
BP42617, Auzeville Tolosane
31326 Castanet Tolosan, France

Fred Hebard
The American Chestnut Foundation
Research Farms,
14005 Glenbrook Avenue
Meadowview
VA 24361, USA

Nathalie Isabel
Ressources naturelles Canada
Service canadien des forêts
Centre de foresterie des Laurentides
1055 rue du P.E.P.S.
C.P. 10380 succ. Sainte-Foy
Québec G1V 4C7, Canada

M. Nurul Islam-Faridi
USDA - Forest Service
Southern Research Station
Southern Institute of Forest Genetics
Forest Tree Molecular
Cytogenetics Lab
Texas A&M University, College Station
TX 77843, USA

Matias Kirst
School of Forest Resources
and Conservation
363 Newins-Ziegler Hall
University of Florida, Gainesville
FL 32611-0410, USA

Hans-Peter Koelewijn
Alterra Wageningen UR
Centre for Ecosystem Studies
P.O. Box 47
6700 AA Wageningen
The Netherlands

Teiji Kondo
Forest Tree Breeding Center
3809-1 Ishi, Juo, Hitachi
Ibaraki 319-1301, Japan
kontei@affrc.go.jp

Antoine Kremer
UMR Biodiversité Gènes
& Communautés, INRA
69 Route d'Arcachon
33612 Cestas, France
antoine.kremer@pierroton.inra.fr

K. V. Krutovsky
Department of Forest Science
Texas A&M University
College Station
TX 77843-2135, USA

Thomas L. Kubisiak
USDA - Forest Service
Southern Research Station
Southern Institute of Forest Genetics
23332 Highway 67, Saucier
MS 39574, USA

S. Kumar
Ensis-Genetics, Ensis
Private Bag 3020
Rotorua, New Zealand

Noritsugu Kuramoto
Kyushu Breeding Office
Forest Treee Breeding Center
2320-5 Suya, Koshi
Kumamoto 861-1102, Japan

Stefano Leonardi
Dipartimento di Scienze Ambientali
Università di Parma
Parco Area delle Scienze 11/A
43100 Parma, Italy

Christian Lexer
Jodrell Laboratory
Royal Botanic Gardens
Kew Richmond, Surrey
TW9 3DS, UK

N. de María
CIFOR - INIA
Genética y Ecofisiología Forestal
& Unidad Mixta de Genómica
y Ecofisiología Forestal INIA-UPM
Departamento de Sistemas
y Recursos Forestales,
Carretera de La Coruña km 7
28040 Madrid, Spain

Cristina M. Marques
RAIZ-Direcçao de Investigacao
Florestal ITQB II Av. Republica
Apartado 127
2781-901 Oeiras, Portugal

Charles H. Michler
USDA Forest Service, Hardwood Tree
Improvement and Regeneration Center
at Purdue University
715 West State Street, West Lafayette
IN 47907-2061, USA
michler@purdue.edu

Alexander A. Myburg
Department of Genetics, Forestry
and Agricultural Biotechnology
Institute (FABI),
University of Pretoria
Pretoria 0002, South Africa
zander.myburg@fabi.up.ac.za

C. Dana Nelson
USDA - Forest Service
Southern Research Station
Southern Institute of Forest Genetics
23332 Mississippi 67, Saucier
MS 39574, USA

J. Paiva
INRA, UMR BIOGECO,
69 route d'Arcachon
33610 Cestas, France

Betty Pelgas
Centre de recherche en biologie
forestière
Université Laval Sainte-Foy
Québec G1K 7P4, Canada

Daniel G. Peterson
Department of Plant and Soil Sciences
117 Dorman Hall, Box 9555
Mississippi State University
Mississippi State
MS 39762, USA

Paula M. Pijut
USDA Forest Service, Hardwood Tree
Improvement and Regeneration Center
at Purdue University
715 West State Street, West Lafayette
IN 47907-2061, USA

Christophe Plomion
INRA, UMR BIOGECO
69 route d'Arcachon
33610 Cestas, France
plomion@pierroton.inra.fr

Ilga Porth
Austrian Research Centre
2444 Seibersdorf, Austria

D. Pot
Coffee Genomic Team, CIRAD,
UMR PIA 1096
34398 Montpellier cedex 5, France

Brad M. Potts
School of Plant Science, and CRC
for Forestry
University of Tasmania
Private Bag 55, Hobart
Tasmania 7001, Australia

D. Prat
Université Claude Bernard - Lyon 1
EA 3731 Génome et Evolution
des Plantes Supérieures, Bât F.A. Forel
6 rue Raphaël Dubois
69622 Villeurbanne Cedex, France

Anne M. Rae
Centre for Novel Agricultural Products
Department of Biology (Area 7)
University of York, PO Box 373
York, YO10 5YW, UK

Kermit Ritland
Department of Forest Sciences
University of British Columbia
Vancouver, British Columbia
V6T1Z4, Canada
kermit.ritland@ubc.ca

Maricela Rodríguez-Acosta
Herbario y Jardín Botánico, BUAP.
Edificio 76 Planta baja
Ciudad Universitaria
Puebla, Pue. CP 72590, Mexico

Jeanne Romero-Severson
Department of Biological Sciences
University of Notre Dame
Notre Dame, IN 46556, USA

Guy Roussel
UMR Biodiversité Gènes
& Communautés, INRA
69 Route d'Arcachon
33612 Cestas, France

Dainis Rungis
Department of Forest Sciences
University of British Columbia
Vancouver, British Columbia
V6T1Z4, Canada
and
Current address:
LVMI Silava
111 Rigas st
Salaspils
LV-2169, Latvia

O. Savolainen
Department of Biology
PL 3000
90014, University of Oulu, Finland

Marta Scalfi
Dipartimento di Scienze Ambientali
Università di Parma
Parco Area delle Scienze 11/A
43100 Parma, Italy

Caroline Scotti-Saintagne
UMR Ecologie des Forêts de Guyane
INRA, Campus agronomique BP 709
Avenue de France
97387 Kourou, French Guyana

F. Sebastiani
Department of Agricultural
Biotechnology, University of Florence
Via della Lastruccia 14
50019 Sesto Fiorentino, Florence, Italy

Paul Sisco
The American Chestnut Foundation
One Oak Plaza, Suite 308, Asheville
NC 28801, USA

Nathaniel R. Street
School of Biological Sciences
University of Southampton
Bassett Crescent East
SO16 7PX, UK

A. Traore
The Schatz Center for Tree Molecular
Genetics, Pennsylvania State University
University Park
PA 16802, USA

G. G. Vendramin
Plant Genetics Institute
National Research Council
Via Madonna del Piano 10
50019 Sesto Fiorentino, Florence, Italy

Fiorella Villani
Istituto per l'Agroselvicoltura
CNR, V. le G. Marconi,
2 - 05010, Porano, Italy

P. L. Wilcox
Cellwall Biotechnology Centre, Scion
Private Bag 3020
Rotorua, New Zealand

Keith E. Woeste
USDA Forest Service, Hardwood Tree
Improvement and Regeneration Center
at Purdue University
715 West State Street, West Lafayette
IN 47907-2061, USA

Abbreviations

ABA	Abscisic acid
AFLP	Amplified fragment length polymorphism
AGE	Agarose gel electrophoresis
ANOVA	Analysis of variance
ARS	Agricultural Research Service (USDA))
ASO	Allele-specific oligo
BAC	Bacterial artificial chromosome
bp	Base pairs
CAPS	Cleaved amplified polymorphic sequence
cDNA	Complementary DNA
CF	Cot filtration
CFS-LFC	Canadian Forest Service-Laurentian Forestry Center
CIM	Composite interval mapping
cM	CentiMorgan
COS	Conserved orthologous set
cpDNA	Chloroplast DNA
dCAPS	Derived cleaved amplified polymorphic sequence
DGGE	Denaturing gradient gel electrophoresis
DNA	Deoxyribonucleic acid
ELP	Expression length polymorphism
EMBL	European Molecular Biology Laboratory
eQTL	Expression quantitative trait locus
EST	Expressed sequence tag
ESTP	Expressed sequence tag polymorphism
FAS	Foreign Agricultural Service (USDA)
FISH	Fluorescent in situ hybridization
GM	Genetically modified
GO	Gene ontology
GUS	β-Glucuronidase
H	Haplotypic diversity
h	Heterozygosity
h^2	Heritability (in narrow sense)
HMM	Hidden Markov model
EUCAGEN	*Eucalyptus* Genome Network
IM	Interval mapping
INDELs	Insertions/deletions
ISSR	Inter-simple sequence repeat
ITS	Internally transcribed spacer
IUFRO	International Union of Forest Research Organizations
kbp	Kilobase pairs (1000 base pairs)
LD	Linkage disequilibrium
LG	Linkage group
LINE	Long interspersed nuclear element
LOD	Logarithm of odds ratio
LSO	Locus-specific oligo

MAB	Marker-assisted breeding
MAS	Marker-assisted selection
Mbp	Megabase pairs (1,000,000 base pairs)
MIM	Multiple interval mapping
MITE	Miniature inverted-repeat transposable element
MM	Multiple marker analysis
mmol	Millimoles
mRNA	Messenger RNA
MS	Mass spectrometry
mtDNA	Mitochondrial DNA
NIL	Near-isogenic line
NRCS	National Resources Conservation Service (USDA)
NTS	Nontranscribed spacer
ORF	Open reading frame
PCR	Polymerase chain reaction
pg	Picogram
PIC	Polymorphism information content
qPCR	Quantitative PCR
QRCPE	*Quercus robur* crown preferentially expressed
qRT-PCR	Quantitative reverse transcription-PCR
QTDT	Quantitative transmission disequilibrium test
QTL	Quantitative trait locus
QTN	Quantitative trait nucleotide
RAPD	Random(ly) amplified polymorphic DNA
rbcL	Chloroplast rubisco large subunit
rDNA	Ribosomal DNA
RFLP	Restriction fragment length polymorphism
RIL	Recombinant inbred line
RNA	Ribonucleic acid
RT-PCR	Reverse transcription-PCR
RuBP	Ribulose 1,5-bisphosphate
SAGE	Serial analysis of gene expression
SAMPL	Selectively amplified microsatellite polymorphic locus
SCAR	Sequence characterized amplified region
SE	Somatic embryogenesis
SM	Single marker analysis
SNP	Single nucleotide polymorphism
SRWC	Short rotation woody crops
SSCP	Single strand conformational polymorphism
SSH	Suppression subtractive hybridization
SSR	Simple sequence repeat
sss	Single-strand specific
STS	Sequence tagged site
TGGE	Thermal gradient gel electrophoresis
TILLING	Targeting induced local lesions in genomes
2D-PAGE	Two-dimensional polyacrylamide gel electrophoresis
USDA	United States Department of Agriculture
UTR	Untranslated region

1 *Populus* Trees

Anne M. Rae[1], Nathaniel R. Street[2], and Maricela Rodríguez-Acosta[3]

[1] Centre for Novel Agricultural Products, Department of Biology (Area 7), University of York, PO Box 373, York, YO10 5YW, UK
e-mail: amr502@york.ac.uk
[2] School of Biological Sciences, University of Southampton, Bassett Crescent East, SO16 7PX, UK
[3] Herbario y Jardín Botánico, BUAP. Edificio 76 Planta baja, Ciudad Universitaria Puebla, Pue. CP 72590, Mexico

1.1 Introduction

The genus *Populus*, which includes poplars, cottonwoods and aspens, is widely distributed over the northern hemisphere. Their rapid growth lends to their use as a source of fuel, fiber, lumber, windbreaks and protective stands to prevent soil erosion. Their ease of vegetative propagation means that they have been closely associated with agriculture since before the Middle Ages in the Near and Middle East. In recent years, poplars have received increasing attention as a renewable source of biomass for energy and short-fiber furnish for papermaking.

The genus also has an important role in the natural populations of their native habitat, particularly riparian ecosystems. Poplars are dioecious and wind-pollinated and produce large amounts of pollen and small, cotton-tufted seed that is dispersed by wind and water in early summer. Capable of rapidly invading disturbed sites, many species occupy habitats in the dynamic environment of riverside floodplains where they form a key component of riparian forests (Braatne et al. 1996). Others, such as the aspens, commonly colonize upland areas after intense, stand-replacing fires (Burns and Honkala 1990). All poplars also have the capacity to reproduce asexually, mostly by sprouting from the root collar of killed trees or from abscised or broken branches that become embedded in the soil.

Traditionally, *Populus* and *Salix* have been regarded as the only two members of the family Salicaceae. However, recent views include many of the genera previously assigned to Flacourtiaceae, giving more than 50, largely tropical, genera. Poplars, cottonwoods, or aspens are the common names for species of the genus *Populus*, recognized as the most abundant woody plant genus in temperate forests around the world. Some estimates put the number of species as high as 70 or more, and there are a number of naturally occurring hybrids. Many species are found in cultivation as well as numerous hybrids and cultivars (Hillier and Coombes 2002).

Populus spp. are fast-growing deciduous trees with alternate leaves. The flowers are extremely reduced and borne in catkins and trees are usually dioecious. For this reason *Populus* spp. have a tendency to hybridize in their natural habitat and especially in cultivation (US Environmental Protection Agency 1999; Wyckoff and Zasada 2003). The extensive interspecific hybridization and high morphological diversity in this group pose difficulties in identifying taxonomic units for comparative evolutionary studies and systematics (Hamzeh and Dayanandan 2004). *P. lasiocarpa* is one of the species that breeds true as it can be monoecious. Because of the extent of hybridization, and the fact that seeds are generally viable for only a short time, most *Populus* trees are propagated vegetatively by cuttings. Good rootability is a property of the members of two sections: Aigeiros and especially Tacahamaca.

The morphology and physiology in *Populus* species vary between geographically distinct populations and are strongly linked with the environment. It follows that any responses to stress will depend not only on the species but also on the particular genotype (Weber et al. 1985; Rogers et al. 1989; Braatne et al. 1992; Dunlap and Stettler 2001; Rowland et al. 2001). Dunlap et al. (1993), for example, looked at poplars growing in different river valleys and found that there was variation of physiological processes like photosynthesis both within and between locations.

Genome Mapping and Molecular Breeding in Plants, Volume 7
Forest Trees
C. Kole (Ed.)

1.1.1 Taxonomic Classification

Kingdom Plantae
  Subkingdom Tracheobionta (Vascular plants)
    Superdivision Spermatophyta (Seed plants)
      Division Magnoliophyta (Flowering plants)
        Class Magnoliopsida (Dicotyledons)
          Subclass Dileniidae
            Order Salicales
              Family Salicaceae (Willow family)
                Genus *Populus* L. (cottonwood)

Bean (1976) describes five sections in the genus *Populus*, from which section Turanga, represented by *P. euphratica*, is the least cultivated as it is a species that is difficult to grow, native to North Africa through southwest and central Asia to China. The rest are mentioned briefly as follows, as they include distinctive features:

Section Populus (Leuce) includes the white and grey poplars characterized by a woolly leaf abaxial surface. It includes *P. alba*, *P.* x *canescens*, and *P. tomentosa*, all native to the Old World, and the Aspens, with glabrous leaves, noted for their restless movement. This subsection includes *P. grandidentata*, *P. tremula*, and *P. tremuloides*. All of these have poor rootability.

Section Leucoides includes a combination of American species such as *P. heterophylla* and Asiatic species such as *P. lasiocarpa* and *P. wilsonii*. The main characteristic of these species is the large and leathery leaves.

Section Tacamahaca, known as the balsam poplars, is a very distinctive group with fragrant buds and scented leaves for which it is named. The leaves are usually whitish and waxy on the underside. This section includes *P. angustifolia*, *P. balsamifera*, *P. koreana*, *P. laurifolia*, *P. maximowiczii*, *P. simonii*, and *P. trichocarpa*. A distinctive characteristic is that all the members of this section are easily propagated by hardwood cuttings or by suckers.

Section Aigeiros, known as Black poplars, includes plants with green leaves on both sides and a large petiole, nearly always in movement. This section is distributed in North America, Europe, and Western Asia. Of all the species in this group, *P. deltoides* is the only one that cannot be propagated easily by hardwood cuttings.

Although *Populus* trees share several floral features with *Salix*, molecular data show a clear separation between the two genera; the most consistent and evolutionarily significant being their mode of pollination. Wind pollination in poplars, combined with their dioecy, has led to natural hybridization, allowing interspecies gene exchange. The outbreeding habit in combination with effective seed dispersal effects the genetic variation within *Populus* species. Patterns of isozyme variation across several species in over 30 enzyme systems show a recent theme of little differentiation among populations, with over 90% of the variation being within populations. Polymorphisms within populations are also common in morphological and phenological traits.

Fossil records show evidence that ancestral poplars were widespread across North America. Members of section Abaso were the sole poplars in North America until the late Eocene, when the precursors of other sections appeared and the first Eurasian poplar fossils are recorded (Collinson 1992).

1.1.2 Botany

Populus species are single-stemmed, deciduous (or semievergreen) trees mostly spread clonally by means of rootborne sucker shoots (soboliferous). They are among the fastest growing temperate trees, a quality tied ecologically to their role as vegetational pioneers as well as functionally to their heterophyllous growth habit. Poplar shoots continue to grow after bud burst by initiating, expanding, and maturing leaves (neoformed or late leaves) throughout the growing season. Cessation of growth and bud formation is induced by photoperiod in some poplars (Pauley and Perry 1954). Preformed and neoformed leaves often differ considerably in texture, shape, and toothing, with preformed leaves often more taxonomically diagnostic than neoformed leaves. Neoformed leaves are relatively convergent among unrelated poplar species, with the exception of the unique lobed leaves of *P. alba* and its hybrids.

The separate male and female trees flower before leaf emergence in spring (except in some subtropical species) from specialized buds containing preformed inflorescences, enabling wind pollination before canopy closure. The capsules and their airborne seeds, which have a readily detached coma of cottony hairs, mature with or after the overwintered preformed leaves.

Poplars have been recognized as a group since very early times and have a unique combination of characteristics that distinguish them from all other genera of plants. The defining features are primarily in the reproductive structures. The flowers are borne in pendent racemes (catkins, aments) that vary in flower number and density among poplar species. Among temperate trees with female catkins, only poplar and willow have seeds with a coma of cottony hairs on parietal placentas in thin-walled capsules. Individual flowers of both catkin types are subtended by thin bracteoles that fall as the catkins elongate during flowering. The caduceus bracteoles distinguish the catkins of poplars from those of willows. There are no ordinary petals or sepals, but 5 to 60 stamens or solitary pistils are borne on a more or less expanded floral disk.

Overwintering vegetative and reproductive buds are covered by several bud scales. The buds are covered with exudates that are rich in a variety of hydrophobic organic chemicals (Greenaway and Whatley 1990) that are thought to be involved in winter hardiness.

The rapid, nearly continuous growth during the favorable season results in a light diffuse-porous wood structure. Poplar wood differs from that of willow in that it has homocellular rays as opposed to the heterocellular rays. The wood differs from many similar light-colored woods in that it lacks some more specialized features like the marginal parenchyma or the spiral thickening and multi-seriate rays. The bark remains thin for a longer period than in most trees.

The leaves are alternate, stipulate, and petiolate with the petiole often transversely flattened and simple with glandular teeth along the margin and often with glands at the junction of the blade and petiole. Poplars generally have a high soil moisture requirement explaining the growth of most species along floodplains of rivers or lake shores.

1.1.3 Poplars as Crops

Rapid growth is the hallmark of poplars. It derives from a growth system that starts with the elongation of a preformed shoot from its bud and then continues to initiate and expand shoot segments and leaves throughout the growing season. The wood is diffuse-porous, light in weight, and yet capable of building trees of 40 m height in less than 20 years.

Several of these features have made poplars attractive to humans since ancient times. Today, poplar is cultivated worldwide in plantations for pulp and paper, veneer, excelsior (packing material), engineered wood products (e.g., oriented strandboard), lumber, and energy. Grown at a commercial scale under intensive culture for 6- to 8-year rotations, production rates with hybrid poplar can be as high as 17 to 30 Mg/ha/year of dry woody biomass (Zsuffa et al. 1996), comparable to the biomass produced by row crops such as corn. Historically, poplar has been widely used in windbreaks and for erosion control. Most recently, poplars have proven to be effective in the phytoremediation of environmental toxins (Flathman and Lanza 1998) and as bioindicators for ozone pollution in the environment (Jepsen 1994).

Poplars have three main properties that make them excellent for short-rotation intensive-culture management: rapid juvenile growth, immediate response to cultural practices, and their coppicing property (Bradshaw et al. 2000). The hybrids of section Tacamahaca are considered to have particularly high water-use efficiency (Mazzoleni and Dickmann 1988). Hybrids are more drought tolerant than native cottonwoods (Hinckley et al. 1989; Wyckoff and Zasada 2003), and this characteristic makes them attractive as a crop to supplement the diminishing supply of natural hardwoods.

An important feature in poplar hybrids and species, linked to their high productivity, is their high rate of stomatal conductance (g_{max} are near 600 mmol m^{-2} s^{-1}), which suggests the transpiration of large volumes of water (US Environmental Protection Agency 1999). Therefore, considerable attention has been given to the study of water regulation in several species, hybrids and cultivars that exhibit a wide range of variation (Reich 1984; Schulte and Hinckley 1987; Schulte et al. 1987; Ceulemans et al. 1988; Mazzoleni and Dickmann 1988; Tschaplinski and Blake 1989b; Bassman and Zwier 1991; Dickmann et al. 1992; Liu and Dickmann 1992, 1996; Dunlap et al. 1993; Tschaplinski et al. 1994; Ridolfi et al. 1996). Important work has been carried out on growth and biomass rates in large-scale poplar plantations (Tschaplinski et al. 1998b; Rae et al. 2004).

The commercial planting of poplars date from half a century ago, and it occurs mainly in America, Canada, Europe, and China. In the UK, the best region in which to grow poplars is the southern half of England, where the best conditions prevail for sustaining high growth rates. It has been shown that *Populus* does not like competition, spacing at planting range from 10 to 24 feet (ca. 3 to 7 m) but no more than 26 feet (~8 m).

Intense culture and study started in the early 1970s, shortly after the US Department of Energy (DOE) embraced the concept of SRWC (short-rotation woody crops), as a way of supplying biomass for conversion to liquid transportation fuels. At present, *Populus* is the main choice of a group of model species for SRWC that includes sycamore (*Platanus occidentalis* L.), silver maple (*Acer saccharum* Marshall), and hybrid willow (*Salix* spp.), all suitable for the Pacific Northwest, north central, and southeastern regions of the USA. Genetic improvement programs, silvicultural studies, and basic research were initiated and continue today throughout the world (Tuskan 1998), and poplar is currently accepted as a model tree in forestry (Bradshaw et al. 2000; Taylor 2002).

In addition to their high biomass yields, hybrid poplars are more drought tolerant than native cottonwoods (Tschaplinski et al. 1998a; Wyckoff and Zasada 2002). Because of this there are major selection and breeding programs, with the objectives of extending crops to marginal agricultural lands and supplementing the diminishing supply of natural hardwoods. One of the most successful hybrids created is the intersectional cross between *P. trichocarpa* and *P. deltoides*, which belong to *Populus* sections Tacamahaca and Aigeiros, respectively. The first of these contrasting species was selected from a maritime and wet climate on the Pacific coast, the second from a more continental and less humid area in the eastern United States. These two species show marked differences in both stomatal behavior and photosynthetic patterns (Hinckley et al. 1989; Dunlap et al. 1993).

Whole-plant and leaf studies have revealed that hybrid poplars close their stomata rapidly in response to atmospheric and soil water deficit, resulting in lower transpiration rates and a greater drought resistance compared to the parental species, a feature that allows hybrids to maintain higher leaf areas for longer periods (US Environmental Protection Agency 1999). Consequently the question of how poplars tolerate increasing drought stress without an impact on productivity has been raised and investigated by several authors. This issue is of particular relevance to increasing the use of marginal agricultural land and in achieving an increase in the energy conversion to liquid transportation fuel in temperate regions (Mazzoleni and Dickmann 1988; Tschaplinski and Blake 1989a, b; Tschaplinski et al. 1994, 1998b, 1999).

1.1.4 Economic Importance

Poplars' wood is soft, rather woolly in texture, pale in color, and inodorous. The commercial uses are as various as food containers, wagon bottoms, wheelbarrows, and colliery tubs, and they are popularly used as matches. Because of its low flammability the wood is suitable for floors of oast and for brake blocks. Uses for small logs include fiberboard, wood chipboard, and pulp. Among the most commonly planted cultivars are *P. eugenei*, *P. gelrica*, *P. robusta*, and *P. serotina*, all forms of the hybrid *P.* x *canadensis* (Bean 1976). Poplars are widely used in the pulp and paper industry, for regeneration on disturbed lands, and for biofuel (US Environmental Protection Agency 1999; Wyckoff and Zasada 2003).

1.1.5 Diseases

According to Bean (1976), the number of cultivars is high and they are mainly chosen for their fast growth, straight stems, and resistance to diseases. Among the most problematic diseases that attack poplars in cultivation are bacterial canker, caused by the bacterium *Aplanobacter populi* that produces cankers on the branches and main stem, generally diminishing the value of the timber; *Marssonina brunnea*, which causes the most serious foliage disease; *M. populinigrae*, which is the cause of death of branches of Lombardy poplar, defoliation, and dieback, more often occurring in the lower crown; *Melampora* species (rust), which cover the leaves with small orange pustules and can cause defoliation when abundant. Another fungus attacking poplars is *Dothichiza populnea* that can be confused with the symptoms of bacterial canker. *Cytospora chrysosperma* attacks dead wood.

1.1.6 Breeding

Poplars can be bred in the greenhouse on detached female branches with pollen that can be stored for several years. Each pollination can yield hundreds of seeds within 4 to 8 weeks. Seeds germinate within 24 h and give rise to 1- to 2-m-tall seedlings by the end of the same year. Few, if any, trees can match such efficiency. Poplar species within the same section, and many of the species from different sections, can be hybridized. Because all members of the genus are diploid ($2n = 38$), hybrids are usually fertile and can generate F_2 and backcross progenies that segregate for a wide range of traits. F_1 hybrids often show heterosis in growth and associated characteristics that make them attractive for commercial use (Zsuffa et al. 1996).

Most poplars will not flower earlier than 4 years of age, and many will take twice that long. The long generation interval is an impediment to practical breeding and selection and the development of informative pedigrees. It limits the applicability of conventional experimental genetic techniques such as induced mutagenesis, since producing homozygotes from heterozygous mutants is impracticably slow.

Poplars are dioecious, so self-pollinations cannot be done (with a very few hermaphroditic exceptions). Classical genetic tools, such as inbred lines, cannot be produced rapidly enough to be useful.

Most poplar breeding programs have relied upon more or less random interspecific hybridization to produce large progeny arrays, followed by very intense clonal selection to identify the best clones out of perhaps 10,000 seedlings. While demonstrably effective at producing remarkable genetic gains in short order, the scarcity of well-planned, long-term, sustained breeding programs and associated breeding materials hampers genetic studies. For example, multigeneration pedigrees are the basis for genetic mapping experiments designed to identify quantitative trait loci (QTLs) affecting important tree phenotypes, yet few such pedigrees exist.

Delayed flowering represents a lost opportunity for tree geneticists to provide suitable pedigrees for physiological studies and is currently the single most limiting factor in the conventional genetic improvement of forest trees. The development of methods for inducible early flowering, either by chemical/physical treatment or genetic engineering, will make it possible to decrease the generation interval for purposes of breeding, while also preventing excessive flowering from siphoning photosynthate away from wood formation in production plantations. While some progress has been made in this area with poplars (Weigel and Nilsson 1995), much remains to be done (Rottman et al. 2000).

1.1.7 Genetics and Breeding Programs

Members of the genus *Populus* are becoming more important because of their suitability for genetic and environmental studies of carbon sequestration as they are some of the fastest-growing trees in the world (Stettler et al. 1996; Tuskan et al. 2002). Their ease of propagation and the intensive work in physiology, biochemistry, agronomy, and genomics around the world have led several authors to regard them as excellent models for forest trees (Bradshaw and Stettler 1995; Taylor 2002; Tuskan 2003). Genetic studies in *Populus* have advanced rapidly. *P. trichocarpa* is the first tree species with its complete genome sequenced in the United States [Tuskan et al. 1998, 2002; US Department of Energy (DOE) 2002] and has an estimated 40,000 to 50,000 coding genes and a genome only four times the size of *Arabidopsis*.

Genomics in species of *Populus* is also advanced, for example, the study of *P. tremuloides* at the University of Agricultural Sciences, in Sweden, where more than 100,000 expressed sequence tags (ESTs) have been obtained by sequencing more than 14 tissue-specific cDNA libraries from *P. tremula* × *tremuloides*. These ESTs have been used as a base to build a cDNA microarray consisting of ca 25,000 probes (Nilsson 2002), and recently the poplar root transcriptome has been profiled where 7,000 ESTs were analyzed (Kohler et al. 2003).

In order to understand the genetics of adaptation, there is a growing body of knowledge about the physiological and molecular genetics of adaptive traits, with an increasing interest in predicting the genetic response of populations to changing climates and a trend toward incorporating adaptive as well as economic traits in breeding programs (Aitken and Adams 1996; Bradshaw and Strauss 2001). Now, the physiological study of *Populus* needs to be accompanied with investigation at a molecular level. Undoubtedly, genomics and transcriptomics will help to unravel the relationships between genotypic, environmental, and adaptive traits.

It has been reported that the hybrids between *P. trichocarpa* and *P. deltoides* in Washington and Oregon in the United States had a volume growth two or three times that of the best growing parent species (Heilman and Stettler 1985; Stettler et al. 1988; Dickman et al. 1992). Exploiting the phenomenon of heterosis requires either a simple method for generating large numbers of hybrid seedlings or the ability to vegetatively (i.e., clonally) propagate desirable hybrid genotypes. Nine species of poplar – *P. angustifolia, P. balsamifera, P. deltoides, P. euphratica, P. fremontii, P. tremula, P. tremuloides, P. tomentosa* and *P. trichocarpa* – are the most used in experiments.

In order to design and implement genetic improvement strategies of interspecific hybrids in forest trees, basic information is needed on the genetics of variation in commercially important traits (Bradshaw and Grattapaglia 1994). Identifying loci contributing to genetic variance in the F_2 or backcross generations makes it feasible to identify loci responsible for F_1 heterosis, and this has been done in *Populus*. The contribution (positive or negative) and mode of action of each parental QTL allele to the phenotype of the hybrid may be determined and used to evaluate the merits of various long-term breeding plans (Bradshaw and Grattapaglia 1994). QTL mapping in *Populus* also aims to give insights into the genetic basis of hybrid vigor (Bradshaw and Grattapaglia 1994; Bradshaw and Stettler 1995).

Interspecific hybrids of poplars have proven to be particularly amenable to map-based QTL analysis. Firstly, they are fast-growing trees and the commercially important phenotypes are revealed soon after planting, while the problem of poor juvenile-mature phenotypic correlations is minimized. Secondly, they grow in a relatively homogeneous environment, resembling agriculture more than traditional extensive forestry. This reduces the nongenetic variance components of complex traits like stem volume growth and facilitates detection of the genetic basis of phenotypic variance. Thirdly, clonal propagation allows precise estimation of the broad-sense heritability and an accurate assessment of the magnitude of the effect of QTLs on the genetic variance (Bradshaw and Grattapaglia 1994).

The physical and genetic maps of *Populus* serve several important purposes: as a starting point for map-based gene cloning, as a means to determine the extent of colinearity between the *Populus* genome and that of the completely sequenced genome of *Arabidopsis* (thus making effective use of the growing knowledge of *Arabidopsis* genome structure and function), and as a scaffold for the eventual annotation of the entire *Populus* genome.

EST microarrays will allow poplar researchers to couple physiological measurements with gene expression profiles, illuminating the genes, biochemical pathways, and cellular processes that are affected by a given environmental perturbation or transgene.

Extensive collaboration between poplar geneticists, physiologists and pathologists has set a solid scientific foundation for joint efforts in the future. Shared genetic materials, genetically informative two- and three-generation pedigrees, DNA-based genetic markers, common field measurement protocols, clonal plantation trials supported by industry/government/academic partnerships, and funding from multiagency grants have established a poplar research network of proven productivity. Working groups under the aegis of the International Poplar Commission (IPC) of the Food and Agriculture Organization (FAO) of the United Nations and of the International Union of Forestry Research Organizations (IUFRO), as well as several university/industrial research cooperatives, help to coordinate the work and the exchange of information.

1.1.8 *Populus* Genome

The first investigation of the *Populus* genome was made in 1921, in which the haploid chromosome number was erroneously reported as four (Graf 1921). By 1924, it became clear that the base chromosome number in *Populus* was 19, based on observations in seven species (Harrison 1924). Since then, various works have revealed that all *Populus* species exist in the diploid form with $2n = 38$ (Smith 1943), with occasional cases of triploid or tetraploid genets (Einspahr et al. 1963; Bradshaw and Stettler 1993). Based on cytology studies, Van Dillewijin (1940) hypothesized that the ancestral chromosome number of *Populus* was eight. However, because *Populus* chromosomes are mostly small and lacking in distinctive morphological features (Smith 1943), there is scant information on chiasma frequencies and chromosomal dynamics to substantiate this claim.

The haploid genome size of *Populus* is 550 million base pairs (bp) (Bradshaw and Stettler 1993). The nuclear genome is relatively small (2C = 1.2 pg) (Dhillon 1987; Bradshaw and Stettler 1993), only 4 times larger than the genome of the model plant *Arabidopsis* and 40 times smaller than the genomes of conifers such as loblolly pine. The small poplar genome simplifies gene cloning, Southern blotting, and other standard molecular genetics techniques. The physical:genetic distance ratio for poplar is close to 200 kb/cM, almost identical to *Arabidopsis*, making *Populus* an attractive target for map-based (positional) cloning of genes.

1.1.9
Populus as a Model System

The woody genus *Populus* offers extremely important potential as a commercially grown tree, but it is also of tremendous value as the model tree, due to its relatively small genome size, with a highly developed number of molecular genetic maps, combined with an ability for easy genetic transformation. *Populus* may be propagated vegetatively, making mapping populations immortal and easing the ability to produce large amounts of clonal material for experiments (Taylor 2002). Hybridization occurs routinely, and in this respect *Populus* has many similarities to *Arabidopsis*. The development of large EST collections and microarray analysis, the best available globally for any tree, and the availability of mapping pedigrees for QTL detection secure *Populus* as the ideal subject for further exploitation. Of crucial importance is the availability of the poplar genome sequence released this year (http://genome.jgi-psf.org/Poptr1_1/Poptr1_1.home.html).

Much work is being carried out in this genus to manipulate the natural variation of *Populus* to enable breeding for traits such as increased yield and improved response to biotic and abiotic stresses. The application of molecular tools has been exploited to understand the genetic resources of the genus of what is now seen as the model tree species.

1.2
Construction of Genetic Linkage Maps

The earliest work in molecular genetics of poplar, as was the case with most organisms, was focused on mutations of genes with large effects on the appearance. A major disadvantage with this is the limited number of markers available as a great deal of genetic variation is not identifiable at the phenotypic level. In addition to this, mutations in single genes is not easy in forest trees because of the difficulty of producing near-isogenic lines or clones differing in the mutant gene only. With the advent of molecular markers at the DNA sequence level came an increase in the number of polymorphic markers available for mapping purposes. In forest trees, where long generation time and high genetic load impede progress by traditional quantitative and classical genetic approaches, molecular genetics has much to offer.

1.2.1
Molecular Markers

Ideal molecular markers for linkage-map construction require that they be evenly distributed throughout the genome with selectively neutral behavior. Previously it was considered necessary to have codominant inheritance so that homozygous and heterozygous states could be discriminated, but recently mapping techniques in outbred poplars does not require this criterion.

Isozyme and Allozyme Markers

Allozyme analysis is based on the correlation between differences in mobility in an electric field due to differences in amino acid sequence, protein structure, and kinetic properties. (Murphy et al. 1990). This analysis is relatively easy to perform, inexpensive, and technically accessible, resulting typically in codominant markers. However, there is a limited number of loci that can be analyzed as the level of genetic polymorphism within coding sequences is relatively low and may be active only at specific physiological stages or in specific tissue.

Early molecular work in poplar relied on the use of allozymes as polymorphic markers. Such work revealed little linkage between markers (Hyun et al. 1987; Rajora 1990; Liu and Furnier 1993). This is likely to be due to the relatively small number of allozyme markers assayed across the 19 linkage groups (LGs).

RFLPs

Restriction fragment length polymorphism (RFLP) markers, which are based on polymorphisms observed after DNA digestion with restriction enzymes, are codominant. Although high levels of polymorphisms can be revealed, RFLP analysis is technically demanding, requires relatively large amounts of DNA, and is limited to the availability of libraries with useful probes. RFLPs have been used in *Populus* to study inter- and intraspecific variation (Keim et al. 1989) and for the production of genetic linkage maps (Bradshaw and Stettler 1993; Liu and Furnier 1993).

Sequence tagged site (STS) markers can be obtained by converting RFLP markers into PCR-based markers by use of the terminal RFLP probe sequence as a primer resulting in codominant length polymorphisms. When the sequenced clones are cDNA, the marker is known as an expressed sequence tag (EST). Bradshaw et al. (1994) generated a genetic linkage map of hybrid poplar composed of RFLP, RAPD, and STS markers.

SSRs

Microsatellites, or simple sequence repeats (SSRs), are tandem nucleotide repeats, polymorphic for the number of repeats. These markers are abundant and widely distributed in plant genomes. They are codominant and have been reported to have the highest polymorphism information content (PIC) of any marker system; however, the effort required to obtain the SSR flanking sequences is a drawback. Their discovery typically involves hybridization to create SSR-enriched genomic libraries followed by sequencing of selected clones and primer design based on 5' and 3' flanking sequence from microsatellite-containing fragments (Karagyozov et al. 1993). In poplar it is possible to take advantage of the growing EST and genomic databases now publicly available to use a more general computational molecular biology method, often referred to as an in silico approach, to identify repeats and their flanking regions (Tuskan et al. 2004). In *Populus* several hundred SSRs have been identified using various approaches for *P. nigra* L. (van der Schoot et al. 2000; Smulders et al. 2001), *P. tremuloides* Michx. (Dayanandan et al. 1998), and *P. trichocarpa* Torr. & A. Gray (Frewen et al. 2000). These SSRs have been derived from enriched genomic libraries and are predominantly simple di- and trinucleotide repeats.

Tuskan et al. (2004) identified and developed SSRs isolated from rapid shallow sequencing of total genomic DNA from *P. trichocarpa* and from selected bacterial artificial chromosomes (BAC) clones known to contain expressed sequences (Stirling et al. 2003). Approximately 23% of the 1,536 genomic clones and 48% of the 768 BAC subclones contained an SSR. 26.6% of the sequences contained multiple SSR motifs in complex or compound repeat structures. A survey of the genome sequence database revealed very similar proportional distribution, indicating that the limited rapid, shallow sequencing effort is representative of genome-wide patterns. The resulting SSRs developed from *P. trichocarpa* were shown to have high utility throughout the *Populus* genus and also in *Salix*, with amplification rates in excess of 70% for all *Populus* species tested. These SSRs have been used alongside other marker types to create genetic linkage maps (Yin et al 2004; Tchaplinski et al 2006).

SSR primer sequence and amplification information is publicly available for research use on the Internet: the *Populus* Molecular Genetics Cooperative (PMGC, GCPM; http://www.ornl.gov/sci/ipgc/ssr_resource.htm), Oak Ridge National Laboratory (ORPM; Tuskan et al 2004), and the Center for Plant Breeding and Reproduction Research (WPMS) (Van der Schoot et al. 2000). SSRs have also been useful for aligning genetic linkage maps with the physical sequence of poplar (Rae et al 2006; Street et al 2006).

RAPDs

Random amplified polymorphic DNA (RAPD) is generated by PCR amplification using arbitrary primers resulting, typically, in between 5 and 20 loci per experiment. This analysis detects single nucleotide changes, deletions and insertions within the primer annealing site. PCR-based marker analysis requires less DNA for the assay and is relatively simple, quick, and inexpensive; however, the resulting markers are dominant unless they are cloned and sequenced for conversion to a sequence characterized amplified region (SCAR) that can be used as codominant markers. Also, they have been found to be difficult to reproduce due to mismatch annealing of the random primer to the DNA. RAPDs have been used in conjunction with other molecular markers to produce linkage maps in poplar (Bradshaw et al. 1994), and they have been used to establish associations between markers and fungal rust resistance (Villar et al. 1996).

AFLPs

Amplified fragment length polymorphism (AFLP) markers were developed at Keygene (Wageningen, The Netherlands) by Vos et al. (1995). These markers assay the presence/absence of restriction enzyme sites and sequence polymorphisms adjacent to these sites.

AFLPs have the advantage over RAPDs in that a higher number of loci can be analyzed per experiment. For example, Cerverea et al. (2001) reported 24 AFLP primer combinations yielding a total of 653 segregating loci, an average of 27 loci per primer combination. There was considerable variation in the number of polymorphic AFLP markers revealed by different primer combinations, ranging from 11 to 51 markers. Pure *P. deltoides* and *P. trichocarpa* have AFLP heterozygosity levels around 20 to 30%. Markers can be codominant if the appropriate equipment and software are used to analyze the gels. It is possible to identify homo- or heterozygous loci (Cervera et al. 1996b) providing more information than dominant markers. AFLPs give highly reproducible banding patterns due to highly specific annealing of the primers to complementary adapter oligonucleotides. Associations between AFLP markers and resistance to fungal rust have been reported in *Populus* (Cervera et al. 1996a), and AFLP markers have been used alongside SSRs to produce high-density linkage maps (Cervera et al. 1996b; Yin et al. 2004)

1.2.2 Linkage Maps

Poplar, being dioecious and long lived, has made the development of linkage maps daunting. The first report of a linkage map for *Populus* was by Liu and Furnier (1993) using a selection of allozymes and RFLP markers in five F_1 crosses of trembling aspen. Work on 97 F_1 individuals from one of the crosses resulted in the identification of 14 LGs based on 3 allozyme and 54 RFLP markers, with a total map distance of 664 cM. This initial study showed that genetic mapping was possible in an F_1 generation of a highly heterozygous *Populus* species.

The first report of a linkage map identifying all 19 chromosomes of *Populus* was produced by Bradshaw et al. (1993). In recent years numerous linkage maps using a variety of mapping pedigrees have been reported.

Mapping Methods and Pedigrees

Much work in the early 1990s focused on the three-generation interspecific cross produced by Bradshaw et al. (1993) between *P. trichocarpa* and *P. deltoides*. Two full-sib individuals from the resulting cross were mated to produce an "inbred F_2" generation, family 331. Mapping in this pedigree was carried out in the same manner as for an inbred F_2 pedigree, assuming that there was enough marker variation between species that the heterozygous grandparents differed for alleles and could be treated as inbred lines. The F_2 generation was treated as though there were three possible genotypes that could occur at any locus: homozygous for parent 1; homozygous for parent 2; or heterozygous, segregating 1:2:1.

Later work on this pedigree modified the mapping procedure specifically for an outbred population structure so that a sex-averaged framework map for the F_2 was produced (Sewell et al. unpublished data; Tschaplinski et al. 2006). Fully informative markers (i.e., a marker that was heterozygous for different alleles in each of the F_1 parents) were preferentially chosen when available. A related interspecific cross using the same maternal *P. trichocarpa* but different male *P. deltoides* grandparent was produced by Frewen et al. (2000), but separate grandparental maps were constructed consisting mainly of dominant markers linked in coupling phase; however, synteny between the two repulsion phase LGs was determined by the placement of codominant AFLPs, microsatellite markers, and candidate genes.

More recently, linkage maps have been produced from F_1 or backcross pedigrees using the pseudotestcross strategy (Grattapaglia and Sederoff 1994). This strategy is mainly based on selection of single-copy polymorphic markers heterozygous in one parent and homozygous null in the other parent and therefore segregating into 1:1 ratio in their F_1 progeny as in a testcross. The term "two-way pseudotestcross" to define this mapping strategy is generally used to describe the two independent genetic linkage maps that are constructed by analyzing the cosegregation of markers in each progenitor. Examples of backcross pedigrees used to map in this way were reported by Wu et al. (2000), Yin et al. (2002, 2004), and Zhang et al. (2004). Cervera et al. (2001) used the pseudotestcross strategy for two half-sib related F_1 mapping pedigrees sharing the same female *P. deltoides* parent, resulting in linkage maps for the three-parental species.

Map Coverage

The first map reported to identify all 19 chromosomes of *Populus* was constructed by Bradshaw et al. (1994), containing 343 markers using RFLPs, STSs, and RAPDs. Marker genotypes were determined for as few as 26 and as many as 90 F_2 trees. MAPMAKER software resulted in the grouping of 35 LGs. The total map distance contained within the largest 19 LGs was 1,261 cM. Using the method put forward by Hubert et al. (1988), they estimated the genome length to be between 2,400 and 2,800 cM, and therefore concluded that their linkage map had ca. 50% coverage of the genome.

Using the same *P. trichocarpa* grandparent, Frewen et al. (2001) developed linkage maps for a related F_2 interspecific cross. Instead of producing a sex-averaged map for the pedigree, the high percentage of dominant markers and the paucity of linkage information from dominant markers linked in repulsion led to the construction of two linkage maps representing the grandmaternal and grandpaternal maps. The *P. trichocarpa* map spanned a distance of 2,002 cM, covering 77% of the estimated genome length (Bradshaw et al. 1994) with an average marker interval of 13.6 cM, while the *P. deltoides* map was reported as 1,778 cM in length, covering 68% of the genome with an average marker spacing of 12.3 cM. There were 26 LGs found in the *P. trichocarpa* map and 24 in the *P. deltoides* map. Because the haploid chromosome number is 19 in *Populus*, some of these LGs represented different sections of the same chromosomes.

More recently a number of genetic linkage maps with improved genome coverage have been produced mainly using backcross or F_1 pedigrees and the pseudo-testcross strategy (Grattapaglia and Sederoff 1994).

Wu et al. (2000) constructed an integrated genetic map for a backcross population derived from two selected *P. deltoides* clones using AFLP markers, using the two-way pseudo-testcross configurations of the markers (testcross markers) heterozygous in one parent and null in the other. By using the markers segregating in both parents (intercross markers) as bridges, the two parent-specific genetic maps were aligned. A number of nonparental heteroduplex markers were detected resulting from the PCR amplification of two DNA segments that had a high degree of homology to one another but differing in their nucleotide sequences. These heteroduplex markers served as bridges to generate an integrated map that included 19 major LGs and 24 minor groups. The 19 major LGs cover a total of 2,927 cM, with an average spacing between two markers of 23.3 cM.

Yin et al. (2001) constructed RAPD-based linkage maps for an interspecific cross between two species, *P. adenopoda* and *P. alba*, based on a double pseudo-testcross strategy. In the female parent, *P. adenopoda*, 82 markers were grouped in 19 different LGs (553 cM), whereas in the male parent, *P. alba*, 197 markers established a much more complete framework map with an observed genome length of 2,300 cM covering 87% of the total *P. alba* genome.

Cervera et al. (2001) constructed linkage maps for *P. deltoides*, *P. nigra*, and *P. trichocarpa* by analyzing progeny of two controlled crosses sharing the same female *P. deltoides* parent. The two-way pseudo-testcross mapping strategy was used to construct the maps. AFLP markers that segregated as 1:1 were used to form the four parental maps. SSR and STS markers were used to align homoeologous groups between the maps and to merge LGs within the individual maps. Linkage analysis and alignment of the homoeologous groups resulted in 566 markers distributed over 19 groups for *P. deltoides* covering 86% of the genome, 339 markers distributed over 19 groups for *P. trichocarpa* covering 73%, and 369 markers distributed over 28 groups for *P. nigra* covering 61%. Several tests for randomness showed that the AFLP markers were randomly distributed over the genome.

Yin et al. (2002) reported molecular genetic linkage maps for an interspecific hybrid population produced by crosses between *P. deltoides* (mother) and *P. euramericana* (father), which is a natural hybrid of *P. deltoides* (grandmother) and *P. nigra* (grandfather). A mixed set of the testcross markers, nonparental RAPD markers, and codominant AFLP markers was used to construct two linkage maps, one based on the *P. deltoides* genome and the other based on *P. euramericana*. The two maps showed nearly complete coverage of the genome, spanning 3,801 and 3,452 cM, respectively.

Zhang et al. (2004) reported the construction of AFLP genetic linkage maps for a hybrid pedigree derived from an interspecific backcross between the female hybrid clone *Populus tomentosa* × *P. bolleana* and the male clone *P. tomentosa*. A total of 782 polymorphic fragments were obtained using 49 primer combinations. Six hundred and thirty two of these fragments segregated into an 1:1 ratio, indicating that these DNA polymorphisms are heterozygous in one parent and null in the other. In the male *P. tomentosa* framework map, 218 markers were aligned in 19 major LGs. The linked loci spanned ca. 2,683 cM (87% coverage) of the poplar genome, with an average distance of 12.3 cM between adjacent markers. For female *P. tomentosa* × *P. bolleana*, the analysis revealed 144 loci, which were mapped to 19 major LGs and covered about 1,956 cM (77% coverage), with an average distance of 13.6 cM between adjacent markers.

Jorge et al. (2005) used an F_1 cross between *P. deltoides* and *P. trichocarpa* to construct linkage maps using RFLP, STS, RAPD, AFLP, and SSR markers with a subset of 90 genotypes using a pseudo-testcross strategy that resulted in a *P. deltoides* map of 2,803 cM and a 2,740 cM map for *P. trichocarpa*. Eight LGs could not be linked between the female and male parental maps. The authors reported that SSR and AFLP genotyping was extended to 253 supplementary genotypes, but they could not publish the map resulting from them.

Several other linkage maps are under construction such as a more complete map for the Bradshaw et al. (1994) *P. trichocarpa* × *P. deltoides* F_2 cross (Tscaplinski et al. 2006; Sewell et al. unpublished), a *P. alba* × *P. alba* F_1 cross (Scarascia-Mugnozza, pers. comm.), and a *P. nigra* × *P. nirga* F_1 cross (Scarascia-Mugnozza, pers. comm.). A summary of published linkage maps is shown in Table 1.

A number of these maps report linkage map lengths greater than that estimated as the complete length reported by Bradshaw et al. (1994). This may be a genuine effect from using different mapping pedigrees, as discrepancies may be due to differences in genome coverage, the choice of mapping function, and differences in recombination rates in the parents of the crosses (Plomion and O'Malley 1996; Echt and Nelson 1997; Remington et al. 1999). There is a tendency for angiosperm females to display higher recombination rates than males (De Vicente and Tanksley 1991; Ganal et al. 1995), and hybrids might be expected to have suppressed recombination compared to pure species because of differentiation of the homologous chromosomes of the parental species (Jackson 1985; Tenhoopen et al. 1996; Chetelat et al. 2000).

However, it is possible that these extended lengths are due to genotyping error. Errors inflate the number of apparent recombinations and expand map distances (Harald et al. 2000). This is especially severe when markers are tightly linked since a misordered marker with genotyping error is most likely to be interpreted as a double crossover in regions with high marker density. For example, it has been shown that a 3% error rate in genotyping can double the genetic map length (Brzustowicz et al. 1993). Double crossovers and possibly misscored individuals or loci can be identified by specific commands in various mapping software packages (e.g., Lincoln and Lander 1992). Therefore, a comparison of map lengths with and without error detection enabled gives some indication of the level of error in the data set. Another indication of the level of genotyping error is the number of markers that cannot be properly mapped.

Possibly the most complete and dense linkage map is that published by Yin et al. (2004) for an interspecific backcross between a hybrid *P. trichocarpa* × *P. deltoides*, backcrossed to a pure *P. deltoides*. This map includes 544 markers mapped onto 19 LGs, with all markers displaying internally consistent linkage patterns. The combined length of the 19 LGs of the maternal parent was 2,313.9 cM with the error detection function of Mapmaker enabled, and 2,564.3 cM with error detection off. The average number of recombinations observed in the progeny was 24.789, which corresponds to a genome size estimate of 2,478.9 cM. The genome length was estimated to be between 2,300 and 2,500 cM, based both on the observed number of crossovers in the maternal haplotypes and the total observed map length. Genome coverage was estimated to be greater than 99.9% at 20 cM per marker. This genetic linkage map provides the most comprehensive view of the *Populus* genome reported to date and will prove invaluable for future inquiries into the structural and functional genomics, evolutionary biology, and genetic improvement of this ecologically important model species.

Segregation Distortion

Liu and Furnier (1993) reported finding no evidence for segregation distortion in the first reported linkage map of *Populus*; however, a number of more recent studies have reported segregation distortion (Bradshaw and Stettler 1994; Yin et al 2004). This could be due to the limited number of available markers for this earlier map.

Reasons for skewed segregation ratios of molecular markers are still not well understood but are generally believed to be related to genetic factors such as chromosome loss and structural rearrangements (Williams et al. 1995; Kuang et al. 1999), genetic isolating mechanisms (Zamir and Tadmor 1986), the presence of an allele for pollen lethality (Bradshaw and Stettler 1994), or gametic selection (Zamir et al. 1982), as well as other nonbiological factors such as sampling in finite mapping populations or scoring errors (Plomion et al. 1995).

Bradshaw and Stettler (1994) detected distortion of expected Mendelian segregation ratios in the three-generation interspecific *P. trichocarpa* and *P. deltoides* cross. An RFLP linkage map was constructed around a single locus showing severe skewing of segregation ratio against F_2 trees carrying the *P. trichocarpa* allele in homozygous form. Several hypotheses for the mechanism of segregation distortion at this locus were tested, including directional chromosome loss, segregation of a pollen lethal allele, conflicts between genetic factors that isolate the parental species, and inbreeding depression as a result of genetic load.

Table 1. Summary of linkage maps produced in *Populus* species

Species	Pedigree	No individuals	Markers type	No markers	Total genetic distance (cM)	Reference
P. trichocarpa × *P. deltoides*	F_2	90	RFLP, STSs, RAPDs	343	1,261	Bradshaw et al. (1994)
P. trichocarpa	F_2	55	AFLP, SSR	147	2,002	Frewen et al. (2001)
P. deltoides				145	1,778	
P. deltoides × *P.deltoides*	BC_1	93	AFLP	198	2,927	Wu et al. (2000)
P. adenopoda	F_1		RAPD	82	533	Yin et al. (2001)
P. alba				197	23	
P. deltoides	F_1	127	AFLP, STS SSR	566	2,178	Cervera et al. (2001)
P. nigra				339	2,356	
P. deltoides		105		566	1,626	
P. trichocarpa				369	192	
P. deltoides	F_1	93	RAPD, AFLP	560	3,801	Yin et al. (2002)
P. euramericana					3,452	
P. tomentosa	F_1	120	AFLP	218	2,683	Zhang et al. (2004)
P. bolleana				144	1,956	
P. deltoides	F_1	253	RFLP, STS, RAPD, AFLP, SSR	200	2,803	Jorge et al. (2005)
P. trichocarpa				191	274	
P. trichocarpa × *P. deltoides*	BC_1	180	SSR, AFLP	544	2,313	Yin et al (2004)
P. deltoides						

A recessive lethal allele, *lth*, inherited from the *P. trichocarpa* parent, was found to be tightly linked to the RFLP marker and to cause embryo and seedling mortality.

Cervera et al. (2001) reported that markers cosegregating with a *Melampsora larici-populina* resistance gene showed a significant deviation in *P. deltoids* due to missing genotypes resulting from the death of susceptible trees.

Yin et al. (2004) detected some markers exhibiting segregation distortion that occured largely in two LGs. In fact, LG IV showed 92.1% of the chromosome was distorted with predominant alleles from *P. trichocarpa*. The distorted region of LG IV contains a locus conferring resistance against the leaf rust pathogen *Melampsora* × *columbiana* (Stirling et al. 2001; Yin et al. unpublished) and the distorted region of LG XIX contains another locus conferring resistance against *Melampsora larici-populina* (Zhang et al. 2001; Yin et al. unpublished results). The researchers hypothesized that divergent selection has occurred on chromosomal scales among the parental species used to create this pedigree and explored the evolutionary implications.

Comparative Maps

Early maps that used anonymous and dominant markers, such as RAPDs and AFLPs, left little scope for alignment or comparative mapping between different pedigrees. Some work on later linkage maps using codominant markers, such as SSRs, has allowed the alignment of these maps. The publicly available SSRs used in many recent linkage maps has allowed alignment of these maps, and LGs have been assigned the names used for the map produced by Cervera et al. (2001). In addition to this, the SSR primer sequences can now be blasted against the genome sequence so linkage maps can be aligned to the physical sequence (Street et al 2006; Rae et al 2006).

1.3 Detection of Quantitative Trait Loci

Fundamental to the breeding of poplar is the understanding of the genetic control of quantitative traits, such as biomass yield and disease resistance. Such traits are often a function of many internal plant processes and their interactions with the environment. These traits are usually complex, requiring knowledge of quantitative genetics to resolve their action. Elucidation of QTLs is useful not only for analyzing important agronomic traits for breeding purposes but also for understanding fundamental aspects of genetic control and about the nature of the QTLs, such as genome position, what they do, and how they act and interact, and can provide a good starting point for future studies on individual genes and genomic regions or for focusing on the inheritance and evolution of specific traits of interest.

QTL analysis involves looking for associations between the trait and molecular marker alleles segregating in a population. This requires accurate data and effective statistical software. Most QTL analyses in plants involve populations derived from pure lines, and several different approaches have been developed to associate QTLs with molecular markers (Kearsey and Farquhar 1998). The difficulties of QTL discovery in *Populus* can never be underestimated due to their outbreeding and dioecious nature. The quantitative genetics of outbreeding species is complicated by their heterozygosity so that up to four alleles may be segregating at each locus.

Early work to identify QTLs in *Populus* used similar methods to those used for pure breeding inbred lines, making the assumption that, in the case of an interspecific F_2 pedigree, the grandparents showed enough variation between species to be treated as purebred parents. An example of this is the work by Wu et al. (1997) on leaf traits in an F_2 pedigree between *P. trichocarpa* and *P. deltoides*. The species differed enough both in genetic polymorphisms used to construct a linkage map and in the leaf traits being analyzed to assume that heterozygous alleles within the grandparental species differed between species. This method assumes that most of the genetic variance for the traits of interest is partitioned between the grandparental species rather than among individuals within the species, but there are disadvantages of treating the parental clones as inbred lines as it fails to take into account that *Populus* clones are highly heterozygous. The advantage of treating the two grandparental species as inbred lines is that any marker that is polymorphic between the two species is informative for QTL mapping under the "simplified" model, and such "fixed" polymorphisms are more abundant than the multiallelic markers necessary for a more thorough analysis of QTL inheritance (Bradshaw and Grattapaglia 1994).

In the same way that the methods for construction of genetic linkage maps have been developed for outbred pedigrees, so have the methods for QTL mapping, particularly with the use of F_1 or backcross pedigrees using the pseudo-testcross approach put forward by Grattapaglia and Sederoff (1994).

1.3.1 Disease

Infection by rust fungus has a devastating impact on poplar plantations worldwide. Leaf rust caused by the fungus *Melampsora larici-populina* causes premature defoliation and can reduce growth by more than 20%. Trees defoliated early in the growing season become more susceptible to secondary pathogen infection and environmental stress. Therefore, durable resistance to rust pathogens is one of the main objectives of poplar breeding in Europe. Both qualitative and quantitative forms of *M. larici-populina* resistance have been described often with qualitative resistance inherited from the North American species *P. deltoides.*

Early works studied the qualitative resistance of poplar. Molecular markers were linked to simple traits for the disease resistance provided by a single gene with the discrete phenotypes for resistant and susceptible. For example, Villar et al. (1996) identified RAPD markers involved in qualitative resistance to rust fungus, *Melumpsora larici-populina.* Cervera et al. (1996a) identified three AFLP markers tightly linked to the locus for rust resistance for three races of *M. larici-populina* in the hybrid progeny derived from an interspecific cross between a resistant *P. deltoides* female and a susceptible *P. nigra* male, using both high-density marker analysis and bulked segregant analysis (BSA).

Newcombe (1998) used the three-generation *P. trichocarpa* × *P. deltoides* hybrid poplar pedigree to investigate the genetic control of resistance to an isolate of *M. medusae* sp. *deltoidae.* Necrotic flecking and rust severity were evaluated over 2 years in field and growth-room experiments and a laboratory assay of leaf-discs. Necrotic flecking in the field and growth-room experiments was found to be governed by a single, dominant gene inherited from the *P. trichocarpa* grandparent, using both nonparametric and QTL analyses of rust severity in the field and growth-room experiments. In contrast, the leaf-disc assay did not support a simple genetic interpretation.

Later works focused on the quantitative trait of rust tolerance rather than qualitative trait rust resistance. Jorge et al. (2005) worked with an F_1 progeny from an interspecific *P. deltoides* × *P. trichocarpa* cross consisting of 343 individuals. The *P. deltoides* parent was attributed to have quantitative and qualitative resistance, while the *P. trichocarpa* had quantitative resistance only (Dowkiw and Bastien 2004). Using a pseudo-testcross strategy, composite interval mapping (CIM) (Zeng 1993; Jansen and Stam 1994) was employed for QTL analysis so that the presence of the qualitative resistance could be fixed as a major genetic cofactor, so as to minimize the negative effect in the detection of other minor QTLs. Two QTLs with broad-spectrum effects were identified, one inherited from *P. deltoides* and one from *P. trichocarpa.* An additional seven QTLs with limited and specific effects were also detected.

Works have also been carried out to understand the genetic control of resistance to leaf spot caused by *Septoria populicola.* Hybrid F_1 clones of *P. trichocarpa* × *P. deltoides* are usually intermediate in disease phenotype between their susceptible *P. trichocarpa* and resistant *P. deltoides* parents. Newcombe and Bradshaw (1996) used the three-generation *P. trichocarpa* × *P. deltoides* pedigree to evaluate the percentage of leaves affected by leaf spot. QTL analysis revealed that a two-QTL model explained 68.3% and 61.2%, and 71.9% and 70.3%, of phenotypic and genetic variance, respectively, in the F_2 generation over the 2 years.

1.3.2 Stem and Growth Traits

Much early work focused on the identification of QTLs for easily measured canopy, stem, and biomass traits. Bradshaw and Stettler (1995) reported QTLs for stem growth and form in an interspecific F_2 pedigree using the inbreeding mapping software MAPMAKER/QTL 1.1 (Lincoln et al. 1992). The study reported several QTLs of large effect, responsible for a large portion of the genetic variance. For example, 44% of the genetic variance in stem volume was controlled by just two QTLs. The authors suggested that, instead of the expected polygenic model (the combined action of environmental effects and multiple genes of small and cumulative effects), the traits studied were under oligogenic control (few QTLs with relatively large effect). This is backed up by QTL analysis on canopy structure (Wu and Stettler 1998). However, later studies using the same pedigree reported a larger number of detected QTLs with smaller effects for stem growth (Rae et al. personal communication). The earlier study was carried out with a considerably smaller sample size (90 F_2 individuals compared to 210) and so was less likely to detect QTLs of small effect. Also later studies have made use of outbred mapping methods (Grattapaglia et al. 1996; Knott et al. 1997) and software such as QTLexpress (Seaton et al. 2002).

Wullschleger et al. (2005) reported 31 QTLs for distribution of biomass between stems, branches, leaves, and coarse and fine roots in the interspecific backcross of a hybrid *P. trichocarpa* × *P. deltoides*, backcrossed to a pure *P. deltoides*. The high-density linkage map produced by Yin et al. (2004) was analyzed using QTL cartographer. The percentage phenotypic variation explained by single QTLs ranged from 7.5 to 18.3%.

Allometric analysis in this study highlighted that as trees increased in age, biomass was preferentially distributed to stems and branches, while distribution to the roots decreased. This brings about the question as to whether QTLs identified early in the development of long-lived tree species can be taken as a good indication of genetic control later in development.

Wu et al. (2003) presented a statistical model for mapping QTLs underlying age-specific growth rates. A maximum likelihood approach based on the mechanistic relationship between growth rates and ages established by a variety of mathematical functions was developed to provide the estimates of QTL position, growth parameters characterized by QTL effects, and residual variances and covariances. The effects of the identified QTLs for growth were described as a function of age, so that age-specific changes in QTL effects can be readily projected throughout the entire growth process. The QTLs displayed increased effects on growth as trees aged, yet the timing of QTL activation was found to be earlier for stem height than diameter, which is consistent with the ecological viewpoint of canopy competition.

1.3.3 Leaf Growth

A number of QTL analyses have been carried out on leaf growth and development traits due to the great variation in leaf morphology between species and the availability of interspecific hybrids showing high levels of leaf area heterosis. Wu et al. (1997) reported the mapping of several leaf QTLs from the interspecific F_2 pedigree between *P. trichocarpa* and *P. deltoides*. The grandmaternal species is known for its long lanceolate, as opposed to deltoid leaves with short round petioles. The leaves differ in that the *P. trichocarpa* has few large cells with a low stomatal density while *P. deltoides* has many small cells with high stomatal density. The F_1 combine the two leaf cell types to result in heterosis in leaf area, and the F_2 generation shows much variation and transgressive segregation. The variation within this mapping pedigree has been utilized in other QTL mapping studies to understand leaf response to drought (Street et al. 2006) and CO_2 level (Ferris et al. 2002; Rae et al. 2006). Figure 1 shows the alignment of the genetic linkage map and physical sequence used by Rae et al. (2006) to compare the QTL positions to differentially expressed genes under elevated CO_2 conditions.

1.3.4 QTLs and the Environment

QTL analysis has been carried out across different environments and treatments to study the genetic control of tree response to the environment. Ferris et al. (2002) reported QTLs for epidermal cell size and number, and stomatal density, in the hybrid pedigree of *P. trichocarpa* × *P. deltoides* in both ambient and elevated concentrations of CO_2.

Genotype by environment interactions have been studied in a number of QTL mapping experiments in poplar using specially designed software such as multiQTL (www.multiQTL.com; Korol et al. 1998). For example, Rae et al. (personal communication) mapped QTLs for stem growth traits across three contrasting sites and tested for differences in additive genetic effects of QTLs mapped at the three sites. Another approach is the mapping of QTLs for "plasticity parameters" or "response QTLs." An example of this is the use of an additive main effects and multiplicative interaction (AMMI) model (Gauch 1988) to determine genotype by environment (GxE) interactions that can be used as measures of trait plasticity and used to identify QTLs that occur in response to altered atmospheric CO_2 (Rae et al. personal communication).

Similar work by Rae et al. (2006) reported QTLs for a number of leaf-growth traits and senescence when the interspecific F_2 *P. trichocarpa* × *P. deltoides* pedigree was grown in ambient and elevated CO_2 conditions. *P. trichocarpa* showed a greater response to elevated CO_2. In the F_2 generation, leaf development and quality traits, including leaf area, leaf shape, epidermal cell area, stomatal number, specific leaf area,

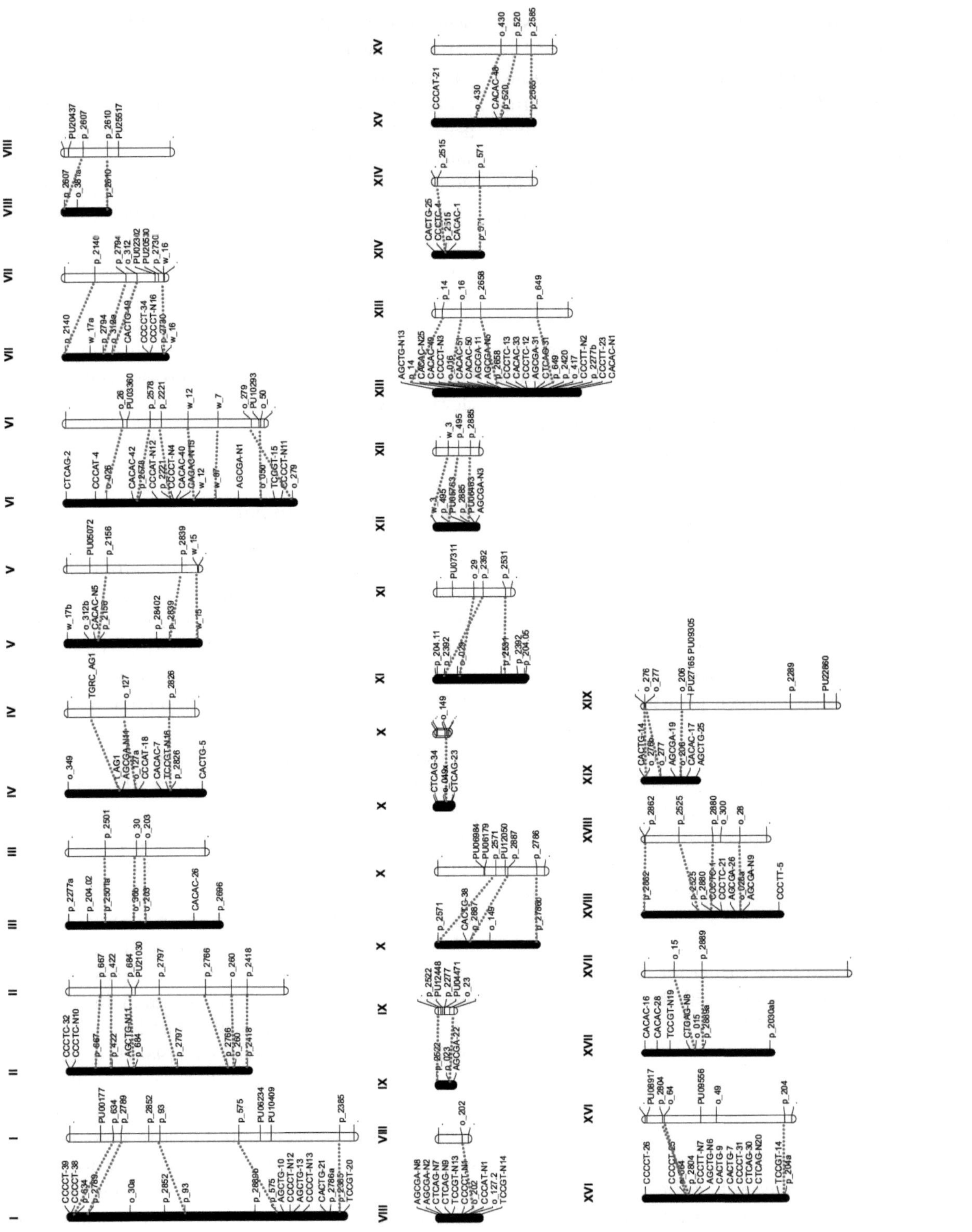

Fig. 1. Linkage map and physical map of *Populus trichocarpa* aligned using microsatellite primer sequences. The linkage map is shown in black showing microsatellites and AFLP markers and the physical sequence is in white showing microsatellites and candidate genes differentially regulated in elevated CO_2

and the canopy senescence index, were sensitive to CO_2 level. Many QTLs for leaf and senescence traits mapped to similar positions under ambient and elevated CO_2 conditions, but some genomic regions showed differential control backed up by the mapping of "response QTLs," which is the percentage difference between trees grown under elevated and control CO_2 conditions calculated for each individual and the results used as a trait for which response QTLs could be mapped. Two candidate genes identified as being differentially expressed under ambient and elevated CO_2 conditions in the grandparents (Taylor et al. 2005) were found to collocate to the regions in which response QTLs were mapped.

The response of *Populus* to drought stress has been studied in several experiments (Monclus et al 2006; Tschaplinski et al. 2006), and QTLs were mapped for such traits as specific leaf area, osmotic potential, and relative growth rate.

Street et al. (2006) utilized different responses of the grandparental species of the F_2 pedigree produced by Bradshaw and Stettler (1994). *P. trichocarpa* comes from the wetland region of the western USA, whereas *P. deltoides* comes from the dryer eastern regions of the USA. The F_2 pedigree has been shown to display transgressive segregation for drought response. This work mapped QTLs for leaf abscission, leaf expansion, photosynthesis, and transpiration when trees were grown under control and drought conditions. In addition to the comparison of QTL location for drought and control trees, response QTLs were also mapped. For each trait, the percentage effect between control and drought was calculated and used as a trait to map QTLs. A number of response QTLs mapped to regions of the genome to which drought QTLs mapped, suggesting that these regions were involved in the control of drought stress response. A transcriptome experiment was run alongside this work to identify genes differentially expressed between F_2 individuals showing the extreme responses to drought stress. SSR markers were used to align the genetic linkage map with the physical sequence of poplar. The degree of differentially expressed genes that collocated to genomic regions identified by QTL analysis was examined. The authors hypothesized that the occurrence of collocation could be due to differences in *cis*-acting elements (promoter sequences) in differentially expressed genes, i.e., the genes may be involved in the control of drought response, but the expression may be regulated in *trans*-acting by changes in a transcription factor regulating the drought response. Work is now under way to map the expression of the genes identified in this study to identify the genes of greatest interest and potential expression QTL (eQTL) mapping as carried out in eucalyptus (Kirst et al. 2004).

1.3.5 Phenology QTLs

Much of the work done to identify candidate genes for mapped QTLs has concentrated on phenological traits, such as bud burst and bud set. The first report of aligning QTLs with putative genes was detailed by Frewen et al. (2000). An interspecific *P. trichocarpa* × *P. deltoides* F_2 was used to map QTLs. Five candidate genes involved in photoperiod perception and transduction were mapped to the linkage map. Two candidate genes were found to collocate with QTLs effecting bud set and bud flush.

Dormancy-related QTLs were also reported to collocate with candidate genes for abscisic acid response signals by Chen et al. (2002)

1.3.6 Metabolite QTLs

Metabolite profiling has gained much attention as a powerful functional genomics tool to unravel gene function. Metabolite levels are controlled by both genetic and environmental factors, enabling Morreel et al. (2006) to consider metabolite concentrations as quantitative traits in a search for metabolite quantitative trait loci (mQTLs) that control their abundance and suggest that QTL analyses of the concentrations of all intermediates in a given biochemical pathway can reveal flux-regulating control points. In contrast to other complex traits, the molecular structure of metabolites and the knowledge of the pathway architecture may already suggest the function of the gene underlying the mQTL. Using the two half-sib F_1 pedigrees and linkage maps described by Ceverea et al. (2001), Morreel et al. (2006) detected four robust mQTLs that control flavonoid levels. A multivariate QTL analysis was used due to its better performance than single-trait analyses in cases of functionally related traits, such as the concentration of pathway intermediates. AFLP and SSR markers from the linkage maps were aligned with the genome sequence.

The chemical structure of the flavonoids, coupled with current knowledge of the pathway architecture and in silico mapping of candidate genes, allowed the tentative assignment of a function to three of these mQTLs. The data indicate that the combination of metabolite profiling with QTL analysis is a valuable tool to identify control points in a complex metabolic pathway of closely related compounds.

1.3.7 Candidate Genes

Candidate genes for QTLs have been suggested using a number of methods such as the collocation of QTLs and genes mapped to the linkage map (Frewen et al. 2000) and alignment of genetic maps with the physical genome sequence using maker sequence information. Candidate genes have then been identified from the literature (Taylor et al. 2006) or transcriptome studies (Rae et al 2006; Street et al 2006).

The release of the full genome sequence for poplar has allowed an estimation of total gene number to be calculated. Based on initial gene model data, it is estimated that the poplar genome consists of around 30,000 genes, but the actual number may be lower due to duplicated regions of the genome containing nonfunctional genes (S. Jansson personal communication). Taking into account the average length of the linkage maps used to map QTLs, it can be estimated that 1 cM distance may contain ca. 15 to 20 genes.

1.4 Advanced Studies

Traditional breeding for *Populus* has mainly focused on the selection of trees for fast growth traits. Work based on the analysis of phenotypes has provided a wealth of information on genetic material through multigeneration pedigrees. Molecular technologies and genetic linkage maps have led to new information for breeding strategies in *Populus*. The release of the complete poplar genome sequence and functional genomic information can be exploited to directly target candidate genes putatively involved in the control of the traits of interest, thus increasing the power of marker-assisted selection (MAS) (Strauss et al. 1992). SSR markers have been shown to provide an ideal bridge for map comparison and direct links to the genomic sequence. Surprisingly, there are no reports yet of actual application of MAS in tree species.

1.4.1 Genomic Resources for Poplar

During the past decade research and resource development in poplar has forged ahead beyond all other forest tree species (Bradshaw et al. 2000; Taylor 2002; Wulshleger et al. 2002; Brunner et al. 2004). As a result, numerous genomic resources are now publicly available that facilitate more rapid candidate-gene selection and collocation with mapped QTLs, for example. Most important among these are the poplar genome sequence, EST collections, and expression microarrays.

1.4.2 The Poplar Genome Sequence

The sequencing of the poplar genome, specifically the female *P. trichocarpa* Nisqually 1, was recently reported by Tuskan et al. (2006) and is hosted by the Joint Genomes Initiative (JGI) at http://genome.jgi-psf.org/Poptr1_1/Poptr1_1.home.html. Genome assembly was achieved through shotgun sequencing combined with genetic mapping to allow chromosome reconstruction with the assembled genome containing 450 Mb and an estimated 45,555 protein-coding loci. Roughly 89% of predicted gene models contained homology to protein sequences held in NCBI, although some 12% (5,248) contained no evidence of homology to known genes in *Arabidopsis thaliana*, suggesting that these may represent tree-specific genes. The reader is referred to Tuskan et al. (2006), and especially the supplementary information available, for further details. A finding of great interest from the genome sequencing project, particularly in relation to the genetic architecture of trait control, was that of the shared Salicaceae genome-duplication event in addition to an older duplication event shared with *A. thaliana* (termed the salicoid and eurosid duplication events by Tuskan et al. 2006). A substantial number of paralogous genes resulting from the more recent salicoid event were identified, and the transcriptional activity and specificity (from a tissue/development/response perspective) of such paralogs must now be examined and considered when interpreting the architecture of trait control. Reexamination of some of our own QTL results suggests that duplicated regions of the genome may be exerting control for some traits (unpublished data). These paralogs also need to be considered carefully when

Library distribution: POPLAR.226

This cluster contains 46 Clones (48 sequences)

Library	Clones	expected	Clones/LibSize
Cambial zone (A + B):	1	2.73	0.000167
Young leaves (C)	1	2.22	0.000206
Flower buds (F)	11	2.97	0.001687
Senescing leaves (I)	4	2.52	0.000723
Apical shoot (K) :	2	2.46	0.000371
Cold stressed leaves (L):	2	1.86	0.000489
Roots (R) :	1	2.64	0.000173
Shoot meristem (T):	11	3.82	0.001313
Male catkins (V):	2	2.23	0.000409
Dormant buds (Q):	2	2.66	0.000342
Female catkins (M) :	1	2.79	0.000163
Petioles (P):	4	2.95	0.000618
Imbibed seeds (S):	3	2.84	0.000482
Virus/fugus-infected leaves (Y):	1	0.64	0.000716

Library distribution: POPLAR.398

This cluster contains 18 Clones (23 sequences)

Library	Clones	expected	Clones/LibSize
Senescing leaves (I)	14	0.99	0.002530
Bark (N):	1	0.87	0.000204
Male catkins (V):	1	0.87	0.000205
Female catkins (M) :	1	1.09	0.000163
Petioles (P):	1	1.15	0.000155

Fig. 2. Library distributions of two EST sequences annotated as ACC oxidase in PoplarDB (http://www.populus.db.umu.se). Such digital northerns provide evidence that ESTs such as these have become functionally differentiated within the *Populus* genome

designing single nucleotide polymorphism (SNP) or other sequence-based gene markers to ensure specificity to only one gene copy. Roughly a third of salicoid duplicated genes represented in *Populus* EST libraries show evidence of differentiated expression, a finding that was supported by examination of the expression patterns of duplicated genes using DNA microarrays (Tuskan et al. 2006).

The availability of a genome sequence makes some gene-finding strategies at least partially superfluous and redundant. For example, chromosome walking and map-based cloning could be considered as offering little purpose as the order and location of virtually all genes are already known and large-scale differences in gene order and content are not expected between members of a genus or even family. Evidence for the near-perfect colinearity and synteny between both fellow members of the *Populus* genus and between *Salix* and *Populus* has been provided by a number of QTL studies using families produced from interspecific crosses, including that used for the chromosome reconstruction of the genome sequence.

Marker development is also greatly facilitated by the genome sequence with sequence repeats being visually displayed within the JGI genome viewer and the obvious ability to develop primers for SNP genotyping. SNP marker development offers a rapid means of adding candidate genes to genetic maps for subsequent QTL mapping, and the availability of intron-exon boundary information means that markers spanning both can be designed to maximize the possibility of SNP detection. Of course, it is also true that exon-specific markers can also be designed for RT-PCR analysis of transcript levels for expression QTL mapping (see below) and for transcript-trait correlation analysis.

1.4.3 EST Resources

A number of extensive EST collections have been developed from a range of poplar species (Table 2). These have facilitated much research as well as playing an important role in the sequence assembly and annotation of poplar (Tuskan et al. 2006). The most advanced EST resource is that available in Populusdb from the Umeå Plant Science Centre (UPSC), which offers extensive search options (annotation, homologs, library, BLAST) as well as the ability to view the library distribution of an EST of interest – something that offers potential insight into gene function and that can be used to produce a digital northern (for example, Street et al. 2006 used digital northerns to determine that genes induced by drought stress show greater similarity to genes associated with winter dormancy than with cold stress). Such library distribution patterns are extremely useful when identifying candidate genes for genomic and genetic studies such as QTL-candidate gene collocation (discussed below). For example, examination of ESTs annotated as ACC oxidase in Populusdb reveals that many appear to have tissue-specific expression patterns, and such knowledge drastically increases the ability to select a functionally relevant candidate (compare clusters POPLAR.3983 and POPLAR.2266) (Fig. 2).

These EST libraries have been constructed from a wide range of library types representing many poplar species, as well as a wide variety of tissue types. In general, there is a bias toward woody tissues, especially differentiating xylem, but across all the collections most tissue types are represented, and a range of treatment conditions have also been used to enrich libraries for genes associated with both abiotic (particularly the *P. euphratica* libraries detailed in Brosche et al. 2005) and biotic (Ralph et al. 2006) stress responses.

1.4.4 Expression Microarrays

The use of microarrays to examine the transcriptional response of organisms caused something of a paradigm shift in biology, and this trend has held true in poplar. There are now a number of microarray resources available for poplar representing nearly the full cross-section of array technologies. The first developed microarray resource was derived from the UPSC EST collection described in Sterky et al. (2004). The Populusdb EST database was used to identify a unigene set of representative EST clones for spotting as cDNA onto glass microarrays. The first generation of developed arrays (referred to as POP1) contained 13,490 clones selected from 7 of the 19 libraries contained in Populusdb. A second-generation array (POP2) was then constructed containing 24,912 spotted clones selected from all 19 libraries and representing ca. 40% of all predicted gene models in the poplar genome (Street et al. 2006). These arrays have been used to examine a range of processes including wood formation (Schrader et al. 2004; Moreau et al. 2005), response to abiotic infection (Smith et al. 2004), gene-expression changes associated with autumnal senenscence (Andersson et al. 2004), and response to biotic factors including the effects of elevated atmospheric [CO_2] (Taylor et al. 2005; Druart et al. 2006) and drought stress (Street et al. 2006). Another cDNA array resource has been developed at INRA that was initially spotted as a nylon membrane macroarray and used to examine the poplar root transcriptome (Kohler et al. 2003) and elevated [CO_2] and [O_3] responses (Gupta et al. 2005). This macroarray was produced from the EST clones represented in PoplarDB (Kohler et al. 2003). These clones have now been donated to the PICME project (www.picme.at) and are available spotted alongside the EST collection

Table 2. Sources of EST sequence data for Populus

EST Resource	# ESTs	Website address	Key reference
Populusdb	121,495	www.populus.db.umu.se/	Sterky et al. (2004)
PoplarDB	20,005	http://mycor.nancy.inra.fr/PoplarDB/	Kohler et al. (2003) Brosche et al. (2005)
AspenDB	5,410	http://aspendb.mtu.edu/	Ranjan et al. (2004)
DFCI Poplar Gene Index	371,518	http://compbio.dfci.harvard.edu/tgi/cgi-bin/tgi/gimain.pl?gudb=poplar	Amalgamation of multiple EST resources
Treenomix	~90,000	http://www.treenomix.ca/cDNA-sequencing/ No public database, ESTs submitted to GenBank	Ralph et al. (2006)
Arborea	11,591	http://www.arborea.ulaval.ca/research_results/est_sequencing_in_poplar/	
GenBank	376,600	As of 26/10/2006 (search term 'populus')	http://www.ncbi.nlm.nih.gov/
Sputnik	137,728	http://sputnik.btk.fi/	

described in Brosche et al. (2005) as a cDNA glass-based microarray representing an estimated 9,500 genes (F. Martin personal communication). There are additionally two short oligomer microarrays available from Affymetrix (www.affymetrix.com) and Nimblegen (www.nimblegen.com). These are both whole-genome arrays with oligo probe sequences having been designed based on predicted genemodel sequences from V1.1 of the genome (Tuskan et al. 2006) as well as EST sequences (in the case of Nimblegen, these are detailed in Tuskan et al. 2006 and for Affymetrix EST sequences stored by TIGR [see DFCI entry in Table 2]). A notable gap in the availability of microarrays is the lack of an oligo expression array, although a 70-mer genomewide array is currently in development (N. Street personal comm.).

As well as the physical availability of microarrays, access to existing microarray data is a key genomics resource, as any user of Genevestigator (https://www.genevestigator.ethz.ch) or other such services will attest. Beyond actual access to the intensity values for arrays, a means of analyzing and interpreting these is also crucial if microarrays are to be exploited to their full potential. In the case of poplar, a microarray resource for the community has been established in the form of UPSC-BASE (www.upscbase.db.umu.se). This is a combined Web-based GUI resource integrating the open-source BASE (http://base.thep.lu.se) project for the storage of microarray data alongside the implementation, normalization, analysis, interrogation, and visualization methods, and the annotation and functional classification information stored in Populusdb. Public access is offered for all published microarray data, and it is intended that this resource be developed to serve the poplar community.

1.4.5 Applying Genomic Resources for Candidate-Gene Discovery

Microarrays are being used for an ever expanding range of applications including the mapping of expression QTL (eQTL), such as was demonstrated in Kirst et al. (2005) in a backcross *Eucalyptus* family. Street et al. (2006) proposed an alternative use for microarrays within an eQTL context whereby they compared the transcriptome in response to drought stress of a highly sensitive and tolerant set of genotypes from an F_2 interspecific population to identify what can be termed gene expression markers (GEMs) (West et al. 2006). They proposed the subsequent mapping of GEMs as eQTL using RT-PCR as this reduces the use of microarrays significantly, making the approach cheaper and statistically less demanding, although such an approach does limit the potential insight offered by mapping eQTL for noncandidate genes. The use of eQTL mapping represents a potentially productive method of elucidating the genetic architecture of complex trait control and can answer questions as to whether the control of a trait response is the result of a *cis*-acting polymorphism within a structural gene or due to a *trans*-acting factor. In the case of a *trans-* effect, mapping eQTLs for multiple genes with collocating QTLs may identify the locating of the *trans-* acting factor. There are additional uses to which microarrays can be put if they are used for hybridizing genomic DNA rather than cDNA (Hazen and Kay 2003). The genomic DNA of the parents of a mapping population can be hybridized to expression arrays in order to identify single feature polymorphisms (SFPs) (Hazen and Kay 2003). If SFPs are identified, the approach can be extended to use for bulked segregant analysis (BSA) by forming two pools of DNA from groups of genotypes lying at contrasting ends of a phenotypic trait distribution, hybridizing these pools, and subsequently identifying polymorphism between them. This has the potential to produce a set of polymorphisms that densely cover a genome and can subsequently be used to map the location of the causal polymorphism with reasonable resolution. Such an approach has recently been demonstrated in *Arabidopsis thaliana* (Borevitz et al. 2003; Singer et al. 2006) and yeast (Brauer et al. 2006; Gresham et al. 2006) but has yet to be extended to forest trees. The approach is most applicable to use on short oligomer arrays such as those made by Affymetrix and Nimblegen.

The above methods of eQTL mapping and array-assisted BSA have particular appeal for use in poplar as they help to overcome or reduce the inherent limitations that often plague bridging the QTL-gene barrier in forest trees. Traditional gene-mapping approaches are hard to apply in forest trees due to the problems of producing and maintaining mapping families of adequate size to represent enough recombination events to ensure adequate mapping resolution. Forest tree species are not only problematic due to their long generation times and the cultural demands of planting and maintaining populations but also because their typically outbreeding nature, with resulting inbreeding depression, limits the type of populations that can be constructed – NILs and RILs are the jealous desire of many a tree geneticist. Genomic resources enable new approaches to be taken to attempt to identify genes containing the causal polymorphisms identified by QTL location. EST databases and microarray analysis both represent powerful means of identifying candidate genes for which the potential collocation to QTL can be examined. In the case of EST libraries, candidate-gene selection will be based on a priori knowledge of gene function, whereas microarrays allow hypothesis-independent identification of genes associated with a response trait or developmental process. In the case of poplar, the availability of the physical genome sequence allows for the immediate physical location of the candidate gene to be known. Genetic maps can then be aligned against the physical map, and the collocation of candidate genes on the physical map and QTL on the genetic map can then be examined (Street et al. 2006). Alternatively, the candidates can be sequenced in the parents of the mapping family, and sequence polymorphisms (SNPs, indels, etc.) can be identified and utilized to develop sequence-based markers. The mapping population can then be genotyped for these markers and added to the genetic map used for QTL analysis, and collocation can be examined in this way. This has advantages over the physical-genetic comparison method as it takes into account local recombination frequencies. In such a case, the candidate gene should represent the closest flanking marker to the peak QTL location score in order for it to represent a convincing candidate. This approach is limited due to the large average size of QTL intervals (10 to 30 cM) that often can contain several hundred to a few thousand genes depending on local recombination frequency. For example, Street et al. (2006) acknowledge that the QTL regions within which there are collocating candidate genes contain an average of 433 genes, making the likelihood of collocation by chance relatively high. Another option made available by the genome sequence is that of identifying all genes lying between the flanking markers of a QTL region and the annotation-based selection of candidates from these using a priori knowledge of their function. Ultimately all of these methods fail to provide causal proof, which is ideally provided through functional means such as over- or under-expression analysis in transgenic plants. Poplar has been shown to be highly transformable, with an increasing number of species and hybrids being successfully transformed using a range of technologies (Han et al. 1997; Rottmann et al. 2000; Groover et al. 2004; Jing et al. 2004; Meyer et al. 2004; Nowak et al. 2004; Strauss et al. 2004; Busov et al. 2005; Filichkin et al. 2006).

These approaches represent powerful means of identifying candidate genes for the control of traits of commercial importance, such as wood quality, biomass yield, and biotic pest resistance, as well as identifying genes responsible for the vast diversity that exists within the *Populus* genus. The identification of ecologically important genes is likely to become increasingly important as a result of the need to understand plant-level responses to climate change as well as requiring the development of new elite varieties that can achieve commercially viable yields in the context of a changing climate.

Comparative genomics approaches comparing species exposed to selection pressures in contrasting natural ranges with, for example, differing precipitation regimes represent a method of identifying existing genetic adaptations that could be exploited to provide commercial benefit, either through means of transgenic manipulation or directed breeding. It can be expected that the *Salix* genome will be largely colinear with that of *Populus* (Tuskan et al. 2006), which should allow for comparative mapping between the two genera. The recent divergence of *Salix* and *Populus* also means that genomic resources developed in one genus, particularly long-oligo and cDNA microarrays, are suitable for use in both (Sterky et al. 2004), making possible a number of interesting comparative studies across the whole of the Salicaceae. Street et al. (2006) recently showed that this approach has the potential through the use of cDNA microarrays to examine the transcriptional drought response of two contrasting *Populus* species whose natural ranges differ in mean annual precipitation and temperature. They showed that microarrays could identify genes with contrasting responses when the two species are exposed to drought, suggesting that gene-expression control is important in the process of speciation. They additionally showed that large-scale differences in transcriptional response could be identified for drought response within an F_2 population formed from a cross between the two species. Many of the transcriptional differences they identified can be expected to be the result of control exerted by few *trans*-acting elements, and subsequent work to identify these represents an excellent method to identify candidate genes for manipulation to produce trees with a desired response to drought.

References

Aitken S, Adams T (1996) Meeting Report, Western Forest Genetics Association Annual Meeting, Newport, Oregon, 29 July-1 Aug 1996. Dendrome. http://dendrome.ucdavis.edu/Newsletter/wfgareport.html [accessed 8 April 2003]

Andersson A, Keskitalo J, Sjodin A, Bhalerao R, Sterky F, Wissel K, Tandre K, Aspeborg H, Moyle R, Ohmiya Y, Bhalerao R, Brunner A., Gustafsson P, Karlsson J, Lundeberg J, Nilsson O, Sandberg G, Strauss S, Sundberg B, Uhlen M, Jansson S, Nilsson P (2004) A transcriptional timetable of autumn senescence. Genome Biol 5:R24

Bassman JH, Zwier JC (1991) Gas exchange characteristics of *Populus trichocarpa, Populus deltoides* and *Populus trichocarpa* × *P. deltoides* clones. Tree Physiol 8:145–159

Bean WJ (1976) Trees and Shrubs Hardy in the British Isles. 4 vols. Murray, London

Borevitz JO, Liang D, Plouffe D, Chang H-S, Zhu T, Weigel D, Berry CC, Winzeler E, Chory J (2003) Large-scale identification of single-feature polymorphisms in complex genomes. Genome Res 13: 513–523

Braatne JH, Hinckley TM, Stettler RF (1992) Influence of soil water on the physiological and morphological components of plant water balance in *Populus trichocarpa, Populus deltoides* and their F_1 hybrids. Tree Physiol 11:325–229

Braatne JH, Rood SB, Heilman PE (1996) Life history, ecology and conservation of riparian cottonwoods in North America. In: Stettler RF, Bradshaw HD, Heilman PE, Hinckley TM (eds) Biology of *Populus* and its Implications for Management and Conservation. NRC Research Press, National Research Council of Canada, Ottawa, ON, pp 57–86

Bradshaw HD, Stettler RF (1993) Molecular genetics of growth and development in Populus. I. Triploidy in hybrid poplars. Theor Appl Genet 86:301–307

Bradshaw HD, Grattapaglia D (1994) QTL mapping interspecific hybrids of forest trees. For Genet 1 (4):191–196

Bradshaw HD Jr, Stettler RF (1995) Molecular genetics of growth and development in *Populus*. IV. Mapping QTL with large effects on growth, form and phenology traits in a forest tree. Genetics 139:963–973

Bradshaw Jr HD, Strauss SH (2001) Breeding strategies for the 21stcentury: domestication of Poplar. In: Dickmann DI, Isebrands JG, Eckenwalder JH, Richardson J (eds) Poplar Culture in North America. Part B, Chap 14. NRC Research Press, Ottawa, Canada, pp 383–394

Bradshaw HD, Ceulemans R, Davis J, Stettler R (2000) Emerging model systems in plant biology: poplar (*Populus*) as a model forest tree. J Plant Growth Regul 19:306–313

Brauer MJ, Christianson CM, Pai DA, Dunham MJ (2006) Mapping novel traits by array-assisted bulk segregant analysis in *Saccharomyces cerevisiae*. Genetics 173:1813–1816

Brosche M, Vinocur B, Alatalo E, Lamminmaki A, Teichmann T, Ottow E, Djilianov D, Afif D, Bogeat-Triboulot M.-B. Altman A, Polle A, Dreyer E, Rudd S, Paulin L, Auvinen P, Kangasjarvi J (2005) Gene expression and metabolite profiling of *Populus euphratica* growing in the Negev desert. Genome Biol 6:R101

Brunner AM, Busov VB, Strauss SH (2004) Poplar genome sequence: functional genomics in an ecologically dominant plant species. Trends Plant Sci 9:49–56

Brzustowicz LM, Merette C, Xie X, Townsend L, Gilliam TC, Ott J (1993) Molecular and statistical approaches to the detection and correction of errors in genotype databases. Am J Hum Genet 53:1137–1145

Burns RM, Honkala BH (1990) Silvics of North America, Vol 1, Conifers. USDA Forest Service Agriculture Handbook 654, Washington DC, USA: http://www.na.fs.fed.us/spfo/pubs/silvics_manual/table_of_contents.htm [accessed 13 Jul 2001]

Busov VB, Brunner AM, Meilan R, Filichkin S, Ganio L, Gandhi S, Strauss SH (2005) Genetic transformation: a powerful tool for dissection of adaptive traits in trees. New Phytol 167:9–18

Cervera MT, Gusmao J, Steenackers M, Peleman J, Storme V, Vanden Broeck A, Van Montagu M, Boerjan W (1996a) Identification of AFLP molecular markers for resistance against *Melampsora larici-populina* in *Populus*. Theor Appl Genet 93:733–737

Cervera MT, Gusmao J, Steenackers M, Van Gysel A, Van Montagu M, Boerjan W (1996b) Application of AFLP- based molecular markers to breeding *Populus* spp. Plant Growth Regul 20:47–52

Cervera MT, Storme V, Ivens B, Gusmao J, Lui BH, Hostyn V, Slycken JV, Van Montagu M, Boerjan W (2001) Dense genetic linkage maps of three *Populus* species (*Populus deltoides, P. nigra* and *P. trichocarpa*) based on amplified fragment length polymorphism and microsatellite markers. Genetics 158:787–809

Ceulemans R, Impens I, Imler R (1988) Stomatal conductance and stomatal behavior in *Populus* clones and hybrids. Can J Bot 66:1404–1414

Chen THH, Howe GT, Bradshaw HD (2002) Molecular genetic analysis of dormancy-related traits in poplars. Weed Sci 50:232–240

Chetelat RT, Meglic V, Cisneros P (2000) A genetic map of tomato based on BC1Lycopersicon esculentum × Solanum lycopersicoides reveals overall synteny but suppressed recombination between these homologous genomes. Genetics 154:857–867

Collinson ME (1992) The early fossil history of Salicaceae. Proc R Soc Edinburgh Sect B Biol Sci 98:155–167

Dayanandan S, Rajora OP, Bawa KS (1998) Isolation and characterization of microsatellites in trembling aspen (*Populus tremuloides*). Theor Appl Genet 96:950–956

De Vicente MC, Tanksley SD (1991) Recombination around the *Tm2a* and *Mi* resistance genes in different crosses of *Lycopersicon peruvianum*. Theor Appl Genet 83:173–178

Dhillon SS (1987) DNA in tree species. In: Bonga JM, Durzan DJ (eds) Cell and Tissue Culture in Forestry, vol 1. Nijhoff, Dordrecht, pp 298–313

Dickmann DI, Liu Z, Nguyen PV, Pregitzer K (1992) Photosynthesis, water relations, and growth of two hybrid *Populus* genotypes during a severe drought. Can J For Res 22:1094–1106

Dowkiw A, Bastien C (2004) Characterization of two major genetic factors controlling quantitative resistance to Melampsora larici-populina leaf rust in hybrid poplars: Strain specificity, field expression, combined effects, and relationship with a defeated qualitative resistance gene. Phytopathology 94:1358–1367

Druart N, Rodriguez-Buey M, Barron-Gafford G, Sjodin A, Bhalerao R, Hurry V (2006) Molecular targets of elevated [CO_2] in leaves and stems of Populus deltoides: implications for future tree growth and carbon sequestration. Funct Plant Biol 3:121–131

Dunlap JM, Braatne JH, Hinckley TM, Stettler RF (1993) Intraspecific variation in photosynthetic traits of *Populus trichocarpa*. Can J Bot 71:1304–1311

Dunlap JM, Stettler RF (2001) Variation in leaf epidermal and stomatal traits of *Populus trichocarpa* from two transects across the Washington Cascades. Can J Bot 75:528–536

Echt CS, Nelson CD (1997) Linkage mapping and genome length in eastern white pine (*Pinus strobus* L.). Theor Appl Genet 94:1031–1037

Einspahr D, Benson MK, Peckham JR (1963) Natural variation and heritability in triploid aspen. Silvae Genet 12:51–58

Flathman PE, Lanza GR (1998) Phytoremediation: Current views on an emergent green technology. J Soil Contaminat 7:415–432

Ferris R, Long L, Bunn SM, Robinson KM, Bradshaw HD, Rae AM, Taylor G (2002) Leaf stomatal and epidermal cell development: identification of putative quantitative trait loci in relation to elevated carbon dioxide concentration in poplar. Tree Physiol 22:633–640

Filichkin SA, Meilan R, Busov VB, Ma C, Brunner AM, Strauss SH (2006) Alcohol-inducible gene expression in transgenic *Populus*. Plant Cell Rep 25:660–667

Frewen BE, Chen THH, Howe GT, Davis J, Rohde A, Boerjan W, Bradshaw HD (2000) Quantiative trait loci and candidate gene mapping of bud set and bud flush in *Populus*. Genetics 154:837–845

Ganal MW, Simon R, Brommonschenkel S, Arndt M, Phillips MS, Tanksley SD, Kumar A (1995) Genetic mapping of a wide spectrum nematode resistance gene (Hero) against Globodera rostochiensis in tomato. Mol Plant Micr Interact 8:886–891

Graf J (1921) Beitrage zur Kenntnis der Gattung Populus. Beih Bot Central 38:405–434

Grattapaglia D, Sederoff R (1994) Genetic linkage maps of *Eucalyptus grandis* and *Eucalyptus urophylla* using a pseudo-testcross: mapping stategy and RAPD markers. Genetics 137:1121–1137

Grattapaglia D, Bertolucci FLG, Penchel R, Sederoff RR (1996) Genetic mapping of quantitative trait loci controlling growth and wood quality traits in *Eucalyptus grandis* using a maternal half-sib family and RAPD markers. Genetics 144:1205–1214

Greenaway W, Whatley FR (1990) Analysis of phenolics of bud exudate of *Populus angustifolia* by GC-MS. Phytochemistry 29:2551–2554

Gresham D, Ruderfer DM, Pratt SC, Schacherer J, Dunham MJ, Botstein D, Kruglyak L (2006) Genome-wide detection of polymorphisms at nucleotide resolution with a single DNA microarray. Science 311:1932–1936

Groover A, Fontana JR, Dupper G, Ma CP, Martienssen R, Strauss S, Meilan R (2004) Gene and enhancer trap tagging of vascular-expressed genes in poplar trees. Plant Physiol 134:1742–1751

Gupta P, Duplessis S, White H, Karnosky DF, Martin F, Podila GK (2005) Gene expression patterns of trembling aspen trees following long-term exposure to interacting elevated CO_2 and tropospheric O_3. New Phytol 167:129–142

Hamzeh M, Dayanandan S (2004) Phylogeny of *Populus* (Salicaceae) based on nucleotide sequences of chloroplast TRNT-TRNF region and nuclear rDNA Am J Bot 91:1398–140

Han KH, Gordon MP, Strauss SH (1997) High-frequency transformation of cottonwoods (genus *Populus*) by *Agrobacterium rhizogenes*. Can J For Res 27:464–470

Harald H, Goring H, Terwilliger JD (2000) Linkage analysis in the presence of errors. II. Marker-locus genotyping errors modelled with hypercomplex recombination fractions. Am J Hum Genet 66:1107–1118

Harrison JWH (1924) A preliminary account of the chromosomes and chromosome behavior in the Salicaceae. Ann Bot 38:361–378

Hazen SP, Kay SA (2003) Gene arrays are not just for measuring gene expression. Trends Plant Sci 8:413–416

Hinckley TM, Ceulemans R, Dunlap JM, et al (1989) Physiological, morphological, and anatomical components of hybrid vigor in *Populus*. In: Krebb KH, Richter H, Hinckley TM (eds) Structural and Funtional Responses to Environmental Stresses. SPB, The Hague, pp 119–217

Heilman PE, Stettler RF (1985) Genetic variation and productivity of *Populus trichocarpa* and its hybrids. II. Biomass production in a 4-year plantation. Can J For Res 15:384–388

Hillier J, Coombes AJ (eds) (2002) The Hillier Manual of Trees and Shrubs. David & Charles, Devon, UK

Hulbert SH, Ilott TW, Legg EJ, Lincoln SE, Lander ES, Michelmore RW (1998) Genetic analysis of the fungus, *Bremia lactucae*, using restriction fragment length polymorphisms. Genetics 120:947–958

Hyun JO, Rajora OP, Zsuffa L (1987) Inheritance and linkage of isozymes in *Populus tremuloides* (Michx). Genome 29:384–388

Jackson RC (1985) Genomic differentiation and its effect on gene flow. Syst Bot 10:391–404

Jansen RC, Stam P (1994) High resolution of quantitative traits into multiple loci via interval mapping. Genetics 136:1447–1455

Jepsen E (1994) Ozone and acid deposition gradients and biomonitoring site selection in Wisconsin. In: Proc 16th Int Meeting on Air Pollution Effects on Forest Ecosystems, Fredericton Canada, 7-9 Sept 1994, p 23

Jing ZP, Gallardo F, Pascual MB, Sampalo R, Romero J, de Navarra AT, Canovas FM (2004) Improved growth in a field trial of transgenic hybrid poplar overexpressing glutamine synthetase. New Phytol 164:137–145

Jorge V, Dowkiw A, Faivre-Rampant P, Bastien C (2005) Genetic architecture of qualitative and quantitative Melampsora larici-populina leaf rust resistance in hybrid poplar: genetic mapping and QTL detection. New Phytol 167:113–127

Karagyozov L, Kalcheva ID, Chapman VM (1993) Construction of random small-insert genomic libraries highly enriched for simple sequence repeats. Nucleic Acids Res 21:3911–3912

Kearsey MJ, Farquhar AGL (1998) QTL analysis in plants; where are we now? Heredity 80:137–142

Keim P, Paige KN, Whitham TG, Lark KG (1989) Genetic analysis of an interspecific hybrid swarm of Populus: occurrence of uni-directional introgression. Genetics 123:557–565

Kirst M, Basten CJ, Myburg AA, Zeng Z-B, Sederoff RR (2005) Genetic architecture of transcript-level variation in differentiating xylem of a eucalyptus hybrid. Genetics 169:2295–2303

Knott SA, Neale DB, Sewell MM, Haley CS (1997) Multiple marker mapping of quantitative trait loci in an outbred pedigree of loblolly pine. Theor Appl Genet 94:810–820

Kohler A, Delaruelle C, Martin D, Encelot N, Martin F (2003) The poplar root transcriptome: analysis of 7000 expressed sequence tags. FEBS Lett 542:37–41

Korol AB, Ronin YI, Nevo E (1998) Approximate analysis of QTL-Environment Interaction with no limits on the number of environments. Genetics 148:2015–2028

Kuang H, Richardson T, Carson S, Wilcox P, Bongarten B (1999) Genetic analysis of inbreeding depression in plus tree 850.55 of Pinus radiata D. Don. I. Genetic map with distorted markers. Theor Appl Genet 98:697–703

Lincoln S, Lander E (1992) Systematic detection of errors in genetic linkage data. Genomics 14:604–610

Lincoln S, Daly M, Lander E (1992) Constructing genetic maps with MAPMARKER/EXP 3.0, Whitehead Institute Technical report, 3rd edn

Liu Z, Dickmann DI (1992) Responses of two hybrid *Populus* clones to flooding, drought, and nitrogen availability. II. Gas exchange and water relations. Can J Bot 71:927–938

Liu Z, Furnier GR (1993) Inheritance and linkage of allozymes and RFLPs in trembling aspen. J Hered 84:419–424

Liu Z, Dickmann DI (1996) Effects of water and nitrogen interaction on net photosynthesis, stomatal conductance and water-use efficiency in two hybrid poplar clones. Physiol Planta 97 (3):507-512

Mazzoleni S, Dickmann DI (1988) Differential physiological and morphological responses of two hybrid *Populus* clones to water stress. Tree Physiol 4:61–70

Meyer S, Nowak K, Sharma VK, Schulze J, Mendel RR, Hansch R (2004) Vectors for RNAi technology in poplar. Plant Biol 6:100–103

Moreau C, Aksenov N, Lorenzo M, Segerman B, Funk C, Nilsson P, Jansson S, Tuominen H (2005) A genomic approach to investigate developmental cell death in woody tissues of *Populus* trees. Genome Biol 6:R34

Morreel K, Goeminne G, Storme V, Sterck L, Ralph J, Coppieters W, Breyne P, Steenackers M, Georges M, Messens E, Boerjan W (2006) Genetical metabolomics of flavonoid biosynthesis in Populus: a case study. Plant J 47:224–237

Murphy RW, Sites JW, Buth DG, Haufler CH (1990) Proteins I: isozyme electrophoresis. In: Hillis DM, Moritz (eds) Molecular Systematics. Sinauer, Sunderland, pp 45–126

Newcombe G, Bradshaw HD (1996) Quantitative trait loci conferring resistance in hybrid poplar to *Septoria populicola* the cause of leaf spot. Can J For Res 26:1943–1950

Newcombe G (1998) Association of *Mmd1*, a majpor gene for resistance to *Melampsora medusae* f.sp. *deltoidae*, with quantitative traits in poplar rust. Phytopathology 88:114–121

Nilsson O (2002) The Swedish *Populus* EST project. In: Plant, Animal and Microbe Genomes X Conf, San Diego, W127

Nowak K, Luniak N, Meyer S, Schulze J, Mendel RR, Hansch R (2004) Fluorescent proteins in poplar: a useful tool to study promoter function and protein localization. Plant Biol 6:65–73

Pauley SS, Perry TO (1954) Ecotypic variation of the photoperiodic response in *Populus.* J Arnold Arbor 35:167–188

Plomion C, O'Malley DM, Durel CE (1995) Genomic analysis in maritime pine (Pins -pinaster) – comparison of 2 RAPD maps using selfed and open-pollinated seeds of the same individual. Theor Appl Genet 90:1028–1034

Plomion C, O'Malley DM (1996) Recombination rate differences for pollen parents and seed parents in Pinus pinaster. Heredity 77:341–350

Rae AM, Robinson KM, Street NR, Taylor G (2004) Morphological and physiological traits influencing biomass productivity in short rotation coppice poplar. Can J For Res 34:1488–1498

Rae AM, Ferris R, Tallis MJ, Taylor G (2006) Elucidating genomic regions determining enhanced leaf growth and delayed senescence in elevated CO_2. Plant Cell Environ 29:1730–1741

Rajora O (1989) Characterization of 43 *Populus nigra* L. clones representing selections, cultivars and botanical varieties based on their multilocus allozyme genotypes. Euphytica 43:197–206

Ralph S, Oddy C, Cooper D, Yueh H, Jancsik S, Kolosova N, Philippe RN, Aeschliman D, White R, Huber D, Ritland CE, Benoit F, Rigby T, Nantel A, Butterfield YSN, Kirkpatrick R, Chun E, Liu J, Palmquist D, Wynhoven B, Stott J, Yang G, Barber S, Holt RA, Siddiqui A, Jones SJM, Marra MA, Ellis BE, Douglas CJ, Ritland K, Bohlmann J (2006) Genomics of hybrid poplar (*Populus trichocarpa* × *deltoides*) interacting with forest tent caterpillars (*Malacosoma disstria*): normalized and full-length cDNA libraries, expressed sequence tags, and a cDNA microarray for the study of insect-induced defences in poplar. Mol Ecol 15:1275–1297

Ranjan R, Kao Y, Jiang H, Joshi CP, Harding SA, Tsai C (2004) Suppression subtractive hybridization-mediated transcriptome analysis from multiple tissues of aspen (*Populus tremuloides*) altered in phenylpropanoid metabolism. Planta 219:694–704

Reich PB (1984) Leaf stomatal density and diffusive conductance in 3 amphistomatous hybrid poplar cultivars. New Phytol 98:231–240

Remington DL, Whetten RW, Lui BH, O'Malley DM (1999) Construction of an AFLP genetic map with nearly complete genome coverage in *Pinus taeda.* Theor Appl Genet 98:1279–1292

Ridolfi M, Fauveneau P, Label P, Garrec JP, Dreyer E (1996) Responses to water stress in an ABA-unresponsive hybrid poplar (*Populus koreana* × *trichocarpa* cv. 'Peace'). New Phytol 134:445–454

Rogers DL, Stettler RF, Heilman PE (1989) Genetic variation and productivity of *Populus trichocarpa* and its hybrids. III. Structure and pattern of variation in a 3-year field test. Can J For Res 19:372–377

Rottmann WH, Meilan R, Sheppard LA, Brunner AM, Skinner JS, Ma C, Cheng S, Jouanin L, Pilate G, Strauss SH (2000) Diverse effects of overexpression of LEAFY and PTLF, a poplar (*Populus*) homolog of LEAFY/FLORICAULA, in transgenic poplar and Arabidopsis. Plant J 22:235–245

Rowland DL, Beals L, Chaudhry AA, Evans AS, Grodeska L (2001) Physiological, Morphological, and environmental variation among geographically isolated cottonwood (*Populus deltoides*) populations in New Mexico. Western North Am Nat 61 (4):452-462

Schrader J, Moyle R, Bhalerao R, Hertzberg M, Lundeberg J, Nilsson P, Bhalerao RP (2004) Cambial meristem dormancy in trees involves extensive remodelling of the transcriptome. Plant J 40:173–187

Schulte PJ, Hinckley TM (1987) The relationship between guard cell water potential and the aperture of stomata in *Populus.* Plant Cell Environ 10 (4):313-318

Schulte PJ, Hinckley TM, Stettler RF (1987) Stomatal responses of *Populus* to leaf water potential. Can J Bot 65:255–260

Seaton G, Haley CS, Knott SA, Kearsey MJ, Visscher PM (2002) QTL Express: mapping quantitative trait loci in simple and complex pedigrees. Bioinformatics 18:339–340

Singer T, Fan Y, Chang H-S, Zhu T, Hazen SP, Briggs SP (2006) A high-resolution map of Arabidopsis recombinant inbred lines by whole-genome exon array hybridization. Plos Genet 2:e144

Smith EC (1943) A study of cytology and speciation in the genus *Populus* L. J Arnold Arbor 24:275–305

Smith CM, Rodriguez-Buey M, Karlsson J, Campbell MM (2004) The response of the poplar transcriptome to wounding and subsequent infection by a viral pathogen. New Phytol 164:123–136

Smulders MJM, van der Schoot J, Arens P, Vosman B (2001) Trinucleotide repeat microsatellite markers for black poplar (*Populus nigra* L.). Mol Ecol Notes 1:188–190

Sterky F, Bhalerao RR, Unneberg P, Segerman B, Nilsson P, Brunner AM, Charbonnel-Campaa L, Lindvall JJ, Tandre K,

Strauss SH, Sundberg B, Gustafsson P, Uhlen M, Bhalerao RP, Nilsson O, Sandberg G, Karlsson J, Lundeberg J, Jansson S (2004) A *Populus* EST resource for plant functional genomics. Proc Natl Acad Sci USA 101:13951–13956

Stettler RF, Fenn RC, Heilman PE, Stanton BJ (1988) *Populus trichocarpa* × *P. deltoides* hybrids for short rotation culture: variation patterns and 4-year field performance. Can J For Res 18:745–753

Stettler RF, Bradshaw Jr HD, Heilman PE, Hinckley TM (1996) Biology of *Populus* and its Implications for Management and Conservation. NRC Research Press, Ottawa, ON, Canada

Stirling B, Newcombe G, Vrebalov J, Bosdet I, Bradshaw HD (2001) Suppressed recombination around the MXC3 locus, a major gene for resistance to poplar leaf rust. Theor Appl Genet 103:1129–1137

Stirling B, Yang ZK, Gunter LE, Tuskan GA, Bradshaw Jr HD (2003) Comparative sequence analysis between orthologous regions of the *Arabidopsis* and *Populus* genomes reveals substantial synteny and microcollinearity. Can J For Res 33:2245–2251

Strauss SH, Lande R, Namkoong G (1992) Limitations of molecular-marker-aided selection in forest tree breeding. Can J For Res 22:1050–1061

Strauss SH, Brunner AM, Busov VB, Ma C, Meilan R (2004) Ten lessons from 15 years of transgenic *Populus* research. Forestry 77:455–465

Street NR, Skogstrom O, Sjodin A, Tucker J, Rodriguez-Acosta M, Nilsson P, Jansson S, Taylor G (2006) The genetics and genomics of the drought response in *Populus*. Plant J 48:321–341

Taylor G (2002) *Populus*: *Arabidopsis* for Forestry. Do we need a model tree? Ann Bot 90:681–689

Taylor G, Street NR, Tricker PJ, Sjodin A, Graham L, Skogstrom O, Calfapietra C, Scarascia-Mugnozza G, Jansson S (2005) The transcriptome of *Populus* in elevated CO_2. New Phytol 143:1469–8137

Tenhoopen R, Robbins TP, Fransz PF, Montijn BM, Oud O, Gerats AGM, Nanninga N (1996) Localization of T-DNA insertions in petunia by fluorescence in situ hybridization: physical evidence for suppression of recombination. Plant Cell 8:823–830

Tschaplinski TJ, Blake TJ (1989a) Correlation between early root production, carbohydrate metabolism, and subsequent biomass production in hybrid poplar. Can J Bot 67:2168–2174

Tschaplinski TJ, Blake TJ (1989b) Water relations, photosynthetic capacity, and root/shoot partitioning of photosynthate as determinant of productivity in hybrid poplar. Can J Bot 67:1689–1697

Tschaplinski TJ, Tuskan GA, Gunderson CA (1994) Water-stress tolerance of black and eastern cottonwood clones and four hybrid progeny. I. Growth, water relations, and gas exchange. Can J For Res 24:364–371

Tschaplinski TJ, Gebre GM, Shirshac TL (1998a) Osmotic potential of several hardwood species as affected by manipulation of troughfall precipitation in an upland oak forest during a dry year. Tree Physiol 18 (5):291-298

Tschaplinski TJ, Tuskan GA, Gebre GM, Todd DE (1998b) Drought resistance of two hybrid *Populus* clones grown in a large-scale plantation. Tree Physiol 18:645–652

Tschaplinski TJ, Tuskan GA, Gebre GM, Todd DE, Bradshaw HD (1999) Limits of Drought Tolerance in *Populus*. Annual Report. Biofuels Feedstock Development Program. Department of Energy. Oak Ridge National Laboratory, Oak Ridge, TN

Tschaplinski TJ, Tuskan GA, Sewell MM, Gebre GM, Todd DE, Pendley CD (2006) Phenotypic variation and quantitative trait locus identification for osmotic potential in an interspecific hybrid inbred F_2 poplar pedigree in contrasting environments. Tree Physiol 26:595–604

Tuskan GA (1998) Short rotation woody crop supply systems in the United States: what do we know and what do we need to know? Biomass Bioenergy 14:307–315

Tuskan GA, Wullschleger SD, Bradshaw HD, Dalham RC (2002) Sequencing the *Populus* genome: Applications to the energy-related missions of DOE. In: Plant, Animal and Microbe Genomes X Conf, San Diego, W128

Tuskan GA (2003) Limits of Drought Tolerance in *Populus*. US Department of Energy (DOE), Office of Fu els Development trough Oak Ridge National Laboratory. http://bioenergy.ornl.gov/doeofd/96_97summaries/woody.html

Tuskan GA, Gunter LE, Yang ZK, Yin TM, Sewell MM, Difazio SP (2004) Characterisation of micrsatellites revealed by genomic sequencing of *Populus trichoarpa*. Can J For Res 34:85–93

Tuskan GA, DiFazio S, Jansson S, Bohlmann J, Grigoriev I, Hellsten U, Putnam N, Ralph S, Rombauts S, Salamov A, Schein J, Sterck L, Aerts A, Bhalerao RR, Bhalerao RP, Blaudez D, Boerjan W, Brun A, Brunner A, Busov V, Campbell M, Carlson J, Chalot M, Chapman J, Chen GL, Cooper D, Coutinho PM, Couturier J, Covert S, Cronk Q, Cunningham R, Davis J, Degroeve S, Dejardin A, dePamphilis C, Detter J, Dirks B, Dubchak I, Duplessis S, Ehlting J, Ellis B, Gendler K, Goodstein D, Gribskov M, Grimwood J, Groover A, Gunter L, Hamberger B, Heinze B, Helariutta Y, Henrissat B, Holligan D, Holt R, Huang W, Islam-Faridi N, Jones S, Jones-Rhoades M, Jorgensen R, Joshi C, Kangasjarvi J, Karlsson J, Kelleher C, Kirkpatrick R, Kirst M, Kohler A, Kalluri U, Larimer F, Leebens-Mack J, Leple JC, Locascio P, Lou Y, Lucas S, Martin F, Montanini B, Napoli C, Nelson DR, Nelson C, Nieminen K, Nilsson O, Pereda V, Peter G, Philippe R, Pilate G, Poliakov A, Razumovskaya J, Richardson P, Rinaldi C, Ritland K, Rouze P, Ryaboy D, Schmutz J, Schrader J, Segerman B, Shin H, Siddiqui A, Sterky F, Terry A, Tsai CJ, Uberbacher E, Unneberg P, Vahala J, Wall K, Wessler S, Yang G, Yin T, Douglas C, Marra M, Sandberg G, Van de

Peer Y, Rokhsar D (2006) The Genome of black cottonwood, *Populus trichocarpa* (Torr. & Gray). Science 313:1596–1604

US Department of Energy (2002) Bioenergy Feedstock Development Program at Oak Ridge National Laboratory. USA P4. http://bioenergy.ornl.gov/forum/94summer.html

US Environmental Protection Agency (1999) Biological Aspects of Hybrid Poplar Cultivation on Floodplains in Western North America – A Review. 9EPA Document No 910-R-99-002

Van Der Schoot J, Pospiskova M, Vosman B, Smulders MJM (2000) Development and characterization of microsatellite markers in black poplar (*Populus nigra* L.). Theor Appl Genet 101:317–322

Van Dillewijin C (1940) Zytologische studien in der gattung Populus. Genetica 22:131–182

Villar M, Lefevre F, Bradshaw HD, duCros ET (1996) Molecular genetics of rust resistance in poplars (Melampsora laricipopulina Kleb *Populus* sp.) by bulked segregant analysis in a 2x2 factorial mating design. Genetics 143:531–536

Vos P, Hogers R, Bleeker M, Reijens M, Van De Lee T, Hornes M, Frijters A, Pot J, Peleman J, Kuiper M, Zabeau M (1995) AFLP: a new technique for DNA fingerprinting. Nucleic Acids Res 23:4407–4414

Weber JC, Stettler RF, Heilman PE (1985) Genetic variation and productivity of *Populus trichocarpa* and its hybrids. I. Morphology and Phenology of 50 native clones. Can J For Res 15:376–383

Weigel D, Nilsson O (1995) A developmental switch sufficient for flower initiation in diverse plants. Nature 377:495–500

West MAL, van Leeuwen H, Kozik A, Kliebenstein DJ, Doerge RW, St Clair DA, Michelmore RW (2006) High-density haplotyping with microarray-based expression and single feature polymorphism markers in Arabidopsis. Genome Res 16 787-795

Williams CG, Goodman MM, Stuber CW (1995) Comparative combination distances among Zea Mays L. inbreds, wide crosses and interspecific hybrids. Genetics 141:1573–1581

Wu R, Bradshaw HD, Stettler RF (1997) Molecular genetics of growth and development in *Populus* (Salicaceae). 5. Mapping quantitative trait loci affecting leaf variation. Am J Bot 84:143–153

Wu R, Stettler RF (1998) Quantitative genetics of growth and development in *Populus*. III. Phenotypic plasticity of crown structure and function. Heredity 81:299–310

Wu RL, Han YF, Hu JJ, Fang JJ, Li L, Li ML, Zeng ZB (2000) An integrated genetic map of *Populus deltoides* based on amplified fragment length polymorphisms. Theor Appl Genet 100:1249–1256

Wu R, Ma CX, Yang MCK, Chang M, Littell RC, Santra U, Wu S, Yin TM, Huang M, Wang M, Casella G (2003) Quantitative trait loci for growth trajectories in *Populus*. Genet Res 81:51–64

Wullschleger SD, Jansson S, Taylor G (2002) Genomics and forest biology: Populus emerges as the perennial favorite. Plant Cell 14:2651–2655

Wullschleger SD, Yin TM, Difazio SP, Tschaplinski TJ, Gunter LE, Davis MF, Tuskan GA (2005) Phenotypic variation in growth and biomass distribution for twoadvanced-generation pedigrees of hybrid poplar. Can J For Res 35:1779–1789

Wyckoff G, Zasada J (2003) *Populus* L. URL http:/wpsm.net/Populus.pdf . March 18

Yin TM, Huang MR, Wang MX, Zhu LH, Zeng ZB, Wu RL (2001) Preliminary interspecific genetic maps of the Populus genome constructed from RAPD markers. Genome 44:602–609

Yin TM, Zhang XY, Huang MR, Wang MX, Zhuge Q, Zhu LH, Wu RL (2002) Molecular linkage maps of the Populus genome. Genome 45:541–555

Yin TM, DiFazio SP, Gunter LE, Riemenschneider D, Tuskan GA (2004) Large-scale heterospecific segregation distortion in *Populus* revealed by a dense genetic map. Theor Appl Genet 109:451–463

Zamir D, Tadmor Y (1986) Unequel segregation of nuclear genes in plants. Bot Gaz 147:355–358

Zamir D, Tanksley SD, Jones RA (1982) Haploid selection for low-temperature tolerance of tomato pollen. Genetics 101:129–137

Zeng Z (1993) Therorectical basis for separation of multiple linked gene effects in mapping quantitative trait loci. Proc Natl Acad Sci USA 90:10972–10976

Zhang J, Steenackers M, Storme V, Neyrinck S, Van Montagu M, Gerats T, Boerjan W (2001) Fine mapping and identification of nucleotide binding site/leucine-rich repeat sequences at the MER locus in Populus deltoides 'S9-2'. Phytopathology 91:1069–1073

Zhang D, Zhang Z, Yang K, Li B (2004) Genetic mapping in (*Populus tomentosa* × *Populus bolleana*) and *P. tomentosa* Carr. using AFLP markers. Theor Appl Genet 108:657–662

Zsuffa L, Giordano E, Pryor LD, Stettler RF (1996) Trends in poplar culture: some global and regional perspectives. In: Stettler RF, Bradshaw HD, Jr, Heilman PE, Hinckley TM (eds) Biology of *Populus* and its Implications for Management and Conservation. NRC Research Press, Ottawa, ON, Canada, pp 515–539

2 Pines

C. Plomion[1], D. Chagné[2], D. Pot[3], S. Kumar[4], P. L. Wilcox[5], R. D. Burdon[4], D. Prat[6], D. G. Peterson[7], J. Paiva[1], P. Chaumeil[1], G. G. Vendramin[8], F. Sebastiani[9], C. D. Nelson[10], C. S. Echt[10], O. Savolainen[11], T. L. Kubisiak[10], M. T. Cervera[12], N. de María[12], and M. N. Islam-Faridi[13]

[1] INRA, UMR BIOGECO, 69 route d'Arcachon, 33610 Cestas, France
e-mail: plomion@pierroton.inra.fr
[2] HortResearch, Private Bag 11030, Palmerston North, New Zealand
[3] Coffee Genomic Team, CIRAD, UMR PIA 1096, 34398, Montpellier cedex 5, France
[4] Ensis-Genetics, Ensis, Private Bag 3020, Rotorua, New Zealand
[5] Cellwall Biotechnology Centre, Scion, Private Bag 3020, Rotorua, New Zealand
[6] Université Claude Bernard - Lyon 1, EA 3731 Génome et Evolution des Plantes Supérieures, Bât F.A. Forel, 6 rue Raphaël Dubois, 69622, Villeurbanne Cedex, France
[7] Department of Plant and Soil Sciences, 117 Dorman Hall, Box 9555, Mississippi State University, Mississippi State, MS 39762, USA
[8] Plant Genetics Institute, National Research Council, Via Madonna del Piano 10, 50019 Sesto Fiorentino, Florence, Italy
[9] Department of Agricultural Biotechnology, University of Florence, Via della Lastruccia 14, 50019 Sesto Fiorentino, Florence, Italy
[10] USDA - Forest Service, Southern Research Station, Southern Institute of Forest Genetics, 23332 Mississippi 67, Saucier, MS 39574, USA
[11] Department of Biology, PL 3000, 90014, University of Oulu, Finland
[12] CIFOR - INIA, Genética y Ecofisiología Forestal & Unidad Mixta de Genómica y Ecofisiología Forestal INIA-UPM, Departamento de Sistemas y Recursos Forestales, Carretera de La Coruña km 7, 28040 Madrid, Spain
[13] USDA-Forest Service, Southern Research Station, Southern Institute of Forest Genetics, Forest Tree Molecular Cytogenetics Laboratory, Texas A&M University, College Station, TX 77843, USA

2.1 Introduction

2.1.1 History of the Genus

Origin and Distribution

Pinus is the most important genus within the Family *Pinaceae* and also within the gymnosperms by the number of species (109 species recognized by Farjon 2001) and by its contribution to forest ecosystems. All pine species are evergreen trees or shrubs. They are widely distributed in the northern hemisphere, from tropical areas to northern areas in America and Eurasia. Their natural range reaches the equator only in Southeast Asia. In Africa, natural occurrences are confined to the Mediterranean basin. Pines grow at various elevations from sea level (not usual in tropical areas) to highlands. Two main regions of diversity are recorded, the most important one in Central America (43 species found in Mexico) and a secondary one in China. Some species have a very wide natural range (e.g., *P. ponderosa*, *P. sylvestris*). Pines are adapted to a wide range of ecological conditions: from tropical (e.g., *P. merkusii*, *P. kesiya*, *P. tropicalis*), temperate (e.g., *P. pungens*, *P. thunbergii*), and subalpine (e.g., *P. albicaulis*, *P. cembra*) to boreal (e.g., *P. pumila*) climates (Richardson and Rundel 1998, Burdon 2002). They can grow in quite pure stands or in mixed forest with other conifers or broadleaved trees. Some species are especially adapted to forest fires, e.g., *P. banksiana*, in which fire is virtually essential for cone opening and seed dispersal. They can grow in arid conditions, on alluvial plain soils, on sandy soils, on rocky soils, or on marsh soils. Trees of some species can have a very long life as in *P. longaeva* (more than 3,000 years).

Botanical Descriptions

The genus is distinguished from other members of the Pinaceae family by its needlelike secondary leaves, borne commonly in fascicles of 1 to 8 on dwarf shoots, with a fascicle sheath of bud scales. The leaves of pines are of four types encompassing the complete plant development: cotyledons, juvenile leaves, scale leaves

Genome Mapping and Molecular Breeding in Plants, Volume 7
Forest Trees
C. Kole (Ed.)

(cataphylls), and secondary leaves. Cotyledons vary in number from 4 up to 24 in *P. maximartinezii*. There are up to three resin ducts in the cotyledons and either one or two vascular bundles (Farjon 1984). Primary leaves are single, generally helically arranged, and acicular, and they are produced in most species only during the first growth season but for a longer period in a few species. Cataphylls, the nonchlorophyllous primary leaves produced on shoots, occur in an extension of the helical arrangement of the primary leaves and subtend all shoot structures, but they are typically small and subulate or lanceolate. Secondary leaves, the needles, appear by the end of the first growing season, or later in some species. They are the most common pine leaves, permanently green, and metabolically active, ranging in length from 2 to 50 cm (generally smaller ones occur in subalpine or aridity-adapted species). They are borne on dwarf shoots axillary to cataphylls, in fascicles of one (*P. monophylla*, with circular section needles) to eight needles, usual numbers per fascicle being two, three, or five. A fascicle is initially bound together by a basal sheath that may then fall off or persist, but actual leaf fall involves the entire fascicle structure. The number of leaves in the fascicles determines the transverse leaf shape (Farjon and Styles 1997). Stomata are arranged in several longitudinal lines along the entire leaf length. There are generally two or more resin ducts in needles. The number of vascular bundles (one or two as in cotyledons) is the major trait for the identification of the main divisions of genus *Pinus*. The trunk is usually single, erect, and columnar. The branches are grouped into pseudowhorls (often called clusters), at least when young. Bark patterns in pines result mainly from fissuring due to expansion growth and to the formation of scales that eventually fall off.

Pine species are monoecious. Pollen cones are relatively small and soft; these ephemeral structures consist of an axis with many helically arranged microsporophylls. Two microsporangia are attached to the underside of each microsporophyll. These cones open in spring, at least for temperate-climate species, and release large quantities of pollen into the air. Pollen is of the bisaccate type giving it great buoyancy in the air. Pine pollen can be blown over long distances. Its morphology is very similar for all pine species. Seed conelets are found in most species at the ends of new twigs, taking the position of a lateral bud. They are usually located on the higher branches. They consist of scales, the megasporophylls, arranged around an axis; on each megasporophyll lie two separate ovules, each consisting of a cell mass protected by an integument. The micropyle through which the pollen tube penetrates is turned toward the axis. After pollination the seed cone closes its scales by expansion. The seed cone then grows rapidly. Pollen germinates and produces the male gametophyte, with two sperm nuclei. Fertilization takes place later, about one year after pollination in temperate pine species. Seed cone maturation requires one growing season after fertilization for many species, and even a third for some species. In tropical climates the cycle can be shortened because of the lack of winter dormancy. Individual seed cones thus persist for 2, up to even 3, years on the same tree for most pine species. Seeds contain an embryo embedded in the remaining megagametophyte and the seed coat. The seed wings derive from tissue on the adaxial face of the seed scale. Even in species with vestigial or absent wings, remnants of basal wing tissue are present on the seed scale, on the seed, or on both. Seeds are mostly wind dispersed. In some species, birds are important seed-dispersal vectors (*P. albicaulis*). Seed cones are serotinous in some species and open only following exposure to fire (*P. banksiana*).

Systematics and Phylogeny

The genus is divided into subgenera, sections, and subsections. Various classifications have been proposed in this genus since Linnaeus. Recent ones obtain support from DNA phylogenetics to identify related species. Many phylogenetic studies have been carried out in pines. Some of the first studies involved restriction patterns of the chloroplast genome (Strauss and Doerksen 1990; Govindaraju et al. 1992; Krupkin et al. 1996). More recent classifications, including a large number of species, were established from nuclear sequences (ITS, Liston et al. 1999) and chloroplast sequences (*rbcL*: Gernandt et al. 2005), the chloroplast genome being paternally inherited in the genus *Pinus* (Neale and Sederoff 1989). Some studies have also focused on subsets of the genus *Pinus*: subgenus *Pinus* (Geada López et al. 2002), section *Parrya* (Gernandt et al. 2003), and Eurasian species (Wang et al. 1999). Comprehensive classifications of the genus *Pinus* were earlier established by Gaussen (1960) and Van der Burgh (1973) using morphological and anatomical traits. Later, Price et al. (1998) and Gernandt et al. (2005) also included molecular data and identified monophyletic subgenera, sections, and subsections. Some features are consistent, but variations are noticed between the classical and molecular approaches. The main division into two subgenera according to the number (one or two) of leaf vascular bundles has been recognized by these authors with various subgenera names (*Haploxylon* and *Diploxylon*, sometimes called, respectively, soft pines and hard pines, and, more recently, *Strobus* and *Pinus* named from type species as recommended by botanical nomenclature code; http://tolweb.org/tree?group=Pinus&contgroup=Pinaceae). The taxonomic position of a singular species with flat secondary leaves *P. krempfii* is not fully agreed. It has been considered as a third monospecific subgenus by Gaussen (1960), while molecular data place it as a member of the subgenus *Strobus* (Wang et al. 2000; Gernandt et al. 2005). Most species belong to the subgenus *Pinus*. Subgenus *Pinus* species are characterized by thick seed cone scales and persistent fascicle sheaths. The numerous sections proposed by Van der Burgh (1973) for this subgenus have subsequently been grouped into two sections *Pinus* and *Trifoliae* (Gernandt et al. 2005), the latter being called New World diploxylon pines by Price et al. (1998). The section *Trifoliae* consists of American species distributed into subsections *Australes* (septal, internal, or medial needle resin ducts), *Ponderosae* (internal or medial needle resin ducts), and *Contortae* (medial needle resin ducts): most of these species are characterized by three-needled fascicles (Table 1).

Each of these subsections groups two or more previously described subsections. *P. leiophylla* and *P. lumholtzii* are now clustered within subsection *Australes* and are not further differentiated. The section *Pinus* is divided into subsections *Pinus* and *Pinaster*. Species within this section, with few exceptions, grow in Eurasia and northern Africa. Subsection *Pinaster*, characterized by the lack of a spine on the umbo of the cone scale, includes *P. pinaster* as the type. More recently, all of the other species included in the section *Pinaster* as defined by Van den Burgh (1973) were found to cluster within the *Australes* subsection of the genus (Gernandt et al. 2005). Species of the subsection *Pinus* including the type species of the genus, *P. sylvestris*, were previously grouped into a section called *Sylvestres* by Van der Burgh (1973). The subgenus *Strobus* has been divided into two sections: *Quinquefoliae* and *Parrya*. They differ from the sections *Strobus* and *Parrya* of Van der Burgh (1973) and Price et al. (1998) by the transfer of the subsections *Krempfianae* and *Gerardianae* from the section *Parrya* into the section *Quinquefoliae*, which also includes the subsection *Strobus*. The subsection *Strobus* consists of species with five-needled fascicles, thin cone scales, terminal position of spines on seed cone, and several other features absent in the subsections *Krempfianae* and *Gerardianae* of Southeast Asia already differentiated by Van der Burgh (1973) and Price et al. (1998). The three subsections of the *Quinquefoliae* section share a deciduous fascicle sheath. The section *Parrya* consists of the subsections *Cembroides*, *Nelsoniae*, and *Balfourianae*; they share an American distribution, the external position of resin ducts (as in subsections *Krempfianae* and *Gerardianae*), and thick cone scales (again as in subsections *Krempfianae* and *Gerardianae*). Subsection *Nelsoniae* shows persistent fascicle sheath not found in other species of subgenus *Strobus*. Most monophyletic groups cannot be identified from unique morphoanatomical traits.

Two of the 11 subsections consist of American species and Eurasian species. Sections *Pinus* and *Quinquefoliae* have an Asiatic origin according to chloroplast data. The subsection *Strobus* lineage would have then evolved in America before coming back to Eurasia. Few dispersal events to eastern North America have probably occurred to explain the presence there of the limited number of species of the subsection *Pinus*. The development and utilization of low-copy-number nuclear genes (Syring et al. 2005) should provide new insights to solve remaining classification problems. Most ancient pine fossils have been dated to the early Cretaceous (Millar 1998). They have been found in China, North America, and Europe (which was very close to eastern North America at that time). They did not further refine the putative geographic origin of genus *Pinus*.

Hybridization

Interspecific hybridization occurs in pines but is limited mostly to related species within a subsection. Some species such as *P. engelmannii*, *P. jeffreyi*, and *P. ponderosa* are compatible in a number of different combinations (Liston et al. 1999). Natural hybridizations are often indicated by the introgression of the paternally inherited chloroplast genome. Barriers occur at different stages, from the failure of pollen germination to failure at embryogenesis (Ledig 1998). A few species are postulated to have been derived from interspecific hybridization. They also exhibit the highly conserved chromosome number in pines ($2n = 24$). This is the situation for *P. densata* that has been shown to combine nuclear polymorphisms of *P. tabuliformis* and *P. yunnanensis* with the chloroplast genome of the latter (Wang et al. 2001) and probably several other

Table 1. Systematics of the genus *Pinus*, according to Gernandt et al. (2005)

Subgenus	Section	Subsection	Number of species	Best-known species	Distribution
Pinus	*Pinus*	*Pinus*	17	*P. sylvestris, P. kesiya, P. merkusii*	Eurasia, North America
		Pinaster	7	*P. pinaster*	Mediterranean, Asia
	Trifoliae	*Contortae*	4	*P. banksiana, P. contorta*	America
		Australes	26	*P. elliottii, P. radiata, P. taeda*	America
		Ponderosae	17	*P. jeffreyi, P. ponderosa*	America
Strobus	*Parrya*	*Balfourianae*	3	*P. balfouriana*	America
		Cembroides	11	*P. cembroides, P. culminicola*	America
		Nelsoniae	1	*P. nelsonii*	Central America
	Quinquefoliae	*Gerardianae*	3	*P. bungeana P. gerardiana*	Asia
		Krempfianae	1	*P. krempfii*	Asia
		Strobus	21	*P. cembra, P. lambertiana, P. strobus*	America, Eurasia

species. *P. densata* exchanged genes with ancestral populations prior to its isolation with local differenciation (Ma et al. 2006). Combinations of parental traits and selection for adaptation to new conditions favored colonization of new territories by the hybrid species.

2.1.2 Cytogenetics, DNA Content, and Genome Composition

Sax and Sax (1933), Mergen (1958), and Khoshoo (1961) were the earliest to describe the karyotypes of various conifer species. They found that species of the genus *Pinus* were diploid with 24 chromosomes ($2n = 2x = 24$). The chromosomal complements generally consist of 10 or 11 pairs of large homobrachial (metacentric) chromosomes and one or two pairs of smaller heterobrachial (submetacentric) chromosomes (Saylor 1961, 1964, 1972, 1983). Several attempts have been made to construct chromosome-specific karyotypes for various pine species using traditional cytogenetics techniques, viz., C-banding, Giemsa, and fluorescent banding (Borzan and Papes 1978; MacPherson and Filion 1981; Drewry 1982; Saylor 1983; Hizume et al. 1989, 1990). More recently, fluorescent in situ hybridization (FISH) has been utilized in several pine species (Doudrick et al. 1995; Lubaretz et al. 1996; Jacobs et al. 2000; Hizume et al. 2002; Liu et al. 2003; Cai et al. 2006). Doudrick et al. (1995) developed a FISH-based karyotype for *P. elliottii* var. *elliottii* using 18S-25S and 5S rDNA probes and CMA (chromomycin A_3) and DAPI (4',6-diamidino-2-phenylindole) banding that distinguished all 12 homologous pairs of chromosomes. They went further to suggest that the presented karyotype might be useful as a standard or reference karyotype for *Pinus*. Lubaretz et al. (1996) used computer-aided chromosome analysis on the basis of chromosome length, chromosome arm length ratio, and the positions of rDNA (18S-28S and 5S) and telomere (*Arabidopsis*-type telomere repeat sequence or *A*-type TRS) detected with FISH to discriminate three chromosomes of *P. sylvestris*. Hizume et al. (2002) used four probes [45S rDNA, 5S rDNA, PCSR (CMA-band specific repeat), and *A*-type TRS] in developing FISH-based karyotypes for four different pine species (*P. densiflora*, *P. thunbergii*, *P. sylvestris*, and *P. nigra*). Liu et al. (2003) and Cai et al. (2006) used FISH to establish rDNA positions in several species of the *Pinus* and *Strobus* subgenera, respectively. Current work in *P. taeda* (subgenera *Pinus*, section *Pinus*, subsection *Australes*) (Fig. 1) emphasizes an improved chromosome preparation technique (based on Jewell and Islam-Faridi 1994 and Islam-Faridi and Mujeeb-Kazi 1995) and statistical analyses of chromosome arm lengths and FISH signal positions and intensities to develop a reference karyotype and cy-

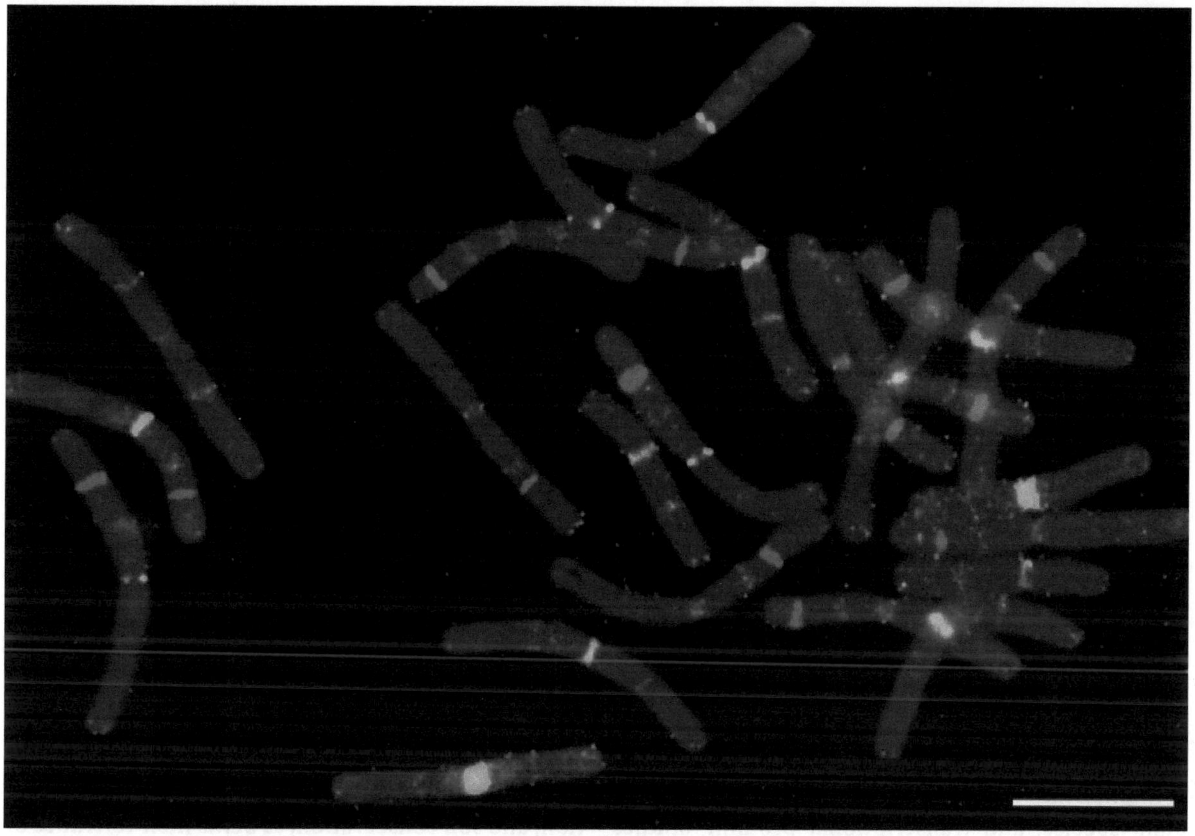

Fig. 1. A fluorescent in situ hybridization (FISH) photomicrograph of *Pinus taeda* metaphase chromosomes showing 23 of the 24 chromosomes (bar = 10 μm). The *red signals* detect the 18S-28S rDNA sites, and the *green signals* detect the *Arabidopsis*-type telomere repeat sequence (*A*-type TRS) sites

togenetic map for use in physical genome mapping in the subsection *Australes* (Islam-Faridi et al. 2003, Islam-Faridi et al. 2007). Comparison of these results with those obtained in other subsections suggests that a subsection-specific karyotype may be required for more robust physical mapping across the entire genus as a whole (Islam-Faridi et al. 2007).

Another feature of the pine genome is its large physical genome size expressed in DNA content. Pines exhibit some of the largest DNA contents per diploid cell in the plant kingdom ranging from ca. 44 pg (*P. banksiana*) to 75 pg (*P. gerardiana*) (1 pg = 960 Mb; Arumuganathan and Earle 1991) based on laser flow cytometry (Grotkopp et al. 2004). For instance, the size of the maritime pine (*P. pinaster*) genome has been estimated to be between 51 and 60 pg/2C (Chagné et al. 2002; Grotkopp at al. 2004), which is about seven times the size of the human genome (7 pg/2C; Morton 1991) and 170-fold larger than the genome of model plant *Arabidopsis thaliana* (0.3 pg/2C). The DNA amount varies according to the subgenus and the section. The genome of subgenus *Pinus* is lower (54.0 pg/2C) than that of subgenus *Strobus* (65.6 pg/2C). This variation has been related to seed mass. Relationships of such variation with ecological conditions, such as drought tolerance, have also been investigated (Wakamiya et al. 1996). Various environmental conditions appear to be related to DNA content, including latitude of range and invasiveness (Grotkopp et al. 2004). An extremely large genome size is common to other gymnosperms (reviewed by Ohri and Khoshoo 1986; Murray 1998; Leitch et al. 2001; Grotkopp at al. 2004).

At the whole-genome level, reassociation kinetics data (i.e., Cot analysis) indicates that 25 to 30% of the pine genome corresponds to low- to single-copy sequences, while 70 to 75% corresponds to highly repeated sequences (Miksche and Hotta 1973; Rake et al. 1980; Kriebel 1985; Peterson et al. 2006). The repetitive sequences of pine have not been studied in much detail. However, it appears that the majority of repetitive DNA consists of repeats of low sequence complexity (Schmidt et al. 2000), retrotransposons (Kamm et al. 1996; Kossack and Kinlaw 1999; Friesen et al. 2001), and 18S-5.8S-25S rDNA genes. As shown by molecular, cytological, and linkage-mapping studies (Friesen et al. 2001; Scotti et al. 2005), most of these repeat sequences are present at multiple loci and are presumably interspersed among other sequences, although they tend to form loose clusters that surround gene-rich islands. The remainder of the genome is composed of low-copy sequences of which an unknown portion are genes. If the pine genome (1C $\sim$ 25 to 30 pg; see above) is similar to Arabidopsis in having about 30,000 expressed genes and an average gene size of 2,000 bp including introns and UTRs (The *Arabidopsis* Genome Initiative 2000), only 0.5% of its genome is likely to be transcribed. In comparison, 54% of the Arabidopsis sequence and 5 to 6% of mammalian genomes are transcribed (Rat Genome Sequencing Project Consortium 2004).

Gene duplication and the formation of complex gene families have been widely cited as a potential cause of the abundance of low-copy DNA in the pine genome. Kinlaw and Neale (1997) suggested that levels of multiplication were greater in conifers than in other plant species, and transcriptional profiling studies have noted surprising levels of transcriptome complexity (Lorenz and Dean 2002). However, considering the relatively low proportion of a conifer genome represented by expressed sequences, it is unlikely that the evolution of multigene families alone can explain the enormous size of pine nuclear genomes. It is possible that a relatively large proportion of low-copy sequences in pine are pseudogenes and/or highly diverged repeat sequences, and indeed there is evidence supporting both possibilities (Elsik and Williams 2000; Rabinowicz et al. 2005).

2.1.3 Economic Importance

Pine species play an especially important role in modern plantation forestry worldwide and now form a large part of both the annual wood harvest and the immature plantation forests that will provide wood in the future (Burdon 2002). Pines enjoy such great popularity because the large number of species allows choice for widely varying site and environmental conditions; the volume of production of some species can be high to very high, even under unfavorable site conditions; they are well suited for reforestation and for simple silviculture (monocultures and clear-felling); their wood is easily processed and utilized for a wide variety of end uses (lumber, pulp and paper, particleboard, etc.); and even if species lack naturally durable heartwood, treatment with preservatives is easy.

Various minor forest products come from pines (Burdon 2002). Some pine plantations, especially *P. pinaster* in the Landes of southwest France, were established largely for resin production. Resin production from pines was also a major economic activity in the southeast of the USA, Mediterranean basin, northern India, and east and Southeast Asia. Resin products are still recovered from chemical pulping of pines. Foliage, in the form of litterfall, has been used for a range of purposes. The bark is used for a variety of purposes, and that of many species has a high tannin content. For a few species with large seeds, notably *P. pinea*, *P. edulis*, and *P. sibirica*, seeds have been a prized item of the human diet. Edible fungi, representing the fruiting bodies of mycorrhizal symbionts, are often collected from pine stands.

Pine species are also used widely for shelter and the interlinked purposes of revegetation, soil stabilization, and soil conservation, as befits their tolerance of exposure and degraded soils in certain species. Most of the widely planted species of pines are used in some degree for one or more of these purposes, often with timber production as a major bonus. Ornamental and festive use is common, with *P. sylvestris* and *P. virginiana* being very widely grown for Christmas trees.

2.1.4 Classical Breeding Objectives

In most pine breeding programs, the setting of goals was originally done subjectively, based on perceptions of the main traits limiting profitability, of the variability and heritability of the traits concerned, and of the genetic correlations (both favorable and adverse) among traits (Burdon 2004). The setting of breeding goals is crucial to any tree-improvement program, and yet most pine tree breeding programs do not have formally defined breeding objectives. Various reasons for this include complexity of the forest processing industry, difficulties in determining the relationships between selection traits and end uses, and long rotation ages that create uncertainty about their use (Apiolaza and Greaves 2001). Earlier attempts of most improvement programs have been to improve growth, form, climatic adaptation, and resistance to certain diseases (Cahalan 1981; Danjon 1995; Paul et al. 1997; Shelbourne et al. 1997).

Most recent efforts have been diverted at improving traits related to pulp and paper and solid-wood products (Shelbourne et al. 1997; Chambers and Borralho 1999; Greaves 1999; Lowe et al. 1999; Pot et al. 2002; Kumar 2004). Wood density and fiber morphology (e.g., tracheid length and coarseness) have been reported to be crucial for pulp yield and quality. Wood properties such as wood density and stiffness are crucial selection traits for the improvement of wood stiffness, while compression wood, spiral grain, and microfibril angle are candidate traits for reducing in-service instability (Shelbourne et al. 1997; Ivkovic et al. 2006). Heartwood, resinous defects, and internal checking are also being considered as selection traits to develop germplasm for appearance-grade products (Shelbourne et al. 1997). There are some situations, especially involving exotic species and disease resistance, etc., where hybrids between selective species are desirable to make genetic improvement in the breeding-objective traits (Hyun 1976; Byun et al. 1989; Blada 1994; Nikles 2000; Shelbourne 2000). Pine species vary widely in their amenability to various forms of vegetative propagation (Hartmann et al. 1990). Ease of propagation was generally a minor criterion for species selection in breeding programs of pines. Currently with improved technology, selection is based more on silvicultural performance and wood quality.

2.1.5 Classical Breeding Achievements

Various selection and deployment strategies are being used for different pine species. Species selection followed by provenance and family-within-provenance selection, and establishment of seed orchards are quite common first steps across various species including *P. radiata* (Falkenhagen 1991; Matziris 1995), *P. oocarpa* (Moura et al. 1998), *P. caribaea* (Zheng et al. 1994), *P. strobus* (Beaulieu et al. 1996), *P. sylvestris* (Quencez and Bastien 2001), *P. pinaster* (Alía et al. 1995, 1997; Danjon 1995), *P. taeda* (Jiang et al. 1999; Lopez-Upton et al. 2000), *P. tecunumanii* (Hodge and Dvorak 1999), and *P. contorta* (Cahalan 1981). In advanced-generation breeding programs, forward selections and/or combined selection (among- and within-family) are the major sources of genetic gain (Wei et al. 1997; Rosvall et al. 1998; Lambeth 2000; Alazard 2001; Olsson et al. 2001; Plomion et al. 2001; Burdon and Kumar 2004). In the species that are easy to propagate, the concept of using clonal replication of individuals within families as a means of genetic testing is being implemented in order to increase the efficiency of genetic improvement. Significant genetic gains from such a strategy have been reported for *P. radiata* (Matheson and Lindgren 1985; Shelbourne 1992) and *P. taeda* (Isik et al. 2004).

Pinus patula, *P. taeda*, and *P. elliottii*, are planted in South Africa in summer rainfall zones, and their average productivity is 15 m^3 ha^{-1} yr^{-1} (Du Toit et al. 1998). Brazil, with its humid summers, largely uses *P. caribaea*, *P. taeda*, and *P. elliottii*. Pines in Brazil produce 8 to 30 m^3 ha^{-1} yr^{-1} on rotations of 20 to 25 years. *P. radiata* is grown as large plantations in the temperate southern hemisphere countries of Australia, Chile, New Zealand, and South Africa. The mean annual increment over 25 years is often 25 to 30 m^3 ha^{-1} yr^{-1} (Lamprecht 1990). However in New Zealand, growth rates of up to 50 m^3 ha^{-1} yr^{-1} have been recorded on the best sites and as low as 11 m^3 ha^{-1} yr^{-1} on very dry sites (Burdon and Miller 1992). The average productivity of *P. pinaster* in southern France is about 10 m^3 ha^{-1} yr^{-1} but could reach 20 to 25 m^3 ha^{-1} yr^{-1} on the best sites. Deployment of genetically improved loblolly pine in the USA has been reported to yield up to about 21 m^3 ha^{-1} yr^{-1} (McKeand et al. 2003). Dhakal et al. (1996) reported a realized gain in volume of about 22% in a slash pine improvement program in the USA. Wood-quality traits that are currently being included, in addition to growth, form, and health traits, in the breeding objectives of various pines species appear to be under moderate to strong genetic control (Burdon and Low 1992; Hannrup et al. 2000; Atwood et al. 2002; Gwaze et al. 2002; Pot et al. 2002; Kumar 2004), and predicted genetic gains from selection appeared to be in the order of about 10% for traits such as wood density and stiffness (Kumar 2004).

Clonal forestry (CLF) represents the large-scale propagation and deployment of selected clones that have been clonally tested. Deployment of tested clones by CLF is being increasingly employed with *P. radiata* in New Zealand (Sorensson and Shelbourne 2005), *P. taeda* in the USA (Stelzer and Goldfarb 1997), and pine hybrids in Australia (Walker et al. 1996). In principle, CLF offers additional genetic gains from capturing nonadditive effects, which are not captured via sexual propagation, plus the benefits of greater uniformity and predictability in performance resulting from a lack of genetic segregation. Genetic gains from CLF have been predicted to be considerably higher than those from family forestry in *P. radiata* (Aimers-Halliday et al. 1997), *P. taeda* (Stelzer and Goldfarb 1997), and *P. strobus* (Park 2002).

2.1.6 Molecular Diversity

The majority of studies aiming to monitor the level and distribution of genetic diversity in *Pinus* natural populations were based on the use of neutral markers. Indeed, molecular markers, such as microsatellites, provided useful information on historical demography and population evolution.

The use of isozyme electrophoresis significantly increased the amount of data on the genetic structure of populations (Petit et al. 2005). These data revealed that pine species had high genetic diversity within populations and only low levels of differentiation among populations. For 28 north temperate pines, genetic differentiation (G_{ST}) averaged 0.076 (Ledig 1998). Exceptions are represented by *P. pinea* (G.G. Vendramin et al. personal communication), *Pinus resinosa*, a species that has a vast range across northeastern North America (Echt et al. 1998; Walter and Epperson 2001), and *P. torreyana* (Ledig and Conkle 1983; Provan et al. 1999) where near absence of variation was observed, and by Mexican pine species where higher differentiation among populations was observed, probably because their natural distributions are more highly fragmented by physiography than those of species at more northerly latitudes. Other pine species with disjunct populations and restricted gene flow also showed higher differentiation among populations: for example, 16 to 27% for *P. radiata* and 22% for *P. muricata* (Wu et al. 1999). On the other hand, experimental evidence indicates that seeds of some pines (e.g., *P. palustris*; Grace et al. 2004) have the potential to disperse greater distances than previously reported, which partly contributes to the low levels of genetic differentiation observed in these species.

In general, the typical distribution of the genetic diversity within and among populations of *Pinus* species is correlated with their mating system and life history (pines, for example, are wind-pollinated and tend to be predominately outcrossing) (Hamrick and Godt 1996; J. Duminil et al. personal communication) and biogeographic history (the distributions of many species have been affected by Pleistocene glacial advances), even if, in some cases, human activities also played a relevant role (e.g., *P. pinaster* in Portugal; Ribeiro et al. 2001).

Self-fertilization in pines generally occurs at a low level (Muona and Harju 1989), and a high outcrossing at the mature seed stage is maintained. There is evidence of selection at the embryonic stage so that the number of inbreds is already low at the seedling stage (Kärkkäinen and Savolainen 1993). Selection after the seedling stage is still severe. This has been shown by observing the survival of selfed seedlings in *P. sylvestris* (Muona et al. 1987) and in *P. leucodermis* (Morgante et al. 1993).

The mode of inheritance has a major effect on the partitioning of genetic diversity in pines, with studies based on maternally inherited markers (transmitted by seeds only) having significantly higher G_{ST} values than those based on paternally or biparentally inherited markers for pine (Burban and Petit 2003; Petit et al. 2005). In fact, the chloroplast and mitochondrial genomes are generally paternally and maternally inherited in pines, respectively (Petit and Vendramin 2006). In pines, G_{ST} is nearly always larger at mitochondrial DNA markers than at chloroplast DNA markers. On the other hand, there is no significant difference between G_{ST} at biparentally inherited markers and at paternally inherited markers in pines (Petit et al. 2005). This is expected considering that both the cpDNA and half the nuclear genomic complement are dispersed by pollen and by seeds, i.e., they use the same vehicles to achieve gene flow.

Chloroplast and mitochondrial DNA markers allow describing phylogeographic structure in pines. While chloroplast DNA generally exhibits the highest diversity, phylogeographic inferences from these markers can be blurred by extensive pollen flow. Low population structure due to extensive pollen flow has been inferred in *P. pinaster* (Burban and Petit 2003), *P. sylvestris* (Robledo-Arnuncio et al. 2005), *P. canariensis* (Gómez et al. 2003), and *P. albicaulis* (Richardson et al. 2002). In contrast, the mitochondrial markers, despite their generally lower level of diversity in pines (but this holds for all conifers; Soranzo et al. 2000; Gugerli et al. 2001), generally provide a clear picture of nonoverlapping areas colonized from different refugia (e.g., in *P. pinaster*; Burban and Petit 2003).

Neutral markers also have allowed investigating spatial genetic structure (SGS) in natural pine populations. A generally weak within-population structure has been described. *P. pinaster* showed a fine-scale structure at the seedling stage with a patch size of ca. 10 m that seems to be produced by restricted seed flow (González-Martínez et al. 2002). Pines with a heavy seed (differences in seed dispersion capability play an important role), such as *P. pinaster*, are expected to have a short dispersal distance, thus producing a fine-scale structure. However, fine-scale structure often does not persist as stands mature. For example, within-population genetic structure in Mediterranean pines may be affected by postdispersal events (e.g., mortality due to the severity of the Mediterranean climate and animal-mediated secondary dispersal during the summer period) that may modify the original spatial structure (González-Martínez et al. 2002). Logging can also play a role in decreasing spatial structuring, as observed in *P. strobus* (Marquardt and Epperson 2004), suggesting that management practices can alter natural spatial patterns, too.

It should be stressed that the presence of fine-scale structure is uncommon in *Pinus* species. Epperson and Allard (1989), studying the spatial pattern of allozyme alleles within *P. contorta* ssp. *latifolia* stands, found a lack of structure in the distributions of most genotypes. Neutral markers by definition do not reflect selective processes and therefore are not used as an indicator of the population adaptive potential to a changing environment (Morin et al. 2004). Recent and well-established markers to detect functional genetic variation are single nucleotide polymorphisms (SNPs), which are particularly useful for finding genes under selection and studying the dynamics of these genes in natural populations.

The availability of high-density markers, such as SNPs, opens the possibility of studying, by association genetics, the molecular basis of complex quantitative traits in natural populations, taking advantage of the fact that genetic markers in close proximity to causal polymorphisms may be in linkage disequilibrium (LD) to them. The magnitude and distribution of LD determine the choice of association mapping methodology. Extension and distribution of LD depend on many factors including population history (e.g., the presence of population bottlenecks or admixture) and the frequency of recombination.

In order to avoid false associations, the optimization of LD mapping requires a detailed knowledge of basic population genetic parameters such as the pattern of nucleotide diversity and LD for each particular species and candidate gene set.

First estimates indicate that nucleotide diversity varies considerably between plant species. Interestingly, the pines (e.g., *P. sylvestris*, Dvornyk et al. 2002; García-Gil et al. 2003; *P. taeda*, Brown et al. 2004; González-Martínez et al. 2006a; *P. pinaster*, Pot et al. 2005a) are not among the most variable species, contradicting expectations from the results obtained using neutral markers and their life history characteristics. First evidences seem to show that broadleaved species (e.g., *Populus*, Ingvarsson 2005; *Quercus*, Table 2.7 in Pot et al. 2005a) display higher levels of nucleotide diversity than pines.

Markers in specific functional regions of the genome need to be statistically analyzed in order to test for the possibility that these regions might have experienced different selective pressures. In unstructured populations, standard neutrality tests might be applied. When variation is structured in populations, a relatively easy approach is the comparison of genetic differentiation estimates, such as Wright's *F*-statistics, among markers tagging a putative gene under selection and neutral markers or expected distributions computed using coalescence theory (see reviews in van Tienderen et al. 2002; Luikart et al. 2003). If population divergence (F_{ST}) is higher for the gene-targeted marker with respect to divergence estimates obtained from random markers, this might indicate divergent selection and local adaptation for the tagged gene (van Tienderen et al. 2002). Pot et al. (2005a) found a high differentiation among populations at *Pp1* (*glycine-rich protein* homolog) gene in *P. pinaster*, higher than at the neutral level. This result is consistent with diversifying selection acting at this locus in this species. On the other hand, the absence of differentiation observed for the gene *CeA3* (*cellulose synthase*), compared with the significant level observed at neutral markers may indicate balancing selection acting on this gene.

Recent studies on pines reveal a rapid decay in LD with physical distance. LD declines very rapidly within 200 to 2,000 bp in *Pinus taeda* (Brown et al. 2004; González-Martínez et al. 2006a), *Pinus sylvestris* (Dvornyk et al. 2002 García-Gil et al. 2003). A rapid decay of LD in pines is consistent with what is expected from outcrossing species with large effective population size.

2.2 Construction of Genetic Maps

In genetics, mapping is defined as the process of deducing schematic representations of DNA. Three types of DNA maps can be constructed depending on the landmarks on which they are based:

- Physical maps, whose highest resolution would be the complete nucleotide sequence of the genome
- Genetic maps, which describe the relative positions of specific DNA markers along the chromosomes, determined on the basis of how often these loci are inherited together
- Cytogenetic maps, a visual appearance of a chromosome when stained and examined under a microscope

To provide a first glimpse of the pine genome, high-resolution genetic maps have been established for several pine species using different types of molecular markers and following different strategies that are reviewed in the following sections.

2.2.1 Development of Molecular Markers in Pines

The construction of a linkage map relies on the availability of enough molecular markers to detect linkage between them. Each type of marker technology has advantages and limitations. Many factors (e.g., polymorphism information content, level of polymorphism exhibited for the mapping progeny, mode of inheritance, genome size) can influence the development of a particular technique and the choice of a marker system for a given purpose (e.g., genetic mapping, quantitative trait loci analysis, survey of genetic diversity, forensic applications). The purpose of this section is to briefly review the different types of molecular marker techniques that have been developed in pines and used for genetic mapping applications. We will not present the details of each technique. Both the review by Cervera et al. (2000a) and the references cited in Table 2 will provide the reader with the necessary information for understanding the scientific basis of each technique.

Isozymes and Proteins

The first markers developed for pine were isozymes. Linkage studies were carried out on more than 10 species for about 15 loci (reviewed by Tulsieram et al. 1992). Conkle (1981) located more loci, but still not enough to cover the pine genome. Proteins revealed by two-dimensional polyacrylamide gel electrophoresis (2D-PAGE; O'Farrell 1975) presented the advantage of being multiplexed compared to isozymes. Importantly, proteins can be easily characterized by mass spectrometry (e.g., Gion et al. 2005) and may be recognizable by sequence similarity to others proteins published in sequence databases, therefore providing functional markers expressed in the tissues analyzed. Two-dimensional protein markers were developed only in *P. pinaster* (reviewed in Cánovas et al. 2004 and Plomion et al. 2004). Although proteins provided physiologically relevant markers to map the expressed genome, this time-consuming technique failed to provide enough markers for genetic application, which requires full genome coverage, such as linkage mapping and quantitative trait loci (QTL) detection.

RFLPs

Restriction fragment length polymorphism (RFLP) markers were developed in the early 1990s for *P. taeda* (Neale and Williams 1991). They offered a sufficient number of markers for high-density genome mapping in pine. However, this labor-intensive and time-consuming technique was only applied to *P. taeda* (Devey et al. 1994) and *P. radiata* (Devey et al. 1996).

RAPDs and AFLPs

In the mid-1990s PCR-based multiplex DNA fingerprinting techniques provided very powerful tools to generate dense linkage maps in a short period of time. Random amplified polymorphic DNA (RAPD; Williams et al. 1990) and then amplified fragment length polymorphism (AFLP; Vos et al. 1995) became the most popular marker technologies in conifers. Despite their biallelic nature and dominant mode of inheritance (which was actually not an issue for mapping with haploid megagametophyte, see Sect. 2.2.2.), these markers tremendously boosted up genetic analysis in most forest tree species including pines (reviewed in Cervera et al. 2000b).

Nuclear Microsatellites

In contrast to other plant species, few polymorphic single-copy nuclear microsatellite markers or simple sequence repeats (SSRs) have been reported in pines (reviewed in Chagné et al. 2004). The genome

Table 2. Characteristics and applications of molecular markers

Feature	Protein markers		DNA markers				
			Hybridization-based	PCR-based			
	Isozyme	2D-PAGE	RFLP	RAPD	AFLP	SSR	SNP
Tissue quantity (g)	0.5–2	1	na	na	na	na	na
DNA quantity (mg)	na	na	10	0.02	0.25–0.5	0.02	0.005–0.02
DNA quality	na	na	high	Medium	Medium	Medium	Medium
Allelism and mode of inheritance	Multiallelic codominant	biallelic, codominant, or dominant	Biallelic codominant	Biallelic dominant	Biallelic dominant	Multiallelic codominant	Biallelic codominant
Number loci analysed per assay	1–5	several tens	1-6	2–30	20–100	1–10 (multiplexing)	1–several hundred (multiplexing)
Polymorphism	Low	Low	Medium	Medium	Medium	High	Medium[a]
Development cost	Low	Low	High	Low	Medium	High	High
Cost per analysis	Low	Medium	High	Low	Medium	Low	Low
Amenable to automation	Low	Low	Low	Medium	Medium	High	High
Technical demand	Low	High	High	Low	Medium	Low	Medium
Reproducibility	High[b]	Medium[b]	High	Low–high[c]	High	High	High
Review	(Müller-Starck 1998)	(Plomion et al. 2004)	(Brettschneider 1998)	(Rafalski 1998)	(Cervera et al. 2000b)	(Echt and Burns 1999)	(Syvanen 2001)
Applications							
Certification	-	-	-	-	+	+++	++
Diversity	+	+	++	+	+	+++	+++
Phylogeny	++	+	++	++	++	++	+++
Mapping	-	-	++	++[d]	++[d]	+++	+++
Comparative mapping	+	+	++	-	-	+++	+++
QTL analysis	+	+	++	+	+	+++	+++
Association studies	+	+	+	-	-	++[e]	+++

[a] Sequence dependent (exon vs UTR and intron)
[b] Under the same environmental conditions, type of tissue, and age
[c] Between laboratories *vs* single laborytory
[d] To construct dense maps
[e] Used as control loci

structure of these species, characterized by a large physical size, with a large amount of repeated sequence (Sect. 2.1.2), has been the main obstacle to the development of useful markers using classical SSR-enriched library approaches (e.g., Auckland et al. 2002; Guevara et al. 2005a; C.S. Echt and C.D. Nelson, unpublished results). In addition, the ancient divergence time between coniferous species (Price et al. 1998) and the complexity of their genomes means that transferability of single-copy SSRs among genera and even within *Pinus* is generally poor, resulting in a large proportion of amplification failure, nonspecific amplification, multibanding patterns, or lack of polymorphism (Echt and Nelson 1997; Mariette et al. 2001). In an attempt to circumvent these genome-related problems, Elsik and Williams (2001) removed most of the repetitive portion of the genome using a DNA reassociation kinetics-based method, and Zhou et al. (2002) targeted the low-copy portion of the genome using an undermethylated region enrichment method. Both approaches yielded remarkable enrichment for useful SSR markers in *P. taeda*. SSRs made from low-copy, undermethylated, and total genomic DNA yielded mappable markers (Nelson et al. 2003; Zhou et al. 2003). *P. taeda* SSRs developed by Elsik and Williams (2001) and Zhou et al. (2002) transferred quite well between American hard pines (Shepherd et al. 2002a) but were shown to be less transferable in the phylogenetically divergent Mediterranean hard pines (Chagné et al. 2004; González-Martínez et al. 2004). Interestingly, perfect trinucleotide SSRs transferred from American to Mediterranean pines better than other motifs (Kutil and Williams 2001). A number of nuclear SSR markers have been developed for *P. radiata*, almost all of which are based on the more frequently polymorphic dinucleotide repeat motifs (Smith and Devey 1994; Fisher et al. 1996, 1998; Devey et al. 2003), and used in a number of applications. Polymorphic chloroplast microsatellite loci have also been identified and applied (Cato and Richardson 1996; Kent and Richardson 1997). More recent SSR discovery efforts have been undertaken in both New Zealand and Australia and have been most commonly applied to fingerprinting (Kent and Richardson 1997; Bell et al. 2004) and QTL mapping applications (Devey et al. 2004a).

EST Polymorphisms

With the availability of sequence data obtained by random sequencing of pine cDNAs (Sect. 2.5.2), there is now a clear trend toward the development of gene-based markers (ESTP: EST polymorphisms). There are basically two groups of technologies used to detect nucleotide polymorphisms (SNPs and INDELs), either based on the knowledge of nucleotide variants or not. Up to now, techniques based on the detection of differences in the DNA stability (denaturating gradient gel electrophoresis, DGGE; Myers et al. 1987), conformation (single-strand conformation polymorphisms, SSCPs; Orita el al. 1989) under specific polyacrylamide gel conditions, or heteroduplex cleavage (TILLING, targeting induced local lesions in genomes; Colbert et al. 2001) have been successfully applied in pines (Plomion et al. 1999; Temesgen et al. 2001; Chagné et al. 2003; Ritland et al. 2006). With the decrease of sequencing costs and the availability of pine cDNA sequences, more targeted and precise approaches are now possible (Pot et al. 2005b). The bioinformatics assembly of ESTs into large contigs (i.e., unigenes) has also made it possible to identify putative SNPs. Le Dantec et al. (2004) identified a set of 1,400 candidate SNPs in *P. pinaster* contigs containing between 4 and 20 sequence reads. This represents a great resource of molecular markers for this species that can be used to map candidate genes, study LD, and develop comparative orthologous sequence markers for comparative genome mapping. In addition to SNPs and INDELs, a large set of microsatellite markers have been developed from *P. taeda* and *P. pinaster* expressed sequence tags (ESTs) (Echt and Burns 1999; Chagné et al. 2004; Echt et al. 2006). These markers present the advantage of bei ng highly polymorphic and located in coding regions.

2.2.2 Haploid- and Diploid-Based Mapping Strategies

The construction of a genetic map requires two components: first, a segregating population (mapping pedigree) derived from a cross between parental trees that are heterozygous for many loci and, second, a set of molecular markers segregating in the progeny according to Mendelian ratios. Linkage map construction is based on the statistical analysis of polymorphic markers in the mapping population, considering that the distance between two loci is related to the probability of observing a recombination event between them. There is a number of mapping software to facilitate automated analysis (http://linkage.rockefeller.edu/soft/list.html).

Haploid- or Half-Sib-Based Mapping Strategy

In conifers, the haploid megagametophyte constitutes an ideal plant material for genetic mapping. This nutritive tissue surrounding the embryo is derived from the same megaspore that gives rise to the maternal gamete. Therefore, it represents a single meiotic event in the parent tree that is genetically equivalent to a maternal gamete. The dominant nature and biallelic mode of inheritance of RAPD and AFLP is not an issue for genetic mapping with haploid megagametophytes. However, quantitative traits can only be measured on half-sib seedlings, limiting the detection of QTLs at the first stages of tree development. Thus, this approach is not applicable to the analysis of QTLs for economically important traits in well-established plantations, unless the megagametophytes were collected and saved, which has generally not been the case.

Diploid- or Full-Sib-Based Mapping Strategy

Different strategies have been followed to construct genetic linkage maps of *Pinus*: the "pseudotestcross strategy," the "F_2 inbred model," and the "three-generation outbred model." The pseudotestcross strategy is mainly based on selection of single-copy polymorphic markers heterozygous in one parent and homozygous null in the other parent and therefore segregating 1:1 in their F_1 progeny as in a testcross. Grattapaglia and Sederoff (1994) introduced the term two-way pseudotestcross to define this mapping strategy, where two independent genetic linkage maps are constructed by analyzing the cosegregation of markers in each progenitor. The efficiency of this strategy, as well as for the haploid mapping strategy, depends on finding individual trees that are heterozygous for many loci, which is quite easy using arbitrarily primed PCR assays (RAPD and AFLP) in highly heterozygous outcrossed tree species such as pines. The F_2 inbred model is based on a three-generation pedigree for which the grandparents are treated as inbred lines. In the F_2 generation, three genotypes occur at any locus – AA, AB, and BB – segregating 1:2:1. The three-generation outbred model (Sewell et al. 1999) is an extension of the pseudotestcross strategy. Within a single outbred pedigree, any given codominant marker will segregate in one of three different ways. When one parent is heterozygous and the other is homozygous, segregation will be 1:1 (i.e., testcross mating type). When both parents are heterozygous, segregation will be either 1:2:1, if both parents have the same genotype (i.e., intercross mating type), or 1:1:1:1, if they have different genotypes (i.e., fully informative mating type). These segregation data are then subdivided into two independent data sets that separately contain the meiotic segregation data from each parent, and independent maps are constructed for each parent. A sex-average map is then constructed using an outbred mapping program, which uses fully informative and intercross markers to serve as common anchor points between each parental data set. Compared to "megagametophyte progeny," full-sibs can be grafted and/or propagated by cuttings, thereby constituting a perpetual population, analogous to recombinant inbred lines in crop plants. The use of such clonally propagated progeny obviously increases the precision of quantitative measurements and therefore enhances the QTL detection power (Bradshaw and Foster 1992).

2.2.3 Genetic Mapping Initiatives in Pines

In this section and Table 3 we summarize what has been done in terms of linkage map construction in the genus *Pinus* with emphasis made on the most studied species. In addition, some maps have been published together with QTL studies and will be found in the references cited in Sect. 2.3.

Maritime Pine

Linkage maps of the Maritime pine (*Pinus pinaster* Ait.) genome were first constructed by analyzing the cosegregation of proteins extracted from megagametophytes collected during the germination of the embryo. Bahrman and Damerval (1989) were the first to report a linkage analysis for 119 protein loci using 56 megagametophytes of a single tree. Extending this approach, Gerber et al. (1993) reported a 65-locus linkage map covering one fourth of the pine genome, using 18 maritime pine trees with an average of 12 megagametophytes per tree. A more conventional pedigree (inbred F_2) was used to map 61 proteins using haploid (Plomion et al. 1995a) and diploid (Plomion et al. 1997; Costa et al. 2000) tissues of the same seedlings. In the latter case, protein loci were found on each chromosome (Thiellement et al. 2001). As stated above, the advance of PCR-based markers has allowed the construction of saturated linkage maps, in a short period of time, with no prior knowledge of DNA sequence (Plomion et al. 1995a, b; Costa et al. 2000).

Table 3. Summary of genetic linkage maps of *Pinus* species

Species pedigree	Number of LGs (+pairs, triplets)	Total number of linked marker loci	RAPD, AFLP	SSR	EST-based	Others	Total genetic distance (cM)	Genome saturation	References
Pinus pinaster									
Haploid (1 tree)	Not estimated	119	–	–	–	119 (2D proteins)	Not estimated	Not estimated	Bahrman and Damerval 1989
Haploid (8 trees)	Not estimated	65	–	–	–	65 (2D proteins)	530	25%	Gerber et al. 1993
Haploid (H12 self)	13(+5)	263	251	–	–	–	1223	90%	Plomion et al. 1995b
Haploid (H12 open pollinated)							1236	90%	
Haploid (H12 self)	12	463	436	–	–	27 (2D proteins)	1860	100%	Plomion et al. 1995a
Haploid (H12self)	11	94	94	–	–	–	1169	Not estimated	Plomion and O'Malley 1996
Diploid (H12 self)	11	94	94	–	–	–	1354	Not estimated	
Haploid (H12 self)	13	398	235+127	–	–	36 (2D protein)	1873	93.4%	Costa et al. 2000
F1 - AFOCEL	12	759	738	14	7	–	1994	Not estimated	Ritter et al. 2002
9.103.3 × 10.159.3 (consensus)	12	620	620	–	–	–	1441	Not estimated	Chagné et al. 2002
9.103.3 × 10.159.3 (consensus)	12	326	276	–	50	–	1638	Not estimated	Chagné et al. 2003
Pinus taeda									
Various pedigrees	Not estimated	–	–	–	–	20 isozymes	Not estimated	Not estimated	Conkle et al. 1981
Base pedigree	20	75	–	–	90	6	Not estimated	Not estimated	Devey et al. 1994
Haploid (Tree 10-5)	16	458	458	–	–	–	1727	97%	Wilcox 1995
Base and QTL pedigrees	18	357	67	–	257	12	1300	76.4%	Sewell et al. 1998
Tree 7-56	12	508	508	–	–	–	1741	100%	Remington et al. 1999
Base and QTL pedigrees	20	265	253	12	–	–	1281	75.4%	Devey et al. 1999
Base pedigree	15	51	–	51	–	–	795	46.7%	Zhou et al. 2003
Pinus sylvestris									
49-2	14	261	261	–	–	–	2638		Yazdani et al. 2003
F1 of P315 × E1101	12(+4)	179	179	–	–	–	1000	50%	Hurme et al. 2000
F1 of AC3065 × Y3038	♀: 12	188	188	–	–	–	1645	98%	Yin et al. 2003
	♂: 12	245	245	–	–	–	1681		
F1 of E635 × E1101	12(+3)	260	194	4	61	–	1314	66–85%	Komulainen et al. 2003

Table 3. (continued)

Species pedigree	Number of LGs (+pairs, triplets)	Total number of linked marker loci	RAPD, AFLP	SSR	EST-based	Others	Total genetic distance (cM)	Genome saturation	References
Pinus radiata									
3-generation	22	208	41	2	165	–	1382	Not estimated	Devey et al. 1996
Haploid (full-sib seed)	14	267	267	–	–	–	1665	93%	Emebiri et al. 1998
S1	19	172	168	4	–	–	1117	56%	Kuang et al. 1999b
Pseudotestcross (Parent 850.055)	20	235	224	11	–	–	1414	85%	Wilcox et al. 2001a
Pseudotestcross (Parent 850.096)	21	194	185	9	–	–	1144	77%	Wilcox et al. 2001a
Two full-sib families	12	311	–	213	98	–	1352	Not estimated	Wilcox et al. 2004
Pinus elliottii									
Tree 8-7	13(+9pairs)	73	73	–	–	–	782	64–75%	Nelson et al. 1993
18-62 × 8-7	17(+12pairs)	129	129	–	–	–	1146	Not estimated	Kubisiak et al. 2000
D4PC40 × D4PC13	15	154	63	–	45	41 (RFLPs) 5 isozyme	1115	Not estimated	Brown et al. 2001
Pinus palustris									
Tree 3-356	16(+6pairs)	133	133	–	–	–	1635	85%	Nelson et al. 1994
P. elliotti and P. palustris hybrids									
F1 of - 3-356 × H-28	♀: 18(+3 pairs)	122	122	–	–	–	1368	81%	Kubisiak et al. 1995
	♂: 13(+6 pairs)	91	91	–	–	–	953	62%	
BC1 of 488 × 18-27	♀: 17	133	133	–	–	–	1338	91%	Weng et al. 2002
	♂:19	83	83	–	–	–	995	81%	
P. elliotti and P. caribea var. hondurensis hybrids									
F1 of 2PEE1-102 × 1PCH1-63	♀: 24	125	117	8	–	–	1548	82%	Shepherd et al. 2003
	♂: 25	155	145	10	–	–	1823	88%	

The two-way pseudotrestcross mapping strategy was used to construct genetic linkage maps of maritime pine using AFLP markers (Chagné et al. 2002; Ritter et al. 2002) as well as AFLP, SAMPL, SSR, and gene-based markers (N. de María and M.T. Cervera, unpublished results). Comparing the total map distance of genetic maps constructed based on haploid and diploid progeny from the same Maritime pine tree, a higher rate (28%) of recombination in the pollen parent was found (Plomion and O'Malley 1996). Such a significant difference between male and female recombination was also reported in other pine species (Moran et al. 1983; Groover et al. 1995; Sewell et al. 1999).

Loblolly Pine

Loblolly pine (*Pinus taeda* L.) has been used extensively for genetic mapping including the development of an early map based on 20 isozyme loci that included five linkage groups (LGs) (Conkle 1981). More recently, maps have been constructed in several pedigreed populations using several types of DNA-based markers. Devey et al. (1994) published the first map utilizing 90 RFLP and six isozyme loci. The map was based on a three-generation pedigree with 95 progeny and revealed 20 LGs. Genomic mapping was successfully used by Wilcox et al. (1996) to define a single gene locus for resistance to an isolate of the fungus (*Cronartium quercuum* f. sp. *fusiforme*) that causes fusiform rust disease. O'Malley et al. (1996) described a RAPD-based map of clone 7-56, a top *P. taeda* parent. This map had been constructed several years prior to publication. The first consensus map was produced by Sewell et al. (1999) combining data from two three-generation pedigrees, including the pedigree used by Devey. RFLP, RAPD, and isozyme markers were placed on the integrated or consensus map containing 357 loci and covering about 1,300 cM of genetic distance on 18 LGs. The first complete genome map was developed by Remington et al. (1999). They utilized haploid megagametophyte samples from an individual mother tree to develop a map based on 508 AFLP markers. This map revealed 12 LGs equaling the basic number of chromosomes and about 1,700 cM of genetic distance. Their analysis suggested that this distance saturated the genome, in slight contrast to an earlier estimate of 2,000 cM based on data from three species of pines (Echt and Nelson 1997). SSR markers developed in radiata pine (*P. radiata*) were used in a comparative mapping project between radiata and loblolly pine (Devey et al. 1999). Of the 20 SSR markers tested, only 9 were mapped in both species; however, these codominant markers along with several codominant RFLP markers were useful in defining homeologous LGs between the species. Temesgen et al. (2001) added 56 ESTP markers to the consensus map developed earlier by Sewell. The DGGE method proved quite useful for assaying ESTP markers and suggested a general method for placing genes on the maps since the markers were developed from expressed sequences. Additional ESTP markers were developed and used to identify anchored reference loci based on their sequence similarity between species and their nature to map to conserved locations in more than one species (Brown et al. 2001). Zhou et al. (2003) mapped 51 SSR markers, covering 795 cM on 15 LGs, in a three-generation pedigree with 118 progeny. The markers, developed from loblolly pine libraries of three types, were not found to be clustered within the genome, further highlighting the value of SSR markers in genome mapping.

Radiata Pine

Over the past 15 years, a range of DNA-marker-based linkage maps have been constructed for this species. Wilcox (1997) briefly reviewed mapping studies undertaken up till that date. Although results were summarized from seven studies involving construction of eight linkage maps, only one map had actually been published by that date, consisting of 208 (mostly) RFLP, SSR, and RAPD markers (Devey et al. 1996, 1999). This map consisted of 22 LGs and covered 1,382 cM. All of the other maps reviewed by Wilcox (1997) were constructed using RAPD markers, either using haploid megagametophytes, or diploid tissues using a pseudotestcross approach. The number of markers used in these studies ranged from 124 to 290, with only one of the maps having LGs equal to the haploid number of chromosomes and the remainder ranging between 14 and 22 LGs. Total map length estimates were undertaken in three studies ranging between 1,978 and 3,000 cM. Subsequent to these earlier studies a number of maps were published, some of which were included in Wilcox's 1997 review. Using 222 RAPD markers to genotype 93 megametophytes, Emebiri et al. (1998) constructed a linkage map that covered 14 LGs and spanned a total distance of 1,665 cM. Kuang et al. (1999a) described a map constructed using megagametophytes of 198 S_1 seeds that had been genotyped with 168 RAPD and four microsatellite markers. The resulting map consisted of 19 LGs, covering 1,116.7 cM, which was estimated to cover 56% of the genome. Because this was constructed using an S_1 family, elevated levels of segregation distortion were observed. Wilcox et al. (2001a) published framework maps of both parents of 93 full-sib progenies based on a total of 429 AFLP, RAPD, and SSR markers. These parent-specific maps were constructed using a pseudotestcross strategy and covered 1,414 and 1,144 cM in 20 and 21 LGs, respectively. These maps have subsequently been added to, using over 300 SSR and EST markers, and have been reduced to 12 LGs (Wilcox et al. 2004). Devey et al. (1999) published a comparative map of radiata and loblolly pine based on RFLP, SSR, and RAPD loci and showed that the highly syntenic nature of *Pinus* applies to these two economically important species. Overall estimates of map length appear to be similar to that of loblolly pine (Wilcox et al. 2001a).

Scots Pine

Early mapping work in *P. sylvestris* has been based on isozyme loci (Rudin and Ekberg 1978; Szmidt and Muona 1989). These maps contained no more than 20 loci. The number of RFLP markers developed for Scots pine had been very low (Karhu et al. 1996). Thus, the next mapping efforts were based on RAPD makers segregating in haploid megagametophytes. Yazdani et al. (1995) mapped 261 markers in 14 LGs in a tree that is part of the breeding program. Hurme et al. (2000) made a low-coverage RAPD map for an F_1 tree that was a result of north × south cross, such that alleles for important quantitative traits were assumed to segregate in the same cross. The map with the best genome coverage, so far, was constructed by Yin et al. (2003). The AFLP mapping in a full-sib family resulted in two maps, for each of the breeding program parents. The map lengths for the two parents, based on about 200 framework markers, were about 1,645 and 1,681 cM for the male and female trees, respectively, with very high estimated genome coverage. Most recently, Komulainen et al. (2003) mapped about 60 gene based markers in the F_1 progeny of a north × south cross. This map also contained markers that had been previously developed for *P. pinaster* (Plomion et al. 1999), and others that had been used for *P. taeda*. Most importantly, the homologous markers allowed defining the correspondence between 12 LGs in the two species. More markers were later added to this map (Pyhäjärvi et al. unpublished).

Longleaf Pine, Slash Pine, Caribbean Pine, and their Hybrids

Longleaf pine (*P. palustris* Mill.), slash pine (*P. elliottii* Engelm. var. *elliottii*), and Caribbean pine (*P. caribaea* Morelet.) are hard pines of subsection *Australes* found along the coastal plains of the southeastern United States, eastern Central America, and the Caribbean islands. Although loblolly pine (*Pinus taeda* L.) has been planted on millions of acres that were once typically occupied by these species, in many situations these pines are proving to be the preferred timber species due to their adaptation to the coastal-plain soils and the associated natural disturbances such as frequent fires and hurricanes (Wahlenberg 1946; Shoulders 1984). A number of genetic maps, consisting primarily of RAPD markers, have been constructed for slash pine, longleaf pine, and their hybrids (Nelson et al. 1993, 1994; Kubisiak et al. 1995), with the main goal being to use these marker maps as a tool for dissecting the inheritance of specific traits of interest and for use in marker-assisted-selection (MAS) strategies within tree-improvement programs (Kubisiak et al. 1997, 2000; Weng et al. 2002). Some markers significantly linked to traits of interest have been converted to more easily scorable markers, such as sequence characterized amplified region (SCAR), to aid selection efforts (Weng et al. 1998). Brown et al. (2001) assembled a genetic linkage map for slash pine using a variety of markers [RAPDs, expressed sequence tag polymorphisms (ESTPs), restriction fragment length polymorphisms (RFLPs), and isozymes]. An additional genetic map for slash pine and one for Caribbean pine, using amplified fragment length polymorphisms (AFLPs) and microsatellite or SSR markers, were created using an F_1 hybrid population (Shepherd et al. 2003). The number of markers mapped and the genetic distances covered by some of the published maps for these species and their hybrids are summarized in Table 3. Unlike loblolly pine, significantly less effort has been focused on comparative mapping across these species. However, studies have shown synteny across slash and longleaf pines using RAPDs (Kubisiak et al. 1995, 1996), slash and loblolly pines using ESTPs as anchored reference loci (Brown et al. 2001), and slash and Caribbean pine using AFLPs (Shepherd et al. 2003). In addition, SSR markers look promising for further comparative analyses across these species (Shepherd et al. 2002a; C.D. Nelson and C.S. Echt, pers. comm.).

Other Pines

Genetic linkage maps have also been constructed in other pine species including *P. brutia* (Kaya and Neale 1995), *P. contorta* (Li and Yeh 2001), *P. edulis* (Travis et al. 1998), *P. massoniana* (Yin et al. 1997), *P. strobus* (Echt and Nelson 1997), *P. thunbergii* (Kondo et al. 2000; Hayashi et al. 2001). These were based mainly on RAPDs and AFLPs.

2.2.4 Genetic vs. Physical Size and Practical Implication

These mapping studies have led to the conclusion that the total genetic distance of the pine genome is around 2,000 cM (Gerber and Rodolphe 1994), i.e., about 167 cM per chromosome. Given a physical size of 25 pg/C, one unit of genetic distance (1 cM) would therefore correspond to 13×10^6nucleotides (13 Mb), while it represents 0.23 Mb in the model plant species *Arabidopsis*! Such high genetic/physical size ratio obviously hampers the characterization of QTLs by fine-mapping and positional cloning approaches. Hence, as will be discussed in Sects. 2.3 and 2.4, the only way for understanding the molecular basis underlying quantitative trait variation is the candidate-gene approach in which genes are identified a priori as likely candidates for the trait of interest and their polymorphisms tested against quantitative trait variation. Interestingly, despite the 56-fold difference between *Pinus* and *Arabidopsis* chromosomes, the number of crossings over per chromosome was found to be highly conserved between both genera: 2 to 4 chiasmata per bivalent (1 chiasma = 50 cM; Ott 1991). A comparison of genome lengths among evolutionary divergent pines found *P. pinaster*, *P. palustrus*, and *P. strobus* to have essentially identical rates of recombination (Echt and Nelson 1997). Thus, genetic mapping studies carried out in pines have clearly demonstrated that the mechanism of crossing over is conserved on a chromosomal basis, and independent of physical map size and the fraction of coding DNA.

2.2.5 Comparative Mapping: Toward the Construction of a Unified Pine Genetic Map

All pine species have the same number of chromosomes (i.e., $n = 12$) as well as a similar genome size. Moreover, they are all diploid, which suggests that their genome might be relatively well conserved among species. Approaches for evaluating genome similarity have used cytogenetics and linkage map comparison. Although cytogenetics can provide a direct idea about the conservation between different genomes (Hizume et al. 2002), most of the interspecies comparisons have been carried out using genetic maps. The same tools have also been developed for other applications such as gene mapping and QTL detection. Comparative genome mapping aims to measure the conservation of gene content (synteny) and order (colinearity) among chromosomes and uses orthologous loci as anchor points between maps. Comparative genome mapping has been successfully used in grasses to explain the genome evolution of cereals (Moore et al. 1995), for choosing a model species for whole-genome sequencing (rice, Ware et al. 2002; http://www.cns.fr/externe/English/Projets/Projet_CC/organisme_CC.html), and for transfering genetic information between related species, such as the position of candidate genes (Schmidt 2002). The lack of genome sequence for a Pinaceae species has made comparative mapping even more important as the primary tool for integrating genetic information across species.

To define the syntenic relationships among phylogenetically related pine species, orthologous markers (i.e., homologous DNA sequences whose divergence follows a speciation event and whose sequence and genome location is conserved between different species) are used. A first example was provided by Devey et al. (1999), who aligned the genetic maps of *P. taeda* and *P. radiata* using RFLPs and SSRs. This first effort was further consolidated in the frame of the Conifer Comparative Genomics Project (http://dendrome.ucdavis.edu/ccgp). Low-copy cDNA PCR-based markers were developed in loblolly pine (Harry et al. 1998; Brown et al. 2001; Temesgen et al. 2001; Krutovsky et al. 2004). These markers showed a relatively good PCR cross-amplification rate between pine species because they target conserved coding regions, showing a relatively high polymorphism rate and a low number of paralogous amplification when PCR primers were chosen carefully. They were used to study the synteny between species belonging to the Family Pinaceae, which included pines, along with other important conifers such as spruces and firs. These markers made it possible to assign LG homologies for 10 out of 12 chromosomes between *P. elliottii* and *P. taeda* (Brown et al. 2001; http://dendrome.ucdavis.edu/

ccgp/compmaps.html), 10 out of 12 between *P. pinaster* and *P. taeda* (Chagné et al. 2003), and 9 out of 12 between *P. sylvestris* and *P. taeda* (Komulainen et al. 2003). About 30 or 40 ESTP markers were proved to be useful in demonstrating large areas of synteny between each species pair. While this comparison was only of low density, these pioneering studies suggested that pine genomes did not show any apparent chromosomal rearrangement. They also provided an indication that gene content and gene order is conserved, as is illustrated for LG 6 in Fig. 2. Current efforts are being expended to add more markers common to *P. radiata* and *P. taeda* (P. Wilcox, personal communication). From an application point of view, these comparisons provide a set of markers that can be used for constructing framework genetic maps of pine species for which maps have not yet been developed.

As more conifer ESTs become available in public databases (329,531 in *Pinus teada*, 132,531 in *Picea glauca*, 27,283 in *Pinus pinaster*, 28,170 in *Picea engelmannii* × *Picea sitchensis*, 80,789 in *Picea sitchensis*, 6,808 in *Pseudotsuga menziesii*, 7,639 in *Cryptomeria japonica:* EMBL 19 March 2006), a computational approach could be used for *in silico* development of putative orthologous EST-based markers (Fulton et al. 2002), as was recently illustrated between loblolly pine and Douglas fir by Krutovsky et al. (2004). Such resources should help to define the precise syntenic relationship across conifers and establish a framework for comparative genomics in Pinaceae.

The alignment of genetic maps of *P. pinaster* and *P. taeda* made it possible to discover putative con-

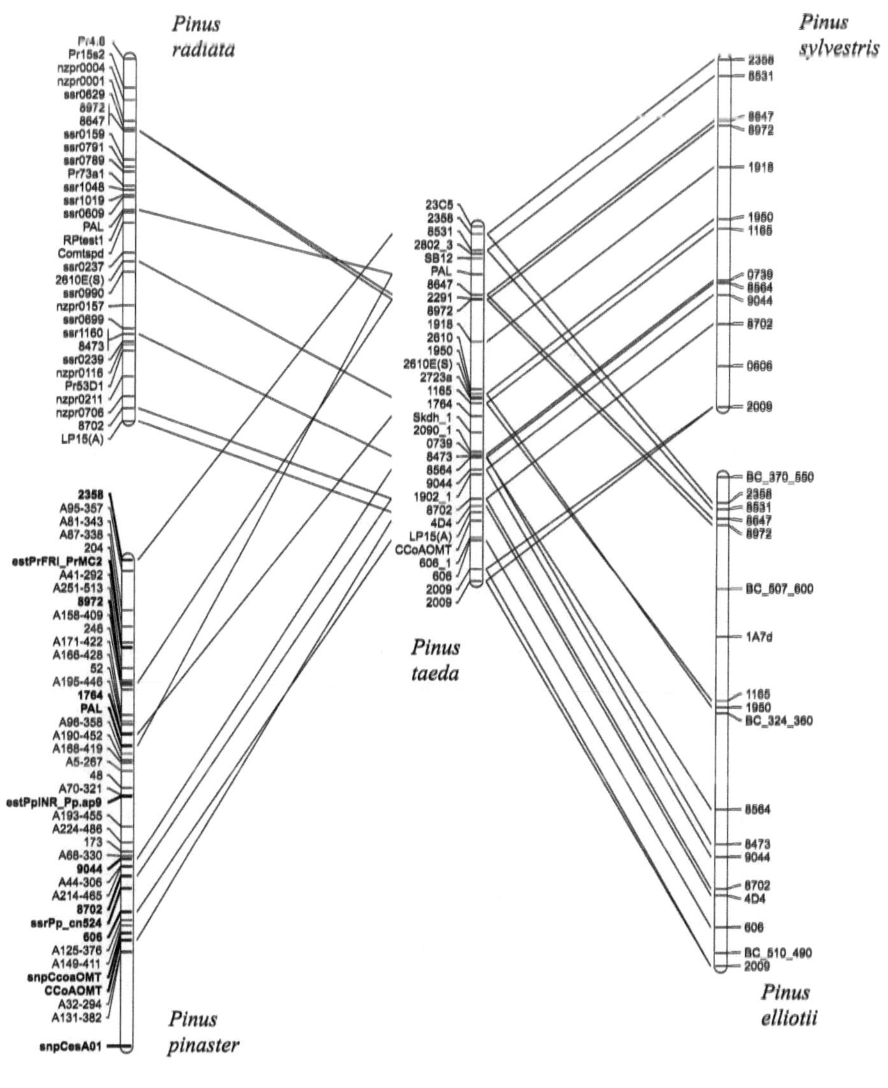

Fig. 2. Synteny in the *Pinus* genus (linkage group 6): alignment of genetic maps of five pine species

served QTLs between the two species (Chagné et al. 2003). Those are QTLs for wood density and wood chemical composition traits located on LGs 3 and 8, respectively. The same observation has been made in *P. radiata* (Telfer et al. 2006). Preliminary studies indicate wood-density QTLs are colocating more frequently than would be expected by chance between these two species (Telfer et al. 2006), and analyses are being extended to other wood property traits. Moreover, candidate genes coding for functions that are linked to wood formation have been mapped in the same regions, which indicates that they may be involved in the molecular control of those traits. These first examples of the application of comparative genome mapping in pines show that comparative genome mapping can be used to verify QTLs across species and that the same genes may be involved in the genetic control of the same traits.

2.3 Genetic Architecture of Complex Traits

Pine tree improvement is hampered by different inherent characteristics: (1) the time needed to reach sexual maturity, (2) the time lag required to evaluate field performance (e.g., growth; Kremer 1992), and (3) in some cases, the cost of phenotyping (e.g., wood-quality-related traits). This makes breeding of these species a slow process compared to that of most common crop plants. In addition, selection of these traits remains imprecise because environmental effects are rather high for most traits of interest. Heritabilities for height, diameter, volume, branching traits, and bole taper, i.e., straightness, are in the range of 0.1 to 0.3, and only slightly higher (0.3 to 0.6) for wood and end-use properties (reviewed by Cornelius 1994). In this context, any tool directed toward selection processes that would improve the evaluation of genetic value and also reduce the generation time would be of considerable value.

Most traits important to forestry, such as biomass production, wood quality, and biotic and abiotic stress resistance are complex quantitative traits. In the theory of quantitative genetics it is assumed that the heredity of a quantitative trait can be ascribed to the additive effects of a large number of genes with small and similar actions, modulated by environment. This assumption has been questioned since the early experiments of Sax (1923) and Thoday (1961), and it is now well known that a small number of segregating loci are involved in the genetic control of quantitative trait variation. These genes act together to provide a quantitative difference and are referred to as quantitative trait loci or QTLs (Geldermann 1975). The basic theory of using genetic markers to detect QTLs was introduced by Sax (1923). Initially, the application of this theory was limited by the lack of available segregating markers; however, rapid advances in DNA marker-based technologies since the 1980s have now made it possible to genoptype hundreds of genetic markers to construct dense genetic linkage maps (Sect. 2.2.3) and further to carry out a comprehensive search of QTLs along the genome. Sewell and Neale (2000) and Guevara et al. (2005b) recently reviewed the science of QTL mapping in pine trees.

In the first part of this section, devoted to the genetic dissection of agronomically important traits, we will present an update of the studies that were carried out in pines. The specificity of pines regarding the different types of populations and strategies used to detect QTLs will be presented. Then, the main results of these studies will be discussed. In the second part, a perspective on the identification of diagnostic markers for pine-tree breeding will be discussed.

2.3.1 Strategy and Methods Used for QTL Detection in Single Family Pedigrees

Pines are characterized by late maturity (longevity), an outbred mating system, and a high genetic variability (Hamrick and Godt 1990). Their outbred mating system and high genetic load have hampered the development of inbred lines, the material of choice from a QTL mapping perspective. In this context, specific populations and statistical methods were developed specially for forest trees, and pines in particular. Until recently, most of the QTL mapping efforts were focused on single family pedigrees. However, given the high rate of polymorphism encountered in pines and the relative lack of stability of QTLs in different genetic backgrounds, methods aimed at validating markers linked to the traits of interest in more complex pedigrees or even in unrelated genotypes are emerging. In this section, we will review the type of population, statistical methods, and main results that have been obtained using simple mapping pedigrees. The limitations of this approach will be underlined, and the use of complex pedigree designs or unrelated populations to unravel the genetic architecture of complex traits will be presented. Finally, the relevance of integrative approaches combining classical QTL studies and transcriptome and proteome analysis, together with studies of molecular evolution, will be highlighted.

Single Family Designs

Two-Generation Full-Sib Design The advantage of naturally high levels of heterozygosity in outbred forest trees can be utilized in a two-generation population structure, where it may be expected that trees chosen as potential parents will likely be heterozygous for some number of QTLs, which will then segregate in the F_1. Typically, "plus trees" are identified and are used as parents of a QTL mapping population. Various studies have taken advantage of this two-generation design to analyze each parent under a pseudotestcross model (Kumar et al. 2000; Lerceteau et al. 2000; Ball 2001; Shepherd et al. 2002b; Weng et al. 2002; Markussen et al. 2003; Yazdani et al. 2003). This model is well suited for dominant markers. However, the main limitation is that the phenotypic effects inherited from each parent are analyzed individually, even though the genetic contribution of each parent simultaneously contributes to the phenotypic variation in the progeny population. Consequently, the genetic information in the four progeny classes of an outbred pedigree is collapsed into only two genotypic classes, thereby reducing the robustness of the analysis. Of course, if codominant markers are used, a consensus map can be built precisely to detect allele effects from both progenitors simultaneously.

Three-Generation Full-Sib Design In the three-generation outbred population structure, two crosses are made among four unrelated grandparents, where each mating pair is selected among individuals displaying divergent phenotypic values for the trait (e.g., Groover et al. 1994). From each grandparental mating, a single phenotypically intermediate individual is chosen as a parent. Presumably, these intermediate parents are heterozygous for both marker and QTL alleles and are potentially heterozygous for different allelic pairs that display a divergent phenotypic effect. This three-generation full-sib structure is typically designed for QTL analysis under an outbred model and has been used extensively (Table 4). Alternatively, Plomion et al. (1996a, b) utilized the selfing ability of *P. pinaster* to design experiments that fit an inbred F_2 model. Although RAPD markers were used, genetic information from progeny and corresponding megagametophytes were utilized to overcome the limitations associated with dominant markers using an F_2 model. It was also proposed to use trans-dominant-linked markers to overcome the problem of dominant markers on the sporophytic phase of F_2 trees (Plomion et al. 1996c).

Two-Generation Half-Sib Design With this population structure, the effects of two maternal QTL alleles are averaged over a large pollen pool (Hurme et al. 2000). This type of structure allows one to test the stability of the effect of the maternal alleles in different genetic backgrounds. In addition to these classical mapping designs, particular populations (F_1S) were especially developed to analyze inbreeding depression in *P. taeda* and *P. radiata* (Kuang et al. 1999a,b; Remington and O'Malley 2000a,b; Williams et al. 2001).

Two-Generation Full-Sib and Half-Sib Design An extension of the pseudotestcross QTL mapping strategy, in which QTLs are defined in a narrow genetic background, Plomion and Durel (1996) show that a "general" value of a "specific" QTL detected in a full-sib family could be easily evaluated, provided that both parents of the full-sib were involved in maternal half-sib (open-polinated or polycross) families. Such two-generation pedigrees are widely available in most pine breeding programs that involve the simultaneous estimation of specific and general combining abilities of selected trees. However, this strategy has never been tested experimentally.

Methods Used for QTL Mapping in Single Family Designs

Regardless of the population structure and size (Beavis 1994), several factors must be considered for successful QTL detection. Statistical methodology significantly influences the accuracy of QTL position and effect estimation. Simple statistical methods such as analysis of variance (ANOVA) have opened the way to the development of more powerful QTL detection methods, integrating information available at multiple markers: interval mapping (IM), composite interval mapping (CIM), and multiple interval mapping (MIM).

The first method, called single marker analysis, proposed by Edwards et al. (1987), is the simplest one. ANOVA is performed with one marker at a time, on the genotypic classes defined by a single marker. This

Table 4. QTL and marker-trait association detected in pines: pedigree structure and methods

Species	Objective	Population (size)[a]	Replications[b]	Clonal replicates	Genetic markers	QTL analysis (analytical model)[c]	Software	Reference
Pinus elliottii	Detection	F1OB (186)	1	No	RAPD	SG (54 : 27 genotypes from each tail), SM, MM	SAS	Kubisiak et al. 2000
Pinus elliottii var elliottii × *Pinus caribaea var hondurensis*	Detection	F1OB (89)	2 sites (60,29)	No	AFLP, SSR	SM, CIM	QTLcartographer	Shepherd et al. 2005
(*Pinus palustris Mill* × *Pinus elliottii Engl*) × *Pinus elliottii Engl*	Detection, stability (age)	F1OB (258)	3 sites (82,83,93)	No	–	SM, IM	SAS, MAPMAKER/QTL	Weng et al. 2002
Pinus sylvestris	Detection	F1OB	1	No	AFLP	IM	Qgene	Lerceteau et al. 2000
	Detection, Stability (time replicate)	F2HS	2	No	RAPD	SG (1994 : 48 vs. 48, 1996 48 vs. 48), SM, Bayesian QTL analysis	SAS, Multimapper/ OUTBRED	Hurme et al. 2000
	Detection	F1OB	1	No	RAPD	SM	SAS	Yazdani et al. 2003
Pinus pinaster	Detection, Stability (Age)	F2S (120)	3 (age)	No	RAPD	IM	MAPMAKER/QTL	Plomion et al. 1996a
	Detection	F2S	1	No	RAPD, AFLP, proteins	CIM	QTLcartographer	Costa 1999
	Detection	F2OB (186)	1	No	AFLP	CIM	MULTIQTL	Brendel et al. 2002
	Detection	F20B (186)	1	No	AFLP, EST	CIM	MULTIQTL	Pot et al. 2005
	Detection	F1OB (80)	1	No	AFLP, SSR, EST	IM	SAS	Markussen et al. 2002
	Detection, stability (age, season)	F20B (186)	1	No	AFLP, EST	CIM	MULTIQTL	Pot 2004
	Detection	F2S (120)	1	No	RAPD	IM	MAPMAKER/QTL	Plomion et al. 1996b

[a] F1OB: two-generation outbred design, F2OB: three-generation outbred design, F2S: three-generation inbred design, F2HS: Two-generation half-sib design, CD: complex design (in this case more information is provided)

[b] When the traits were measured on one site and at one given time, 1 is indicated, otherwise more details are provided

[c] SM: single-marker analysis, MM: multiple-marker analysis (without map information), IM: interval mapping, CIM: composite interval mapping, SG: selective genotyping, BSA: bulk segregant analysis

Table 4. (continued)

Species	Objective	Population (size)[a]	Replications[b]	Clonal replicates	Genetic markers	QTL analysis (analytical model)[c]	Software	Reference
Pinus radiata	Detection, stability (age)	F1OB (174)	1	No	RAPD	SM, BSA (9 vs 9)	-	Emebiri et al. 1997
	Detection, stability (age)	F1OB (80-93)	3 times (age)	No	RAPD, AFLP, SSR	IM, MTM	-	Kumar et al. 2000
	Detection, stability (age)	F1OB (93)	2 times (age)	No	RAPD, AFLP, SSR	Bayesian approach	Splus	Ball 2001
	Detection	CD	1	No	SSR	-	-	Kumar et al. 2004
	Detection, validation, stability (genetic background)	3 unrelated F1OB (400 each)	2 sites	No	RFLP and SSR	SG (Juvenile wood density : 50 vs 50; Diameter at breast height : 100 vs 100), SM	GENSTAT	Devey et al. 2004a
	Detection, validation	Detection: 6 related F2OB (202)	1 site for the detection population	No	RFLP + SSR	SM within family + SM among families	GENSTAT	Devey et al. 2004b
		Verification: 1 F2OB (400)	2 sites for the verification population	Yes				
Pinus taeda	Detection	F20B (Detection population : 172)	6 sites (19 to 35 trees per sites)	No	RFLP	SG, SM	Home made	Groover et al. 1994
	Detection	F20B (Detection population : 172)	6 sites (19 to 35 trees per sites)	No	RFLP	IM	Home made	Knott et al. 1997
	Detection, stability (age, genetic background)	2 F2OB (Base : 95, Detection : 172)	2 sites for BASE (48+47), 6 sites for Detection (19 to 35 trees per sites)	No	RFLP, RAPD, Isozymes	IM	Home made	Kaya et al. 1999
	Detection, stability (age), development of statistical methods	F2OB (91)	5 times (age)	No	SSR	IM	Home made	Gwaze et al. 2003
	Detection, stability (age, season)	F20B (Detection population : 172)	6 sites (19 to 35 trees per sites)	No	RFLP	IM	Home made	Sewell et al. 2000
	Detection, stability (age, season)	F20B (Detection population : 172)	6 sites (19 to 35 trees per sites)	No	RFLP	IM	Home made	Sewell et al. 2002
	Detection, validation, stability (genetic background)	1 F2OB detection (D) : 172, validation (V) : 457	D : 6 sites (19 to 35 trees per sites)	No	RFLP + EST	IM	QTL express	Brown et al. 2003

method suffers from several limitations. First, it does not provide any information regarding the location of the QTLs in the genome, and furthermore the definition of the QTL effect is largely inaccurate given the inability to separate small-effect QTLs with tight linkage from QTLs of large effect but more distant linkage.

In order to improve the efficiency of QTL mapping through the use of genetic map information, Lander and Botstein (1989) developed the IM method. This approach allows QTLs to be detected in the intervals defined by the markers. The IM has been widely used in pines (Groover 1994; Plomion et al. 1996a, b; Knott et al. 1997; Kaya et al. 1999; Costa and Plomion 1999; Costa et al. 2000; Sewell et al. 2000, 2002; Brown et al. 2003). This method is, however, rather limited as QTL detection is made in a linear way, i.e., the same test is applied at each point of the interval without taking into account the results of successive tests.

More recently, statistical approaches have been developed to increase the statistical power of QTL detection and have received increasing attention in pines (Costa 1999; Brendel et al. 2002; Pot et al. 2005b). One of these approaches is CIM, developed by Zeng (1993a, b) and Jansen (1993), which combines IM with multiple regression. Like IM, this method evaluates the presence of a QTL at multiple analysis points across each interlocus interval. However, at each point it also includes in the analysis the effect of one or more markers elsewhere in the genome. However, although IM and CIM brought significant improvements, both methods lack dimensionality: i.e., only a single QTL is being searched at a time.

The method proposed by Kao et al. (1999) and Zeng et al. (2000) called MIM differed from the previous methods through the implementation of multi-QTL models. The selection of the model (i.e., QTL number, position, effect, and interactions) that best fits the data follows an iterative process. After identification of QTL number and position by CIM, the MIM strategy consists of looking for additional QTLs through forward-backward selection cycles while integrating interaction information between the different QTLs. Within each iterative cycle, QTL position and effect are reevaluated. MIM also allows the evaluation of QTL epistasis. However, it is important to note that, although this strategy provides significant improvements over CIM, it does not allow the detection of nonsignificant QTLs at the individual level. Therefore, even if the concept aimed at detecting multiple QTL effects is simple, its implementation is relatively complex given the number of potential QTLs and the resulting number of possible interactions. Thus, the problem is no longer genetic but resides in the ability to test all the possible genetic models and select the one that best fits the observed data (*model selection*). Until now, the MIM algorithm implemented in QTL Cartographer developed by Kao et al. (1999) and Zeng et al. (2000) has not been directly used in pines, but Bayesian approaches, also based on model selection, have been applied in these species (Hurme et al. 2000; Ball 2001).

The simultaneous analysis of multiple correlated traits has also been incorporated into algorithms in order to improve QTL detection efficiency (Jiang and Zeng 1995; Korol et al. 1995, 1998). Although these algorithms have not been specifically used up till now, there is much interest in such algorithms, especially for such traits as annual growth or wood density. Additionally, application of specific genotyping strategies has further maximized the efficiency of QTL mapping. One such strategy, called selective genotyping, has been extensively used in pines (Groover et al. 1994; Hurme et al. 2000; Kubisiak et al. 2000; Devey et al. 2004a). Another strategy, termed bulked segregant analysis (BSA), commonly used to study qualitative traits, has also been applied to analyze quantitative traits (Emebiri et al. 1997).

Besides statistical procedures, it has been clearly shown that the power of QTL detection largely depends upon the quality of the phenotypic assessment. Poor phenotypic assessments, i.e., imprecise measurements, generally result in QTLs with true effects being left undected, but in some circumstances (but to a lesser extent) might even lead to the detection of false QTLs. Without the possibility of developing F_3 populations or recombinant inbred lines to precisely estimate the value of the traits, clonally propagated material has become the material of choice for forest tree geneticists (e.g., Scotti-Saintagne et al. 2004). However, this type of material has only been rarely used in pines (Devey et al. 2004b). This is most likely due to the large capital investments and technical expertise needed to clonally propagate pines and the wide variability noted among specific genotypes in their ability to produce rooted cuttings.

2.3.2 QTL Discovery in Single Family Pedigree Designs

Twenty-six QTL studies aimed at detecting associations between molecular markers and trait variation have been performed in seven pine species. However, most of the efforts have been concentrated in three species, *P. pinaster*, *P. taeda*, and *P. radiata* (Tables 4 and 5). In addition, studies on inbreeding depression have been carried out in *P. radiata* and *P. taeda* (Kuang et al. 1999a, b; Remington and O'Malley 2000a, b; Williams et al. 2001).

Qualitative Traits

Associations between qualitative traits and molecular markers have led to the identification of markers linked to resistance to different rust diseases: pine needle gall midge in *P. thunbergii* (Hayashi et al. 2004), white pine blister rust in *P. lambertiana* Dougl (Devey et al. 1995; Harkins et al. 1998), and fusiform rust disease in *P. taeda* (Wilcox et al. 1996). A major gene controlling the biosynthesis of δ-3 carene has also been mapped in *P. pinaster* (Plomion et al. 1996b). Most of these studies have been based on intraspecific mapping populations. Additionally, two interspecific crosses were used to increase the level of polymorphism and the range of variation for branch architecture in the mapping population (Shepherd et al. 2002b).

Quantitative Traits

Growth has been the most studied trait in QTL mapping studies in pines. To date, a comparable number of studies have been achieved for wood and end-use properties. These two classes of traits have been studied either globally (e.g., height growth, specific gravity) or after decomposition into simpler components (e.g., growth unit, ring density). Traits involving adaptation to the environment have also been studied, including tree response to heavy metal, drought, and cold stress, as well as bud phenology (Costa 1999; Hurme et al. 2000; Kubisiak et al. 2000; Lerceteau et al. 2000; Brendel et al. 2002; Yazdani et al. 2003).

QTL Results

The diversity of population types, population sizes, marker types, QTL detection methods, detection thresholds, and variation of phenotypic trait measurements have made the comparison between QTL experiments a difficult task. However, general observations can be made regarding QTL number, position, phenotypic effect, and stability.

QTL Number and Effect Most QTLs have been detected at a single maturation stage and in a single environment. Therefore, although all the analyzed traits show continuous variation, suggesting polygenic control of the traits, only a limited number of QTLs per trait (Table 4: between 0 and 8) have been detected. This number is in general smaller than in annual crops and has led to a smaller proportion of the phenotypic variation being explained (Table 5). As stated in the chapter on Eucalypts, "the limited power to detect QTL in forest trees compared to crop species may be due to the high environmental and developemental variation in tree plantations, as well as to the small size of the analyzed populations" (mainly between 91 and 200 genotypes, Table 4). However, QTL analyses carried out by Brown et al. (2003) and Devey et al. (2004a) using large mapping populations (>400 individuals) do not support the latest hypothesis. Brown et al. (2003) reported QTL effects that were two- to threefold smaller than those reported by Sewell et al. (2000, 2002) for the same traits. Such divergence likely represents more accurate estimates of the QTL effects owing to the larger segregating population analyzed. These results suggest that most QTL studies performed in pines, with the exception of the analyses carried out by Devey et al. (2004a, b) and Brown et al. (2003), have yielded an overestimation of the QTL effects, which was also suggested by Beavis et al. (1994). Consequently, as underlined by Wilcox et al. (1997), the genetic determinism of most target traits for pine breeding is likely to be explained by small-effect genes, rather than any moderate- to large-effect genes. However, a larger number of QTL experiments with larger pedigree sizes will be required to validate this hypothesis.

Compared to the small number of QTLs detected at any one maturation stage, the analysis of QTL stability along a cambial age or seasonal gradient (Table 4) revealed a significant increase in the number of QTLs detected. For instance, the simultaneous analysis of different maturation stages combined with seasonal variation allowed for the detection of 7 to 23 QTLs for wood-specific gravity, the percentage of latewood, and microfibrilar angle (Sewell et al. 2002). Similar results were reported by Pot (2004), who observed a total of 30 QTLs for wood density, 42 for wood heterogeneity, and 33 for radial growth when measured over several cambial ages.

QTL Stability In their review, Sewell and Neale (2000) pointed out that "before a commitment to marker-aided selection or breeding (MAS/MAB) can be made in tree breeding, QTL that have been detected must be verified in different experiments, as well as in differ-

Table 5. QTLs and marker-trait associations detected in pines: number of QTLs and phenotypic variance explained

Species	Type of trait	Trait (abbreviation used in the article)	#QTls[a]	%Variance explained by each QTL[b]	Note	Reference
Pinus elliottii	Adaptative traits	Aluminium tolerance	3*	15.61[all]		Kubisiak et al. 2000
Pinus elliottii var elliottii × *Pinus caribaea var hondurensis*	Wood-quality traits and growth traits	Average branch angle (AVBRA)	0	–		Sheperd et al. 2002b
		Average number of branch per whorl (AVBRN)	3***	12–18		
		Average branch diameter (AVBRD)	2***	16–17		
		Average whorl spacing (AVWS)	2***	15–19		
		Regularity of whorl spacing (CVWS)	1***	17		
		Stem class (SC)	0	–		
		Number of distinctively large, steep angled branches observed per tree (RAM)	0	–		
		Number of "leaders" (DL)	0	–		
		Bark thickness (AVBT)	2***	11–12		
		Relative bark thickness (RBT)	0	–		
		Trunk height (HT)	3***	13–21		
		Overbark diameter at breast height (OBDBH)	1***	16		
		Underbark diameter at breast height (UBDBH)	1***	17		
		Basic density (BD)	1***	14		
(*Pinus palustris Mill* × *Pinus elliottii Engl*) × *Pinus elliottii Engl*	Growth traits	Total height (month 7,16,29,41)	5[lod2]	3.6–11		Weng et al. 2002
		Stem diameter	6[lod2]	4–10		
		Height increments	5[lod2]	4.3–8.5		
		Diameter increments	5[lod2]	4.2–11		
Pinus sylvestris	Growth, wood-quality, and adaptive traits	Tree height (TH)	3[lod2]	11.5–12.2		Lerceteau et al. 2000
		Trunk diameter at breat height (DBH)	2[lod2]	9.3–15		
		Trunk diameter 0.5m from ground (D0.5)	2[lod2]	12.6–13.3		

[a] * significant at chromosomewide level ($p < 0.05$),
*** significant at genomewide level ($p < 0.05$),
[lodx] lod threshold value provided by authors (IM),
[pv=x] significance level used for single-marker analysis,
♣: unique QTLs, defined as subset of QTLs (suggestive or significant) that map within ca. 15 cM of one to another and have the same general profile for their parental and interaction effects (magnitude and direction of effects)
[b] when the % of phenotypic variance explained by each QTL was not provided, the phenotypic variance explained by all QTLs is indicated [all]

Table 5. (continued)

Species	Type of trait	Trait (abbreviation used in the article)	#QTls[a]	% Variance explained by each QTL[b]	Note	Reference
		Branch diameter of average branch at fourth branch level from terminal bud (BDA)	0	–		
		Branch angle (BA)	1^{lod2}	15.9		
		Basic wood density measured with Pilodyn (PIL)	0	–		
		Frost hardiness, critical temperature giving mean injury (CTm)	2^{lod2}	12.1–21.1		
		Frost hardiness, estimated temperature causing slight injury to 50% of needles (CT50)	2^{lod2}	11.3–22.7		
		Tree volume calculated with DBH (VOL)	2^{lod2}	10.7–16.7		
		Tree volume calculated with D0.5 (VOL0.5)	2^{lod2}	11–14.6		
	Adaptive traits	Bud set	3	2.1–12.7		Hurme et al. 2000
		Frost hardiness	7	3–11.1		
	Adaptive traits	Cold acclimation	$16^{p=0.01}$	$9–19.5^{all}$		Yazdani et al. 2003
	Growth traits	Height growth	$11^{p=0.01}$	$10.8–30.8^{all}$		
Pinus pinaster	Growth traits	Total height week 15	2**	7–12		Plomion et al. 1996a
		Total height week 38	3**	6.2–11.5		
		Total height week 92	1**	10		
		Fertile zone length (LF)_ cycle 5	1**	19.6		
		Fertile zone length (LF) adjusted to NSU_ cycle 5	1**	11.8		
		Number of stem units in the fertile zone (NSU)_ cycle 5	1**	20.4		
		Mean stem unit length (MSUL)_ cycle 5	1**	6		
		Germination date (GERMD)	2**	15–17		
		Hypocotyl length (Lhypo)	2**	9.5–12.7		
		Megagametophyte weight (MW)	2**	6–8.6		
	Adaptive and growth traits	Delta 13C	3*** + 5**	4.7–12.4	pop : id to Pot et al 2006 and Pot 2004	Brendel et al. 2002
		Ring width	1*** + 5**	5.9–18.1		
	Wood-quality and growth traits	Total height	0	–	pop : id to Brendel et al 2002 and Pot 2004	Pot et al. 2006
		Mean density of all the rings	2**	4.6–4.8		
		Density estimated by Pilodyn penetration	2** + 1***	3.7–8.8		
		Density heterogenity	3**	5.1–8		
		Lignin content	7**	5–9		

Table 5. (continued)

Species	Type of trait	Trait (abbreviation used in the article)	#QTLs[a]	%Variance explained by each QTL[b]	Note	Reference
		Water extractives content	2** +1***	5.8–8.8		
		Acetone extractive content	0	–		
		Alpha cellulose content	4**	3.7–7.8		
		Hemicellulose content	3** + 1***	4.5–8.1		
		Lignin composition	3**	5.1–9.4		
		Kraft pulping yield adjusted to Kappa number	1***	8.8		
		Kappa index	3**	4.7–6.4		
		Arithmetic fiber length	2** + 1****	4.9–12.3		
		Weighted fiber length	2** + 1***	6.1–11.9		
		Fibre width	2** + 1***	5.4–8.2		
		Coarseness	2** + 1***	4.4–11.9		
		Curl	1** + 1***	6–10.2		
		Zero span tensile value	5** + 2****	4.9–7.3		
	Wood-quality and growth traits	Alpha cellulose content	5**	5.98–12.24		Markussen et al. 2002
		Lignin content	6**	5.98–13.17		
		Pulp yield	7**	9.65–18.43		
		Extractives content	2**	4.26–13.36		
		CIE brightness	3**	7.76–13.74		
		Mean wood density	8**	6.17–12.95		
		Minimum wood density	3**	5.09–8.58		
		Maximum wood density	6**	7.50–14.51		
		Diameter	7**	5.04–14.51		
		Height growth	3**	4.93–9.78		
	Secondary metabolism	δ3Carene content	1***	26.5		Plomion et al. 1996b
Pinus radiata	Growth traits	Stem growth index	$2^{p=0.01}$	9.23–10.56		Emebiri et al. 1997
	Wood-quality trait	Wood density 1–5 years old	1***	–	Pop identical to Ball 2001, LG3 only, multitrait model analysis	Kumar et al. 2000
		Wood density 6–10 years old	0	–		
		Wood density 14 years old	0	–		
	Wood-quality trait	Wood density 1–5 year-old	1***	–	Pop identical	Ball 2001

Table 5. (continued)

Species	Type of trait	Trait (abbreviation used in the article)	#QTLs[a]	%Variance explained by each QTL[b]	Note	Reference
		Wood density 14 year-old	0	–	to Kumar et al. 2000, LG1 and 3 only	
	Growth and wood-quality traits	Diameter at breast height	2*** _0	–		Kumar et al. 2004
		Stem straightness	2*** _0	–		
		Branching cluster frequency	2*** _0	–		
		Wood density	1*** _0	–		
	Growth and wood-quality traits	Diameter at breast height	Verification: $2^{p=0.05}$ Bridging pop : 0	Verification: 2.2[all]		Devey et al. 2004a
		Juvenile wood density	Verification : $8^{p=0.05}$ Bridging pop : $4^{p=0.05}$	Verification: 14.1[all]		
	Biotic stress tolerance	Resistance to Dothistroma needle blight	$4^{p=0.05}$ in the detection population, $4^{p<0.05}$ in the validation population	detection: not estimated, validation: 1.76–4.8		Devey et al. 2004b
Pinus taeda	Wood-quality trait	Wood specific gravity	$5^{p=0.05}$	23[all]		Groover et al 1994
	Wood-quality trait	Wood specific gravity	4**, 1***	–	re-analysis of Groover et al. 1994	Knott et al. 1997
	Growth traits	Height increment	Base: 6***, detection: 7***	Base: 23.1–30.5[all], detection: 7.3–11.7[all]	Detection population: same as Groover et al 1994, Knott 1997, Sewell 2000, 2002, Brown et al 2003.	Kaya et al. 1999
		Diameter increment	Detection : 8	Detection : 12.5–59.5[all]		
	Growth traits	Total height 2 years	2**	7.9–10.8		Gwaze et al. 2003
		Total height 3 years	1***	10.3		
		Total height 4 years	1***	12.2		
		Total height 5 years	1***	12.2		

Table 5. (continued)

Species	Type of trait	Trait (abbreviation used in the article)	#QTls[a]	%Variance explained by each QTL[b]	Note	Reference
		Total height 10 years	1***	10.5		
		Growth rate	1***	11.3		
	Wood-quality trait	Wood specific gravity (individual rings and composite rings, measured for early and latewood)	23♣	5.4–15.7	Same population as Groover et al. 1994 and Knott et al. 1997	Sewell et al. 2000
		Percentage of latewood (individual rings and composite rings)	16♣	5.5–12.3		
		Average microfibrilar angle (individual rings and composite rings, measured for early and latewood)	7♣	5.4–11.9		
	Wood-quality trait	Cell wall content	8♣	5.3–12.7	Same population as Groover et al. 1994 and Knott et al. 1997	Sewell et al. 2002
	Wood-quality traits	Wood-specific gravity	V : 18♣, U : 5♣	V : 1.7–5.7, U : 1.8–4.4	Detection population: same as Groover et al 1994, Knott et al. 1997, Sewell et al. 2000, 2002	Brown et al. 2003
		Percentage of latewood (individual rings and composite rings)	V : 12♣, U : 5♣	V : 1.7–5.7, U : 1.8–4.4		
		Average microfibrilar angle (individual rings and composite rings, measured for early and latewood)	V : 4♣, U : 2♣	V : 1.7–5.7, U : 1.8–4.4		
		Cell wall content	V : 10♣	V : 1.7–5.7		

ent genetic and environmental backgrounds." Indeed, estimation of QTL stability (position and phenotypic variation explained) is one of the most critical factors if QTL analysis is to be performed for application in MAS. This question has been addressed at different levels, including the stability of QTLs across different developmental stages (ontogenic or cambial age effect), time points during the growing season (seasonal effect), environments, and among diverse genetic backgrounds.

Ontogenic effect: Given the long-lived characteristic of forest trees, it has been questioned whether or not the same genomic regions would control quantitative traits (e.g., annual growth, density) as trees mature. Several experiments have been conducted to answer this question (Plomion et al. 1996a; Emebiri et al. 1997; Kaya et al. 1999; Sewell et al. 2000, 2002; Weng et al. 2002; Brown et al. 2003; Gwaze et al. 2003; Pot 2004). In general, a rather low QTL stability has been noted across different maturation stages for growth (Brown et al. 2003; Pot 2004). In only one case did the authors report on the detection of the same QTL regions at different maturation stages (Gwaze et al. 2003). Conversely, Kaya et al. (1999) did not find a single "common" QTL controlling growth rate through tree development. The same trend – low QTL stability across maturation stages – was observed for wood density, a trait that presents a higher juvenile-mature correlation than growth (e.g., Williams and Megraw 1993, Hannrup and Ekberg 1998). This unexpected result suggests that some of the wood-density QTLs probably reflect the genotypic response to annual climatic variation. Recently, Rozenberg et al. (2002) reported on the alteration of wood-density profiles in response to drought. More interestingly, they showed that the alteration of annual wood-density profiles (the presence of a false late wood ring in the early wood zone) in response to drought was genetically controlled.

Seasonal effect: Sewell et al. (2000, 2002) and Pot (2004) analyzed the seasonal stability of growth and wood-quality QTLs for traits measured in spring (early wood) or summer (late wood). Overall, half of the detected QTLs were specific to one type of wood (early vs. late wood). These results agree with recent transcriptome studies that reveal that different sets of genes are regulated throughout the growing season (Le Provost et al. 2003; Egertsdotter et al. 2004).

Environmental effect: although several studies have been performed in multisite trials (Groover et al. 1994; Knott et al. 1997; Kaya et al. 1999; Sewell et al. 2000, 2002; Shepherd et al. 2002b; Weng et al. 2002; Brown et al. 2003; Devey et al. 2004a, b), in only one study did the authors analyze QTL stability at different sites (Groover et al. 1994).

Genetic effect: a complete understanding of the genetic variability of the traits of interest will rely on the analysis of multiple populations as all the major genes involved in the genetic control of a given trait are unlikely to be polymorphic in a single family. On a more practical side, QTL stability across different genetic backgrounds is a prerequesite to marker-assisted breeding (MAB) in multiparental tree breeding programs. As underlined by Brown et al. (2003), MAB will reveal its full potential under two scenarios: (1) if a genetic marker in full LD with molecular polymorphism causing trait variation at the population level is discovered or (2) if the gene (polymorphism) underlying a QTL is identified. Given their allogamous reproductive system and their recent domestication, pines are characterized by high levels of genetic diversity and low levels of LD (reviewed in González-Martínez et al. 2006b). The combination of these two factors (high diversity and low LD), together with their perennial characteristics (maturation, environmental heterogeneity), is likely to contribute to QTL instability across genotypes. Yet few studies have addressed this important issue in pines. Kaya et al. (1999) did not find any QTL shared between pedigrees, while Brown et al. (2003) and Devey et al. (2004a, b) found only a small fraction of the detected QTLs to be common across different genetic backgrounds.

Multiple-Trait QTL Analysis As stated ealier, multiple-trait QTL analysis will likely become very important for breeding purposes since pine breeding is a multitrait process. In several QTL experiments, more than one trait (e.g., growth, wood quality) was studied. In most cases, colocalizations between QTLs for different traits were observed, which might be expected for highly correlated traits. Possibly more important was the occurrence of multiple colocalizations in the genome, suggesting the effect of pleiotropic genes rather than the existence of physically linked genes controlling different traits. It should be noted that QTL clusters were also observed for traits that were not phenotypically correlated (Brown et al. 2003; Pot et al. 2005b), suggesting strong environmental/developmental effects masking genetic correlation.

2.3.3 Future Direction on QTL Mapping

As underlined in the previous section, the most reliable QTLs – from a breeding perspective – are those that have been consistently detected at different developmental stages, in different environments, and in diverse genetic backgrounds. However, it is important to remind the reader that most of the studies performed so far in pines have been based on single pedigree analysis, and that only a handful of experiments have attempted to validate QTLs across different genetic backgrounds (Brown et al. 2003; Devey et al. 2004a, b). Considering that pine improvement involves the deployment of many families/clones, the genetic stability of marker-trait association is a prerequisite before any extended use of molecular markers is considered in operational breeding programs. There have been two major attempts to identify diagnostic markers for MAB either using neutral markers spanning the genome or selecting candidate genes based on their coincidence with QTLs.

Complex Designs for Detecting Marker-Trait Associations: LD Mapping

As underlined in Guevara et al. (2005b), while QTL identification is based on physical LD generated in one or a few generations of crossing, association mapping or LD mapping takes advantage of events that created associations in the past to find a statistical association between molecular markers and a phenotype on a much finer scale. As presented in Sect. 2.1.6., five studies have provided valuable information regarding the extent of LD in pines. In all cases, it was found that LD extended only over short distances. In *P. taeda*, Brown et al. (2004) and González-Martínez et al. (2006a) revealed a rapid decay of LD within 800 to 2,000 bp in candidate genes for wood quality and drought-stress response. In *P. sylvestris* a rapid decay of LD was also detected by Dvornyk et al. (2003) and García-Gil et al. (2003). The genotyping of a breeding population of *P. radiata* with microsatellites (Kumar et al. 2004 also yielded the same result, i.e., no significant LD was observed between pairs of genetically linked markers, suggesting that LD decreases rapidly with physical distance.

LD mapping is only in its infancy in pines. To our knowledge, only one study has reported the use of molecular markers (SSRs) in unrelated trees (Kumar et al. 2004). Marker-trait associations were analyzed in a *P. radiata* trial consisting of 45 parents (40 males and 5 females). Parental trees were genotyped and the association between parental genotypes and the performance of 200 full-sib-generated families were analyzed according to the strategy used in larch and eucalyptus by Arcade et al. (1996) and Verhaegen et al. (1998), respectively. This analysis allowed the identification of several significant associations between markers and traits. However, it is important to note that none of the marker-trait associations found using full-sib family performance were identified when the parental general combining ability was regressed on the allelic frequencies of the marker. This result suggests that the first associations detected were probably biased due to population structure (only five females), as in larch and eucalyptus.

Pan-genomic LD mapping in pines will require an extremely high marker density, given the low extent of LD as estimated so far within the few genes analyzed. Alternatively, marker-trait association can be performed on selected candidate genes.

The Candidate-Gene Approach

Candidate genes (CGs) can be proposed based on the coincidence between QTLs and known functional genes putatively involved in the genetic control of the trait. Such positional CGs have been described in *P. taeda* (Brown et al. 2003) and *P. pinaster* (Chagné et al. 2003; Pot et al. 2005b) for wood-quality-related traits. In *P. taeda*, colocalizations between genes involved in monolignol biosynthesis (4CL, C4H, C3H, and CcOAOMT) and QTLs for wood density were observed (Brown et al. 2003). In *P. pinaster*, a single candidate gene-QTL colocalization was found between KORRIGAN, a gene involved in the hemicellusose/cellulose biosynthesis, and QTLs for hemicellulose and fiber characteristics.

Other types of analysis can be used to select CGs related to plant adaptation. As underlined by González-Martínez et al. (2006b), "standard neutrality tests applied to DNA sequence variation data can be used to select candidate genes or amino acid sites that are putatively under selection for association mapping." Unusual patterns of nucleotide diversity and/or population differentiation have been detected in *P. taeda* (González-Martínez et al. 2006a), *P. pinaster*, and *P. radiata* (Pot et al. 2005a). Pot et al. (2005a) observed singular patterns of nucleotide diversity in three genes: a glycin-rich protein homolog that was found to be up-regulated in late wood-forming tissue (Le Provost et al. 2003); *CesA3*, a cellulose synthase gene; and *KORRI-*

GAN, a membrane-bound endo-1,4-beta-glucanase involved in cellulose/cellulose biosynthesis. As mentioned before, colocalization between *KORRIGAN* and wood-quality QTLs has been reported (Pot et al. 2005b). In *P. taeda*, although the action of neutral processes cannot be completely ruled out to explain the patterns of nucleotide diversity observed for *CcoA-OMT1*, several characteristics of its nucleotide diversity seem to indicate the action of natural selection on this gene. The colocalization of this gene with a QTL of water use efficiency (Brendel et al. 2002; Pot et al. 2005b) and the differential expression of this protein under different watering regimes (Costa et al. 1998) clearly emphasize the putative role of this gene in wood-trait variation. Association studies should now be used to validate this hypothesis.

2.4 Marker-Assisted Breeding

This section discusses the application of information from DNA polymorphisms in conifer-type tree-improvement programs. Note here that such information includes not only purposefully designed markers but also information obtained directly from (re)sequencing, for which specific markers may not have been designed. Applications of information from DNA polymorphisms fall into four generic areas: audit and quality control, elucidation of genetic phenomena, population management, and selection and breeding. We discuss each of these aspects below, covering both existing and potential applications. Furthermore, we provide examples where appropriate, as well as comments regarding the current status of each of these applications.

2.4.1 Quality Control and Audit

Development of an array of DNA marker systems over the past two decades has provided tree breeding programs with a range of tools to achieve basic aspects of quality control that hitherto could only be addressed with difficulty, if at all. Seedlot and/or clonal fidelity have been the key concerns for both commercial and research applications. Indeed, this area was the earliest commercial application of DNA markers in tree breeding programs and is still the most widespread, at least in coniferous species.

From a commercial perspective, the key objective is assurance of genetic gain by ensuring that seedlots and/or clones are true to intention. The applicability and efficacy of various DNA marker systems is dependent, therefore, on the way in which genetic gain is delivered. In conifers, such delivery can come in a number of different forms (and costs), even within the same breeding program.

Open-Pollinated Seed

Open-pollinated seed from a mixture of selected seed orchard material generally consists of a few dozen maternal parents, with little if any control of the pollen source. This is a common means of seed production, particularly in commercial *Pinus* species in the southeastern United States. In such cases, there may be a need to ensure no contributions from unwanted maternal parents, which can be achieved by genotyping megagametophytes with sufficient markers, assuming maternal parentage information is maintained for all seed. Furthermore, markers could be used to quantify the relative contributions of each maternal parent to individual seedlots if seed counts are not available. A variation on this method of seed production is used in mass pollination techniques, such as supplemental mass pollination or liquid pollination. In these approaches, receptive conelets are pollinated using pollens from selected genotypes by various means, but the conelets themselves are not actually covered to prevent fertilization by unwanted pollens.

Control-Pollinated Seed

Control-pollinated seed, where pollens from selected parents are used to fertilize selected seed parents, in the process excluding pollen from other sources using bags to cover receptive cones and/or undertaking pollinations in contained greenhouse facilities. This is usually undertaken to produce full-sib families or individual half-sib families with known selected pollen parents. Such methods are used in both breeding and commercial production for species such as *P. radiata*. Vegetative propagation, sometimes involving in vitro technologies, can be used to amplify genotypes, particularly for commercial production, largely because of shortages of seed and/or cost of seed production. DNA markers – particularly codominant multiallelic marker systems such as microsatellites – have been developed for such purposes (e.g., Devey et al. 2003). In addition, paternal parentage can sometimes be evaluated via paternally inherited chloroplast markers. Such markers have been developed and applied in *P. radiata* for checking paternal inheritance (Kent and Richardson 1997), although judiciously chosen nuclear DNA markers may also suffice. In general, markers have shown that misidentification is common in breeding programs; for example, Bell et al. (2004) estimated that 2.6% of parents were misidentified in a sample of an Australian breeding population of *P. radiata* and that 8.4% of offspring of ten families were not consistent with expected parentage. These results also indicate that a proportion of misidentification of open-pollinated seedlots is possible. Furthermore, our experience at Scion with putative full-sib families used for gene mapping experiments has revealed very few such families – produced either commercially or by research groups – are completely consistent with expected parentage (unpublished data). DNA markers, particularly codominant marker systems, have therefore been useful in checking and assuring parentage.

Clonal Production

Clonal production is used either for experimental purposes (such as clonal tests) or for mass-propagation of tested clones for clonal forestry. In these cases, genotyping is undertaken to ensure that ramets do represent the desired genotype(s). Similarly, ramets deployed in seed orchards raise the same issue, although sometimes misidentification of parents is detected via parentage testing of the seed obtained via methods described above. For applications where genotype fidelity is needed, "profiling" marker systems such as RAPDs and AFLPs is useful in that they are generally cheaper to both develop and use for this specific application, particularly as the high level of polymorphism revealed overcomes issues associated with dominance. However, in more recent years, fingerprinting kits have tended to utilize codominant markers (see references above), as these have been developed for other purposes (above) and are generally adequate for clonal fingerprinting, particularly if enough marker loci are used (e.g., Kirst et al. 2005b).

DNA markers have, therefore, been developed and utilized extensively in tree breeding programs, for both verifying commercially deployed materials and for ensuring that experimental materials meet the requirements. Even so, some programs still do not universally implement or rely on such genotyping, largely due to the expenses involved.

2.4.2 Elucidation of Genetic Phenomena

All genetic gain ultimately depends on DNA polymorphisms. Knowledge of the nature and effects of the polymorphisms has the potential to generate far more gain than the use of purely phenotypic data – on the parents and/or their progeny. Information from DNA polymorphisms enables the elucidation of phenomena such as understanding of the genetic architecture of trait variation, revelation of population structure and history, and detection of selection fingerprints, all of which can have direct or indirect applications in tree breeding.

The genetic architecture of trait variation can be defined as the frequencies, location, magnitude, and mode(s) of action of quantitative trait loci/nucleotide (QTL/N) effects underpinning quantitative traits. While QTL mapping has been very informative in this regard, the results are relevant only to the pedigree(s) used, not to whole populations. Association genetics may, therefore, be more relevant for understanding the genetic landscape of trait variation in forest trees.

Estimation of Genetic Variance Structures and Heritabilities

DNA markers offer new ways of obtaining some key knowledge that is fundamental to tree improvement regarding genetic parameters, in particular genetic variances, heritabilities, and correlations among traits. Such knowledge informs the breeder not only about the feasibility of breeding for certain traits or combinations thereof, but also how it might be efficiently undertaken.

For obtaining genetic gain, a trait must be both variable and heritable. Genetic correlations between economic traits can be a major constraint if they are adverse, but they can provide the breeder with great opportunities if they are favorable. Even genetic correlations between noneconomic indicator traits and economic ones can be used to great advantage if the correlations are strong and the indicator traits highly heritable. For genetic variances and heritabilities, if they have to be estimated from seed collected in natural stands, then the coefficients of relationship within seed-parent families need to be known. Indeed, the relative contributions of inbreeding as such and finite effective numbers of (unrelated) pollen parents per seed parent represent important information. While in the past isozymes were a valuable tool for providing such information, DNA markers can be much more powerful for the purpose, and with a greater range of applications. This issue of nonrandom components in the mating system can be important with native stands of conifers (Burdon et al. 1992), despite their wind pollination, even if it is less acute than in insect-pollinated species like eucalypts (Hodge et al. 1996). The wind pollination of conifers tends to reduce the nonrandomness of mating, while the conifers' mechanism of archegonial polyembryony probably reduces the inbreeding component still further. For estimating genetic correlations between traits, knowledge of coefficients of relationship within such families is likely to be much less crucial. As we will mention later, a priori family information as such may not be essential for this purpose, but it can be supplemented with information from markers.

Population Structuring and History

The structuring and history of populations influence both the availability of genetic variation for the breeder to exploit and the potential for association genetics to contribute. While large, essentially panmictic populations cannot be expected to have appreciable across-family linkage disequilibrium (LD), cryptic structuring may exist that generates significant disequilibrium that could be useful for breeders. For example, localized population bottlenecks, followed by coalescences, could easily cause this. Such LD could provide valuable clues to "metapopulation" history. Despite wind pollination, various factors can generate population structure in conifers (Mitton 1992). Interesting possibilities of structure exist in populations derived from recent admixtures. In *P. radiata*, the exotic, domesticated "landraces" still have large elements of the wild state. Interestingly, they evidently represent a genetically recent fusion of two of the native populations, Año Nuevo and Monterey (Burdon et al. 1998), which may provide a basis for some admixture disequilibrium.

Polymorphisms revealed by DNA sequence data derived from both genic and nongenic regions can reveal much about the genetic history of those regions. Departures from Hardy-Weinberg equilibrium could reveal the presence of previously undetected genetic phenomena such as the presence/absence of inbreeding. Indeed, genetic variance (and gain) estimates are based on assumptions regarding the relatedness of parents used in genetic tests. Such data can be used to check these assumptions and provide empirical data for more accurate estimates. Similarly, sequence data from genic regions can reveal evidence of selection: for example, balancing selection was detected by Cato et al. (2006b) in a gene associated with wood density and growth rate in *P. radiata*. Krutovsky and Neale (2005) found evidence for selection in three of 18 genes in Douglas-fir. Such evidence – which can be generated on a relatively small subset of genotypes – could be an effective prescreen for genes more likely to be associated with trait variations, although some caveats apply regarding the power to detect the effects of selection (Wright and Gaut 2005).

Assignment of Gene Function

Knowledge of gene function, if acquired, offers the greatest long-term opportunities to capture genetic gain, using either endogenous variation or genes introduced by conventional breeding or genetic engineering. LD mapping and association genetics can provide clues in the search for quantitative trait nucleotides (QTNs) regarding which genes have functional roles in trait variation. While relatively short stretches of disequilibrium (usually hundreds to low thousands of base pairs) represent constraints for breeding applications, a key advantage is the potential for assignment of function – even identification of individual QTNs. However, because of the size of conifer genomes and the lack of genomic sequence, identifying candidate genes will be crucial. Possible approaches are described in the following section and in more detail by Wilcox et al. (2006).

In carefully selected cases, genetic transformation can be used to verify the role of a QTN in generating phenotypic variation. The costs of achieving a transformation, and often the regulatory issues, will demand a highly selective application of this approach. On the other hand, knowledge of gene function may be useful for identifying genes to target for genetic transformation, to create new variations of use for breeders, and to provide commercial cultivars. There are no reports yet of cloning QTLs from tree species, partly due to the large number of candidates within QTL confidence intervals, but also because of the time required for trait expression of transformants arising from complementation studies. Nonetheless, association genetics will be a key step in increasing resolution and, in some cases, identifying putative QTN for further analyses.

2.4.3
Population Management

Pedigree Reconstruction and Detection of Genetic Contamination

Breeding populations represent the "engine room" for capturing the additive gene effects that allow cumulative genetic gain over successive generations, through recurrent cyles of selection, intermating, evaluation, selection, and so on. Maintaining full pedigree has been favored on the grounds that it helps preclude inbreeding and maintains effective population size. However, the expense of maintaining full pedigree can limit the size of the breeding population that can be handled, raising the question of whether larger populations can be handled if pair crossing is not mandatory. Moreover, there may be situations where the breeder has reason to resort to material, e.g., commercial stands, in which pedigree has been sacrificed but which has the advantage of huge numbers (Burdon 1997).

Modern marker technology has major potential for pedigree reconstruction (Lambeth et al. 2001; Kumar et al. 2006), at least in open-pollinated families of known seed parentage. Complete reconstruction of predigree will be more challenging, especially with the wind pollination that characterizes conifers. In maintaining gene resources that underpin breeding populations, controlled crossing is typically far too expensive, yet there may be a call for quantifying and even detecting individual cases of contamination (Burdon and Kumar 2003). Different types of markers may be required for this purpose, and the task may be challenging, but it appears inherently feasible. Pedigree reconstruction is discussed later, in connection with selection.

Tracking and Maintenance of Genetic Diversity

A key element of population management is maintenance of genetic diversity, to give the breeder flexibility in both the short and long terms and to safeguard continuing long-term genetic gain. While pedigree information is an indicator of genetic diversity, it does not provide definitive information in itself. Achieving that poses significant challenges.

The "gold standard" for functional genetic diversity will usually be performance in well-designed and properly located common-garden genetic experiments. However, such experiments are expensive and often slow to deliver results. The use of DNA polymorphisms is clearly much quicker, but such marker diversity will need to be cross-referenced with the functional diversity since the two classes of diversity are not necessarily closely coupled, at least among species (Morgante and Salamini 2003; Paran and Zamir 2003). Components of DNA diversity obviously include percentage of polymorphic genes (in either the coding or regulatory regions), percentage of base pairs that are polymorphic, particularly for QTN that must exist in coding or regulatory regions), and allele frequencies for the polymorphisms. Such information, in conjunction with knowledge of magnitudes of QTN effects, provides a benchmark for monitoring changes in diversity, for any forest trees.

Loss of low-frequency alleles is an obvious manifestation of a decrease in genetic diversity. Paradoxically, abrupt increases in the frequencies of such alleles, as can occur through genetic drift, can be a manifestation of the same phenomenon. While the inherently outbreeding genetic systems of forest trees may be able to cope with significant inbreeding in the wild, through selection for balanced heterozygotes, such mechanisms may be impeded under conditions of artificial breeding. In principle, almost any sort of genetic marker should be able to manifest the losses of alleles or sharp fluctuations in their frequencies. Nevertheless, it seems preferable to know what genes are of particular current or contingent importance and monitor their frequencies. This, however, will depend on knowing the functions of the genes.

Provision for Biotic Crises

While it appears QTNs typically exert minor individual effects in conifers, disease-resistance genes can represent a notable exception (Burdon 2001). Such genes can both have large effects (despite outward appearances of classical quantitative inheritance) (e.g., Kinloch et al. 1970; Wilcox et al. 1996) and be present at low frequencies (R.D. Burdon and P.L. Wilcox personal communication). These genes can also have the feature of gene-for-gene specificities between host genotypes and fungal strains (pathotypes), which has important implications for ensuring durability of disease resistance against pathogen mutations (Burdon 2001). Establishing the nature of such genes can require carefully planned mating between parents and inoculation studies based on single-spore isolates, preferably backed up with genomic studies. Here, as in other areas, comparative genomics can have a major role, at least in studying resistance genes of lower specificity.

The implications of such patterns of genetic variation within both hosts and pathogens, which are called pathosystems, can be major for both population management and selection. Disease resistance can be very important in plantation forestry, especially with exotics that are being grown in the absence of natural or other pathogens. Conserving low-frequency genes of large effect, which are sometimes important in conifers, can pose a threefold challenge. A large population may be needed to find trees with such resistance. The requisite population size can be increased by the desirability of obtaining resistance genes in unrelated pedigrees. It may be further increased by the desirability of combining ("pyramiding") genes in the genetic material that represent diverse mechanisms of resistance to the pathogen, toward ensuring durability of resistance against mutation or genetic shifts in the pathogen population. In implementing this approach, identification of resistance genes will be of enormous help.

The information on forest-tree pathosystems, in general, is still very sketchy, although there are resistance genes identified for some coevolved forest pathosystems (e.g., Wilcox et al. 1996). Nevertheless, what is known indicates that preparations for biotic crises in the form of new fungal diseases should include having very large population resources available. In at least some cases, it is very doubtful whether breeding programs effectively contain such a provision (Burdon and Gea 2006).

2.4.4 Selection and Breeding

Pedigree Reconstruction as an Alternative to Maintaining Pedigree Records

A proposed application (Lambeth et al. 2001) is to use markers in genetic tests to reconstruct pedigrees retrospectively, as opposed to maintaining pedigree information throughout the life of a genetic test. This involves applying pollen mixes of known composition to a range of seed parents, planting the resultant offspring in designed experiments with limited or no maintenance of family information (Kumar et al. 2006). Upon subsequent measurement, pedigree identity is ascertained via DNA markers. Benefits of such an approach include logistical simplicity, cost reductions for breeding and testing, potentially better estimates of genetic parameters and increased gains, and assurance of parentage (see above). Furthermore, fewer financial costs are incurred if experiments are abandoned or lost prior to remeasurement – which does happen in tree-improvement programs. Disadvantages include underrepresentation of some specific families due to either chance or factors affecting fertilization, prevalence of inbreeding in some families distorting breeding-value estimates of those families, difficulties in detecting full-sib-specific combining ability in full-sib families, and genotyping costs. Most of these disadvantages could be overcome, although further research is needed to fully evaluate various possible application scenarios. To date, this approach is currently being evaluated in the context of operational breeding in New Zealand with *Eucalyptus* and *P. radiata*.

Estimation of Heritability Based on Marker-Ascertained Relatedness

Because molecular markers have the potential to estimate relatedness between genotypes, methods to estimate the heritability of phenotypic characteristics have been proposed and evaluated (Andrew et al. 2005; Kumar and Richardson 2005) without the need to have any prior knowledge of genetic relationships (see above). Such methods have the potential benefit of allowing one to obtain information from existing forests (e.g., natural forests or commercial plantations), thereby obviating the need for genetic testing involving progenies, thus speeding up the generation of information, particularly for species where little a priori information is available. Some disadvantages involve the reliability of marker-based estimates, particularly as assumptions need to be either made or checked regarding genetic structure and prior levels of relatedness, as well as the reliability of phenotypic information, particularly if sourced from forests rather than purpose-designed common-garden tests with appropriate controls. These approaches have not been extensively investigated but may have a role. In part, this may be due to the fact that existing approaches are well established and successful, as well as to the existence of infrastructure such as extensive testing involving progenies. Kumar and Richardson (2005) compared phenotype-based with marker-based heritability estimates for wood density in *P. radiata* and found very little difference. Similarly, Andrew et al. (2005) reported nonzero estimates of heritability for a number of foliar defence-related chemicals in *Eucalyptus melliodora*, some of which were consistent with independent estimates based on phenotype (see Andrew et al. 2005 and references therein). Such approaches may be of use also for association genetics, particularly in revealing cryptic population structuring in natural forests, as well as for selecting maximally unrelated individuals for estimating LD.

Within-Family Selection Based on Marker-Trait Associations Derived from Pedigreed QTL Mapping Populations

Such selection is based upon selecting individuals within known full- or half-sib families where marker-trait associations have been previously determined. Genotypes with the desired multilocus genotypic composition are selected from within families. In theory, this can be applied within families in breeding populations and/or production populations. Generic benefits include earlier selection, increased selection intensity, and potentially cheaper selection (Stromberg et al. 1994). In conifer improvement, breeding populations are usually based on a few hundred parent genotypes; thus it is largely impractical and very expensive to detect associations for each genotype independently (Strauss et al. 1992; Johnson et al. 2000). Rather, application of markers is more likely to be restricted to populations where specific families are advanced, as for example in elite populations or in small nucleus- or mainline breeding populations.

A number of situations have been evaluated either for species-specific scenarios or from a wider theoretical perspective. One of the earliest such studies was that by Strauss et al. (1992), who concluded that within-family marker-assisted selection (MAS) was of limited or no value unless a high proportion of additive genetic variation could be explained by markers for traits of low heritability but high value and where selection intensities within families were high compared to that among families. They also concluded that difficult- or expensive-to-measure traits such as wood quality or resistance to certain diseases showed the most potential for MAS. Williams and Neale (1992) evaluated the relative efficiency (RE) of MAS and also concluded RE was greatest where (a) traits were not expressed sufficiently early to enable a reduction in generation length as well as (b) with lower-heritability traits where markers explain a high proportion of additive genetic variability for the traits. However, results from many QTL mapping experiments have indicated that the genetic architecture of quantitatively inherited traits is dominated by genes of small effect (Wilcox et al. 1997; Sewell and Neale 2000; Brown et al. 2003; Devey et al. 2004a, b), thus limiting the opportunity to explain sufficient variation with markers. Wu (2002) evaluated the tradeoff between proportion of variance explained by markers and shortening of selection interval. For traits of moderate heritability (0.2 to 0.4) marker-assisted early selection (MAES) was more efficient than phenotypic selection, if they explained more than 5 to 10% of additive genetic variance and allowed the selection interval to be reduced by half. For later-expressed traits, even less variance explained by markers was necessary for markers to be more efficient than phenotypic selection. Wu (2002) also evaluated a number of other scenarios, including combining phenotypic selection with markers, and found that MAES is only marginally more efficient than phenotypic selection except in cases of relatively low juvenile-mature phenotypic correlations. Given genetic architectures of QTLs explaining a few percent each (Wilcox et al. 1997; Brown et al. 2003; Devey et al. 2004a, b), and limitations for the selection of up to ten unlinked markers per family linked to such QTLs, the effectiveness of early selection is limited for traits with low-moderate age-age phenotypic correlations, particularly once costs of QTL detection are taken into account. Time savings for traits of delayed expression or low juvenile-mature correlation will, however, depend on the prior establishment of QTL/phenotype associations.

All the above studies investigating the application of markers for breeding population advancement have generically shown that while there are scenarios where markers could be used to generate genetic gains, the financial gains could be quite limited. This is in part due to the cost of genotyping as well as the length of time, meaning that additional costs will be incurred. A further limitation stems from the fact that most breeding objectives involve multiple quantitatively inherited traits whose genetic architectures are likely to involve predominantly small-effect genes (see references above). This means limited genetic gains from individual markers. Furthermore, relatively few markers can be used to select within any particular family: for a simple two-genotype-per-marker scenario, it would be necessary to generate and genotype thousands of offspring per family to have sufficient power to generate the optimal ten-marker-locus genotype. If such selection were restricted to a limited range of traits, there would also be the need to augment marker information with phenotypic records to effectively address the breeding objective; therefore the opportunity for earlier selection could be lost. On the other hand, few studies have investigated the efficacy of marker-based or marker-assisted selection

southeastern USA, and *P. pinaster* in Europe. The evidently close synteny among these species promises great opportunities to apply genomic information on any one species to the others. Using these as model species within conifers should allow one to capitalize on the most favorable model features of all the species at once. Similarly, whole-genome sequences of *Populus* and *Eucalyptus* will allow for even more rapid advances in these species.

Should association genetics be utilized by tree breeders, then a number of other requirements need to be addressed. These include:

- Populations and analytical methods for adequate detection of marker-trait association (see Ball 2005),
- Suitable laboratory and bioinformatics capabilities functionally integrated into tree-improvement programs, and
- Development of strategic alliances among appropriate entities to effectively manage resource association genetics programs in a manner that allows operational implementation.

The above issues are discussed in more detail by Wilcox et al. (2006).

2.4.5 Summary of Applications of DNA Polymorphisms in Conifer Breeding

Information from DNA polymorphisms has a wide range of applications for tree improvement, including quality control, pedigree reconstruction, elucidation of genetic phenomenon, monitoring and maintenance of genetic diversity, and selection and breeding based upon polymorphisms associated with trait variation. To date, only some of these applications have been implemented in operational breeding programs – largely those associated with quality control. Some potential applications, such as association genetics and pedigree reconstruction as part of operational testing, are largely in the research and development phase, while others such as within-family MAS have been sufficiently evaluated to identify at least some strategies for implementation. The lack of uptake across the spectrum of potential applications is likely due to cost, which is particularly important in tree-improvement programs, which usually take years to recoup such costs. Nonetheless, technological advances will increase the scope of applications for tree improvement.

2.5 Genomics Resources for the Genus *Pinus*

2.5.1 Efforts Toward Complete Genome Sequencing

BAC Libraries

Large-insert genomic DNA libraries in which each DNA clone is stored and archived individually (i.e., ordered libraries) are a fundamental tool in modern genomics research. The most popular large-insert vector is the bacterial artificial chromosome (BAC), which, despite its name, is not a chromosome but rather a modified bacterial plasmid (F-factor) that can stably carry large inserts (ca. 50 to 400 kb) (Zhang and Wu 2001). BAC libraries have been made for a host of taxa and employed in a variety of applications (reviewed by Zhang and Wu 2001).

Despite the importance of pine and other conifers and the value of BAC libraries to genomic research, BAC resources are extremely limited for gymnosperms. To our knowledge, there are three gymnosperm BAC libraries in existence – all of them for pine. The first pine BAC library, constructed by Islam-Faridi et al. (1998), affords 0.05× coverage of the *P. taeda* genome. A second library, developed by Claros et al. (2004) for *P. pinaster*, provides considerably higher genome coverage (0.32×), although the probability of finding a gene of interest is still relatively low (27%). However, the *P. pinaster* library is reportedly being expanded to 3× (Claros et al. 2004), which would provide 95% probability of finding a given genome sequence (Fig. 3). In September 2004, the US National Science Foundation funded a 3-year project that included construction of a 10× BAC library from the *P. taeda* genotype "7-56" (Peterson et al. 2005) as a primary objective (Fig. 4). When completed, this library will be, to our knowledge, the single largest BAC library ever made with roughly 1.7 to 2 million individual clones. As of this writing, the 7-56 BAC library affords 2× coverage of the pine genome. When the 7-56 library reaches 3× coverage (ca. June 2006), it will be gridded onto BAC macroarrays and screened with STS and molecular genetic markers including those associated with genes of economic importance. The resulting information will be utilized to isolate and sequence intact pine genes. Additionally, marker localization will represent the start of STS-based

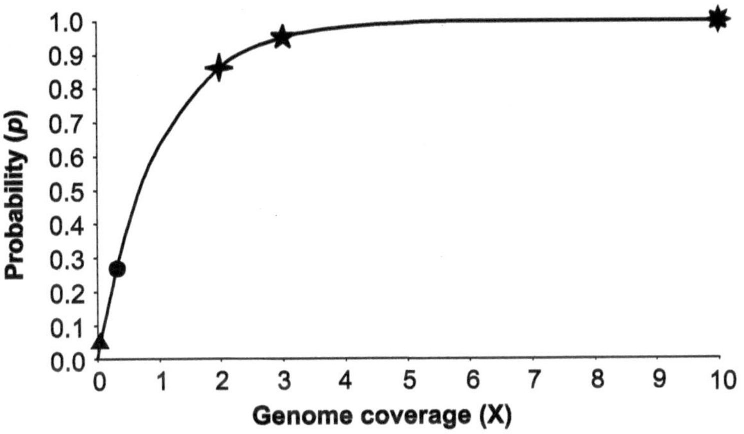

Fig. 3. The relationship between genome coverage (×) provided by a BAC library and the probability (p) of finding a particular sequence of interest in that library. A 0.05× BAC library, like the *P. taeda* BAC library generated by Islam-Faridi et al. (1998), affords only a 4.8% chance ($p = 0.048$) of finding a sequence of interest (*triangle*), while a 0.32× library, like the current *P. pinaster* library (Claros et al. 2004), affords a considerably better chance ($p = 0.27$; circle). The *P. taeda* 7-56 library (Peterson et al. 2006) currently has a size of 2× ($p = 0.86$; *four-point star*) and will be used to isolate genes of interest once it reaches 3× coverage ($p = 0.95$; *five-point star*). At 10× coverage ($p = 0.9999$; *eight-point star*), the *P. taeda* 7-56 BAC library will be of adequate size for essentially any use including genome-wide physical mapping and genome sequencing

physical mapping in pine. Comparison between the *P. pinaster* and *P. taeda* libraries should provide considerable insight into genome evolution within *Pinus*.

Need for Pine Genomic Sequence

Although EST sequencing in pine is relatively advanced (Sect. 2.5.2), there are few genomic sequence data available for pine or any other gymnosperm. As of 6 March 2006, the longest continuous gymnosperm/conifer genomic sequence in GenBank was only 6,884 bp. While ESTs provide the coding regions of many expressed genes, an understanding of gene function and regulation requires knowledge of those noncoding sequences that coordinate gene expression in response to biotic and abiotic cues (e.g. promoters, enhancers, silencers). Such information necessarily comes from sequencing genomic DNA.

As the pine genome is ten times larger than that of maize (the largest plant genome to be the target of a full-scale genome sequencing effort, NSF 2005), it may be a while before whole-genome sequencing of pine becomes a reality. Nonetheless, pine genomics can be greatly advanced by even relatively modest sequencing of genomic DNA, especially sequencing of large continuous pieces of DNA such as BAC inserts. The following are just a few areas of pine genome research that will benefit from more extensive sequencing of genomic DNA.

- *Genome structure*: Very little is known about the structure of any conifer genome, and relationships between genes, repeats, and pseudogenes have only been tangentially explored (e.g., Elsik and Williams 2000; Rabinowicz et al. 2005 ; Peterson et al. 2006). BAC and shotgun sequencing of pine DNA will provide a more detailed understanding of the sequence structure of the pine genome. Such information will be essential in the development of an efficient strategy for sequencing the *gene space* of pine.
- *Differential gene expression*: The lack of genomic sequence for pine has severely limited the study of the noncoding regulatory regions of its genes. Information on these regions is essential if we are to understand differential gene expression in pine and manipulate conifer genes in a useful and controllable manner (No et al. 2000). To determine the general tissue/development/environment specificity of the regulatory sequences for a given gene, the expression profiles of that gene can be examined; since there is already a considerable amount of expression profile data available (e.g., Egertsdotter et al. 2004; Lorenz et al. 2006), associating regulatory sequences with specific tissues and/or developmental events could be initiated immediately after sequencing. Comparison of regulatory sequences from genes with similar expression patterns can be used to gain insight into the ac-

Fig. 4. Constructing the *Pinus taeda* 7-56 BAC library. A Genetix QPixII robot picks 7-56 BAC colonies off of agar in a "Q-Tray." Picked clones are used to inoculate media in 384-well microtiter plates (*left side* of image)

tivation/repression mechanisms underlying major developmental events.

- *Comparative genomics of conifers*: Pine BAC/shotgun sequences and comparative genomic approaches can be utilized to advance the study of other gymnosperm genomes, which will afford considerable insight into the evolution of this important group of organisms. Comparison of orthologous regions of conifer genomes (e.g., through sequencing of BACs and/or BAC contigs recognized by common markers) will provide high-resolution means of investigating conifer sequence evolution (see Stirling et al. 2003 and Yan et al. 2004 for angiosperm examples).
- *Gymnosperm/angiosperm comparisons*: Information garnered from pine genomic sequence should provide nearly limitless opportunities to explore comparative genome evolution between gymnosperms and angiosperms. BAC sequences from pines can be compared with existing plant genome sequences as a means of evaluating macro- and microsynteny as well as sequence conservation/divergence between pines and angiosperms. As pine has been widely used as an outgroup in angiosperm comparative research, pine genomic sequences should enable investigations by many angiosperm research groups throughout the world.
- *Polymorphism discovery and characterization*: EST sequencing has afforded tremendous insight into polymorphisms within the coding sequences of pine genes (e.g., Krutovsky et al. 2004; Le Dantec et al. 2004). However, sequence changes in the noncoding regulatory regions of pine genes may provide as much, if not more, information on phenotypic diversity than coding sequence.
- *Association mapping*: Relatively long DNA stretches produced by BAC sequencing will facilitate genomewide studies of LD, which, in turn, will fuel association mapping projects (e.g., Brown et al. 2004; Krutovsky and Neale 2005).
- *Sequencing adaptive and economic genes of pine*: There is considerable interest in sequencing intact genes of economic importance and/or particular value to conifer/gymnosperm research. Potential target genes include those associated with outcrossing (Williams et al. 2001), embryogenesis (Ciavatta et al. 2001), disease resistance/susceptibility (e.g., Devey et al. 1995; Morse et al. 2004; Kayihan et al. 2005), abiotic stress resistance (e.g., Chang et al. 1996; Dubos and Plomion 2003; Dubos et al. 2003 ; Krutovsky and Neale 2005; Lorenz et al. 2006), and wood properties (Brown et al. 2003; Devey et al. 2004a; Krutovsky and Neale 2005; Pot et al. 2005b). Complete gene sequences can be utilized to explore pine/conifer evolution and facilitate pine improvement through traditional breeding, marker-aide selection, and genetic engineering. Of note is the fact that a BAC containing a target gene may well contain other genes of interest. The study of these "bonus" genes will likely be as important as studying target genes themselves.

- *Exploring the evolution of repeat sequences*: Recently a 454 Life Sciences Genome Sequencer 20 (Margulies et al. 2005) was used to generate 100 Mb of sequence from the *P. taeda* genotype 7-56 (Peterson et al. 2006). This sequence should provide considerable insight into pine genome structure, although the short length of 454 reads (ca. 100 bp each) limits the utility of the sequence. While pseudomolecule contigs have been assembled, 100 Mb affords only 0.046× coverage of the pine genome, and consequently it is likely that any 454 contig of significant length is a conglomeration of sequences in a repeat sequence family rather than an individual element found in native DNA. However, repeat sequences obtained from sequenced BACs and/or contigs assembled from Sanger/capillary shotgun reads can presumably be used as scaffolds on which 454 reads can be aligned. The depth and shape of the resulting alignments should permit detailed analysis of the evolution of major repeat-sequence families in pine.
- *Genetic engineering*: Regulatory sequences obtained from genomic sequences can be attached to reporter genes and used to study promoter/regulator specificities (e.g., No et al. 2000). Such testing will provide insight into pine gene regulation, which will eventually enable insertion of constructs containing pine promoters and genes/alleles of value into conifers. In addition, the genomic sequences can serve as sources of molecular markers that can be utilized in tree improvement through MAS.

Physical Mapping

Given the enormous size of the pine genome and the time and financial resources required to conduct physical mapping research, it is likely that complete physical mapping of the pine genome will be too costly to pursue, at least with present technology. Because DNA sequencing technologies are advancing rapidly (e.g., Margulies et al. 2005), it is probable that complete shotgun sequencing of the pine genome will become affordable before complete physical mapping does. However, physical mapping of gene-rich genomic regions has been utilized to advance understanding of important chromosomal regions in many species (e.g., Folkertsma et al. 1999; Sanchez et al. 1999; Dilbirligi et al. 2004; Barker et al. 2005), and indeed, physical mapping of gene-rich regions is a goal of the research group constructing the *P. taeda* 7-56 BAC library.

Sequencing the Pine Genome

C. Plomion: Will the pine genome be sequenced?
D.G. Peterson: Yes.
C. Plomion: When?
D.G. Peterson: Not this week.

The complete sequencing and assembly of the *Arabidopsis* and rice genomes have afforded considerable insight into the evolution of higher plants including pines. For example, Kirst et al. (2003) compared the genome sequence of *Arabidopsis* with pine ESTs and showed that gymnosperms and angiosperms possess highly similar gene complements. Other plant genomes, most notably poplar, sorghum (*Sorghum bicolor* L.), and maize (*Zea mays* L.), are current subjects of full-scale genome sequence efforts, and the genome sequences of these plants will further facilitate understanding of plant genome evolution and function (Paterson et al. 2005). The poplar sequence, in particular, may provide information that will help advance understanding of wood formation in trees (Tuskan et al. 2004). However, angiosperms and gymnosperms diverged from a common ancestor more than 300 million years ago (Bowe et al. 2000), and consequently the utility of angiosperm sequences in the study of pine and other gymnosperms will be limited. Ultimately, the best means of advancing pine and conifer genomics is complete sequencing of a conifer genome. While it is now theoretically feasible to sequence the genome of any organism, the large, repetitive nature of conifer genomes will likely prevent them from being targets of full-scale genome sequencing for at least a few years. However, the likelihood that the pine genome will be sequenced is high based upon past and present investments made by the NSF (including construction of the *P. taeda* 7-56 BAC library) and other granting agencies and growing worldwide interest in pine as a biofuel/carbon sequestration crop (Jackson and Schlesinger 2004; Perlack et al. 2005).

While whole-genome sequencing may have to wait, sequencing of gene-rich BACs is likely to begin in relatively short order. Additionally, reduced-representation sequencing (RRS) methods (see Peterson 2005 for review) are being used to investigate sequence subsets of the pine genome and are likely to play a role in eventual whole-genome sequencing as well. Two RRS techniques that have been successfully utilized in maize genome exploration are methylation filtration (MF) and Cot filtration (CF)[1] (Whitelaw et al. 2003; Springer et al. 2004).

- MF is based on the preferential hypermethylation of retroelements and hypomethylation of genes observed in some plants (Rabinowicz et al. 1999). In MF, genomic DNA fragments are ligated into a plasmid containing an antibiotic-resistance gene and the resulting recombinant molecules are used to transform a bacterial strain containing enzymes that preferentially cleave hypermethylated DNA. Linearization of a hypermethylated recombinant molecule results in its loss and makes its host cell susceptible to an antibiotic in the growth medium. Consequently, only hypomethylated sequences are successfully cloned. While a useful gene-enrichment tool in maize and sorghum (Rabinowicz et al. 1999; Bedell et al. 2005), MF has proven considerably less effective when applied to the "mega genomes" of pine and wheat (Rabinowicz et al. 2005).
- CF is rooted in the principles of Cot analysis, the study of DNA reassociation in solution (Peterson et al. 2002). When sheared DNA is heated to 100 °C, the two complementary strands of each double helix come apart in a process known as denaturation. If the denatured DNA is slowly cooled, complementary DNA strands find each other and form double helices (duplexes). The rate at which a particular DNA sequence finds a complementary sequence with which to pair is proportional to the number of times that sequence (and hence its complementary sequence) is found within the genome. In other words, repetitive sequences renature more quickly than low-copy sequences. In CF, sheared genomic DNA is denatured and allowed to reassociate for a period of time in which only repetitive DNA sequences are likely to form duplexes. The double-stranded repetitive DNA is then separated from single-stranded low-copy DNA using hydroxyapatite chromatography. The low-copy DNA and/or the repetitive DNA can then be cloned and sequenced (Peterson et al. 2002; Yuan et al. 2003; Lamoureux et al. 2005) or used directly as a pyrosequencing substrate. With regard to the latter option, Peterson et al. (2006) recently sequenced Cot-filtered *P. taeda* 7-56 DNA including isolated highly repetitive, moderately repetitive, and low-copy sequences using the new 454 Life Sciences Genome Sequencer 20 (see Margulies et al. 2005 for details on this instrument). These data are currently being evaluated, although initial results look promising. CF has proven highly effective in wheat, where it provides 19.5-fold enrichment for low-copy DNA (Lamoureux et al. 2005).

[1] Cot filtration was originally called "Cot-Based Cloning and Sequencing" (CBCS; Peterson et al. 2002) and later "high Cot" (HC; Yuan et al. 2003) sequencing . However, the term "Cot filtration" and the acronym CF are becoming more widely used because (a) the acronym "CBCS" can also stand for "clone-by-clone sequencing," (b) "high Cot" is frequently confused with "high copy," and (c) Cot components may be sequenced without prior cloning (Peterson et al. 2006).

2.5.2 Genomic Tools to Identify Genes of Economic and Ecological Interest

Characterization of the Pine Gene Space

A large-scale expressed sequence tag (EST) sequencing project can generate a partial sequence for a large proportion of genes from a given organism (Adams et al. 1991). As only genic sequences are analyzed regardless of genome size, this approach is considered the most cost-efficient strategy of gene discovery for organisms whose genome sequences are not yet available (Rudd 2003), especially for species, such as pines, with exceedingly large genomes primarily composed of nongenic repetitive elements. For this reason, such projects have been initiated for many organisms including the numerous agronomically important plant species (Rounsley et al. 1998). To rapidly scan for all the protein coding genes and to provide a sequence tag for each gene of the pine genome, large collections of ESTs has been developed in *P. taeda* and, to a lesser extent, in *P. pinaster*.

Loblolly pine ESTs Three major EST sequencing projects, mainly funded by the NSF, have been conducted so far in *P. taeda*. A first project, "Genomics of wood formation in Loblolly pine," was concerned with gene discovery in wood formation (http://pinetree.ccgb.umn.edu/). It resulted in ca. 60,000 ESTs placed in the public domain. A second project, entitled "Transcriptome responses to environmental conditions in Loblolly pine roots," aimed at completing this information by targeting especially roots undergoing a variety of biotic and abiotic stresses (http://fungen.org/Projects/Pine/Pine.htm). It resulted in 140,000 additional ESTs. A third project, "Genomics of Loblolly pine embryogenesis," seeks to identify and characterize

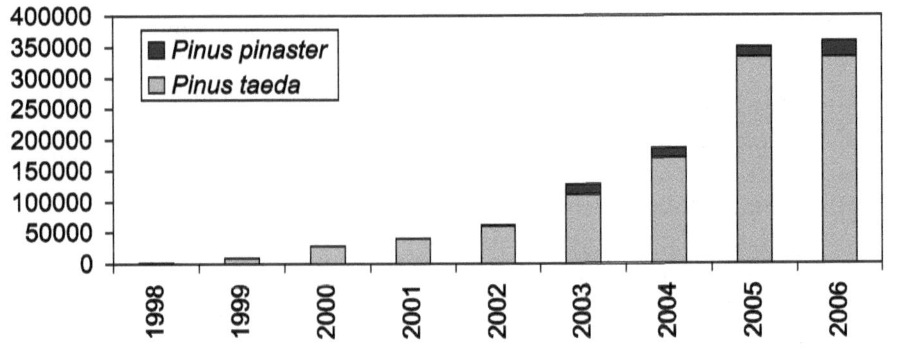

Fig. 5. Pine sequence resources available at EMBL (www.ebi.ac.uk/srs), on 28 February 2006

an additional 85,000 ESTs derived from somatic and zygotic embryos (http://www.tigr.org/tdb/e2k1/pine/index.shtml).

Maritime pine ESTs Early molecular biology studies paved the way for the construction of cDNA libraries from different woody tissues (Cantón et al. 1993; García-Gutiérrez et al. 1995), roots (Dubos and Plomion 2003) and buds (C. Collada, M.A. Guevara, and M.T. Cervera, unpublished results) of maritime pine. Funded by the European Union (GEMINI project), INRA (lignome project), and CNS (ForEST project), 28,000 high-quality sequences have been generated in about equal number from wood-forming tissues, roots, and buds. These ESTs have been made available at the EMBL database and at the following URL: http://cbi.labri.fr/outils/SAM/COMPLETE/. The *P. pinaster* cDNA clones are available for the scientific community, at the Platform for Integrated Clone Management PICME Web site (http://www.picme.at/). A Web-based environment for assembly and annotation of collections of ESTs (Le Provost et al. in prep.) was used to analyze the *P. pinaster* ESTs.

A comprehensive unigene assembled for the user community The number of pine ESTs in public databases has increased dramatically during the past decade (Fig. 5), especially for *P. taeda*, ranking at the 19th position in dbEST with 329,469 ESTs, and to a lesser extent for *P. pinaster* (about 28,000 ESTs). The other pine species, *P. elliottii*, *P. banksiana*, *P. patula*, and *P. sylvestris*, represent less than 2,000 sequences. In 2005, the *Pinus* genus was introduced in the TIGR gene index (Quackenbush et al. 2001), where a comprehensive collection of all publicly available pine ESTs has been clustered into a tentative unigene set of 45,500 genes (http://www.tigr.org/tigr-scripts/tgi/T_index.cgi?species=pinus), representing probably a major fraction of the coding genome of pine. This resource has laid the foundation for future identification, characterization, and cloning of economically important genes in pines and should provide breeders with a wealth of information to significantly enhance and speed up the breeding process.

Besides this resource, a significant number of ESTs have not been released yet in the public domain and are mainly held by private companies. For instance, Strabala (2004) reported that in August 2001, 344,279 ESTs from *P. radiata* were sequenced by AgriGenesis Biosciences Ltd. based in New Zealand.

Microarrays: New Nools to Study the Functioning of the Pine Genome

ESTs represent snapshots of the genes expressed in a given tissue or at a given developmental stage and provide a good representation of the expressed regions of the genome if a sufficient number of different libraries from a range of tissues and developmental stages and after various environmental challenges are analyzed. To this end, several cDNA libraries have been produced from pine roots, needles, differentiating xylem, and buds (see previous section). These genes have enabled the development of microarrays to identify genes whose expression varies in response to various environmental and developmental cues (mainly drought-stress response and wood formation). These studies, which will likely increase in the future, are listed in Table 6 with others that have used other transcriptomic (SAGE, SSH, cDNA-AFLP) and proteomic (2D PAGE MS/MS) approaches to study the functioning of the pine genome.

Table 6. Transcriptome and proteome analysis in pines

Species	Methodology	Study	Tissue	Authors
P. contorta	Microarrays	Adventitious root development	Roots	Brinker et al. 2004
P. halepensis, P. taeda	cDNA library, Northern blot, RT-PCR	Drought response in two pine species	Roots, needles	Sathyan 2005
P. pinaster	cDNA-AFLP, reverse Northern	Drought response	Roots	Dubos and Plomion 2003
P. pinaster	cDNA-AFLP, reverse Northern	Drought response	Needles	Dubos et al. 2003
P. pinaster	cDNA library	Bud phenology	Buds at different developmental stages	Collada, Guevara, and Cerera, unpublished results
P. pinaster	Six SSH cDNA libraries	Wood formation	Early vs. late, juvenile vs. mature wood-forming tissues.	Pacheco et al. 2005
P. pinaster	cDNA-AFLP; reverse Northern; RT-PCR	Wood formation	Early vs. late, opposite vs. compression wood-forming tissues	Le Provost et al. 2003
P. pinaster	cDNA library	Wood formation	Composite xylem tissues	Cantón et al. 2003
P. pinaster	EST sequencing, microarrays, qPCR	Wood formation	Early vs. late, juvenile vs. mature wood-forming tissues and 8 different organs	Paiva 2006
P. pinaster	EST sequencing, macroarrays, qPCR, 2DE MS/MS	Drought response in two ecotypes	Roots	Chaumeil 2006
P. pinaster	2DE, tandem MS	Wood formation	Wood-forming tissues	Gion et al. 2005
P. pinaster	2DE, micro-sequencing	Drought response	Needles	Costa et al. 1998
P. pinaster	2DE, micro-sequencing	Wood formation	compression and opposite wood forming tissues	Plomion et al. 2000
P. pinea	Two SSH cDNA libraries	Cytokinine induction of adventitious buds	Cotyledons	Alonso, Cantón, Ordás, unpublished results
P. sylvestris	cDNA library	Nitrogen and carbon assimilation	Cotyledons	Avila et al. 2000
P. radiata	CAGE	Wood formation	Wood-forming tissues	Cato et al. 2006a
P. radiata	Three SSH cDNA libraries	Adventitious rooting capacity	Rooting competent tissues	Sánchez et al. 2005
P. taeda	Differential screening, northern blot	Drought response	Needles, stems, roots	Chang et al. 1996
P. taeda	Microarrays	Wood formation	Wood-forming tissues	Yang and Loopstra 2005
P. taeda	Microarrays	Drought response and recovery	Needles	Watkinson et al. 2003
P. taeda	Microarrays	Drought response	Needles, rooted cuttings, 2 unrelated genotypes	Heath et al. 2002

Table 7. (continued)

Species	Methodology	Study	Tissue	Authors
P. taeda	Differential display	Response to galled tissues	Healthy and galled stem	Warren and Covert 2004
P. taeda	0.35k microarrays; qRT-PCR	Wood formation	Unlignified early and latewood xylem: cambial region and developing secondary nonlignified xylem	Egertsdotter et al. 2004
P. taeda	cDNA and subtractive libraries; EST sequencing; Northern blot	Wood formation	Compression and opposite wood	Allona et al. 1998
P. taeda	EST sequencing, microarray	Wood formation	Different tissues and organs, including several types of differentiating xylem samples	Whetten et al. 2001
P. taeda	SAGE profiling	Wood formation	Juvenile and mature wood formatting tissues	Lorenz and Dean 2002
P. taeda	cDNA library; EST sequencing, comparative sequence analysis	Wood formation	Six types of wood forming tissues	Kirst et al. 2003
P. taeda	Microarray, qRT-PCR	Identification of genes preferentially expressed in P. taeda	Differentiating xylem, needles, megagametophytes, embryo (growing radicule)	Yang et al. 2004
P. taeda	Comparative sequence analysis, digital profiling	Computational analysis of transcript accumulation in xylem.	Differentt typoes of wood-forming tissues	Pavy et al. 2005
P. taeda	EST sequencing, in silico detection of differentially expressed genes	Drought response	Roots	Lorenz et al. 2005
P. taeda, P. sylvestris, P. abies	Macroarrays	Heterologous arrays	Needles	van Zyl et al. 2002
P. taeda	Microarrays	Disease resistance	Stem (infection court)	Myburg et al. 2006
P. elliottii var. *eliottii*	Microarrays	Disease resistance	Stem (infection court)	Morse et al. 2004

2.6 Future Perspective: Challenges for Molecular Breeding of Pines

Pines offer a great opportunity for molecular breeding given their large size and long generation interval. The large size causes inefficiencies in progeny testing and concomitant decreases in heritability of traits. Molecular-marker information can reduce these inefficiencies and effectively increase heritability, leading to more accurate selections and commensurate increases in genetic gain (Fernando and Grossman 1989; Lande and Thompson 1990; Hospital et al. 1997). The long generation intervals cause delays in obtaining genetic gain. Savings of many years in the breeding cycle can easily be envisioned if selections can be based on DNA marker information collected at a very early age as opposed to phenotypes measured much later (Williams and Neale 1992). However, these gains can only be achieved if breeding can commence very soon after the selections are made. This has traditionally been known as accelerated breeding, and it is a critical, complementary technology (e.g., Bramlett and Burris 1995).

While the opportunity for increased genetic gain per year is great, major challenges to implement molecular breeding include pines' highly heterozygous, outbreeding nature and the presence of low levels of LD in natural populations (Strauss et al. 1992; Brown et al. 2004). Each of these factors limits the opportunity for finding QTLs by association (Kruglyak 1999; Hirschhorn and Daly 2005) or candidate-gene (Pflieger et al. 2001; Neale and Savolainen 2004) mapping approaches, and for extending QTL results from one pedigree to another within a breeding population. However, given very dense genetic maps or good leads about possible candidate genes, these low LD levels allow for the possibility of discovering tight linkages to important genes. These close linkages should be generally useful across the population for predicting genetic value and implementing molecular breeding.

The challenge in moving forward lies in the development of powerful genetic markers and maps and populations providing genetic information on traits of interest (e.g., Georges et al. 1995; Farnir et al. 2002; Laurie et al. 2004). These populations can include both pedigreed and nonpedigreed populations, and both are valid for the application of molecular breeding. Pedigreed populations place and maintain large regions of the genome in LD, allowing for the tracking of most regions with mapped genetic markers (Lander and Botstein 1989). Once certain regions are identified as causing (or at least being predictive of) significant variation in the phenotype, marker alleles for these regions can be selected for two or more successive generations, thereby increasing the frequency of favorable alleles in the population (Edwards and Page 1994; Hospital et al. 2000). One such view of this for pine breeding is termed marker-directed population improvement (MDPI) (Nelson and Echt 2003, 2004).

Nonpedigreed populations provide an opportunity for tracking very small regions across the genome (Hirschhorn and Daly 2005). These are the regions that have remained in LD for a very long time, possibly since the creation of the DNA polymorphism used as the genetic marker or the gene mutation itself. Extremely dense genetic maps are required for mapping QTLs in nonpedigreed populations through a genome-scan analysis (Kruglyak 1999). Testing specific genes known as candidate genes for association to phenotype does not require a genetic map (Pflieger et al. 2001; Neale and Savolainen 2004). Instead, a hypothesis about what gene(s) might affect a trait's phenotype is needed as well as markers that allow detection of alternative alleles of the candidate gene(s) in the population (Krutovsky and Neale 2005).

The greatest opportunities for implementing molecular breeding exist in breeding programs that utilize best testing practices, including good experimental design, careful field site selection and maintenance, and high-quality measurements on all important traits (e.g., Stuber et al. 1999; Tanksley and Nelson 1996). Most of these programs will contain populations with pedigree structure, allowing for moderately dense genetic maps (10 to 20 cM spacing) to be used in various linkage-mapping approaches (Darvasi et al. 1993) including MDPI. In addition, all trees are drawn from a source population where tight linkages will be maintained over short recombination intervals, allowing for mapping by association through a genome-scan or candidate-gene approach (Wu et al. 2002; Lund et al. 2003).

Where no breeding program exists, large random mating populations could be established specifically for association mapping and subsequent marker-assisted selection and breeding. These would most likely use candidate-gene methods, as candidate genes could be selected from first principles or knowledge of related species and then tested for association with phenotype in the target species and population. It would seem that genome-scan approaches will only be viable in very intensively studied species where dense genetic maps will be developed. In all cases, highly efficient DNA isolation and genotyping methods (e.g., Darvasi and Soller 1994; Mosig et al. 2001) and effective database management and bioinformatic tools will be required for cost-effective implementation (Nelson 1997). Expected results include increased genetic gain per generation due to increased effective heritability and with accelerated breeding, increasing the number of generations per unit of time. In addition, traits that are difficult to measure can be improved more readily assuming they are measured and mapped in an early generation.

Acknowledgement. This work was funded in part by (A) ANR awards GENOQB (GNP05013) and DIGENFOR (ANR-05-GPLA-006-01) to C Plomion, (B) NSF award DBI-0421717 to DG Peterson, CD Nelson, and MN Islam-Faridi, (C) award RTA03-213 from the Programa Nacional de Recursos y Tecnologías Agroalimentarias (Ministerio de Educación y Ciencia, MEC) to MT Cervera, (D) National Research Council of Italy (Commessa: "Evoluzione e analisi della diversità genetica in piante forestali") to GG Vendramin, (E) New Zealand Foundation for Research Science and Technology (FRST) contract CO4X0207 to Scion (P.L. Wilcox and R.D. Burdon), (F) the Finnish Funding Agency for Technology and Innovation and by the Biosciences and Environment Research Council of Finland, to O Savolainen.

References

Adams MD, Kelley JM, Gocayne JD, Dubnick M, Polymeropoulos MH, Xiao H, Merril CR, Wu A, Olde B, Moreno RF, Kerlavage AR, McCombie WR, Venter JC (1991) Complementary DNA sequencing: expressed sequence tags and human genome project. Science 252:1651–1656

Aimers-Halliday J, Shelbourne CJA, Hong SO (1997) Issues in developing clonal forestry with *P. radiata*. In: Burdon RD, Moore JM (eds) "IUFRO '97 Genetics of Radiata Pine", Proc IUFRO Conf 1-4 Dec and Workshop 5 Dec, Rotorua, New Zealand. FRI Bull 203:264–272

Alazard P (2001) Genetic gain in seeds from first generation seed orchards of maritime pine. Informations Forests, Afocel, France, No 628

Alía R, Gil L, Pardos JA (1995) Performance of 43 *Pinus pinaster* provenances on 5 locations in Central Spain. Silvae Genet 44:75–81

Alía R, Moro J, Denis JB (1997) Performance of *Pinus pinaster* Ait. provenances in Spain: interpretation of the genotype-environment interaction. Can J For Res 27:1548–1559

Allona I, Quinn M, Shoop I, Swope K, ST-Cyr S, Carlis J, Rield J, Retzel E, Campbell M, Sederoff R, Whetten RW (1998) Analysis of xylem formation in pine by cDNA sequencing. Proc Natl Acad Sci USA 95:9693–9698

Andrew RL, Peakall R, Wallis IR, Wood JT, Knight EJ, Foley WJ (2005) Marker-based quantitative genetics in the wild?: The heritability and genetic correlation of chemical defenses in Eucalyptus. Genetics 171:1989–1998

Apiolaza LA, Greaves BL (2001) Why are most breeders not using economic breeding objectives. In: IUFRO Symp "Developing the Eucalypt for the Future", 10-14 Sept 2001, Valdivia, Chile

Arcade A, Faivre-Rampant P, Le Guerroue B, Paques LE, Prat D (1996) Quantitative traits and genetic markers: analysis of a factorial mating design in larch. In: Ahuja MR, Boerjan W, Neale D (eds) Somatic Cell Genetics and Molecular Genetics of Trees. Kluwer, Dordrecht, pp 211–216

Arumuganathan K, Earle ED (1991) Nuclear DNA content of some important plant species. Plant Mol Biol Rep 9:208–218

Atwood RA, White TL, Huber DA (2002) Genetic parameters and gains for growth and wood properties in Florida source loblolly pine in the southeastern United States. Can J For Res 32:1025–1038

Auckland L, Bui T, Zhou Y, Shepherd M, Williams CG (2002) Conifer Microsatellite Handbook. Corporate Press, College Station, TX

Avila C, Muñoz-Chapuli R, Plomion C, Frigerio J-M, Cánovas FM (2000) Two genes encoding distinct cytosolic glutamine synthetases are closely located in the pine genome. FEBS Lett 477:237–243

Bahrman N, Damerval C (1989) Linkage relationships of loci controlling protein amounts in maritime pine (*Pinus pinaster* Ait). Heredity 63:267–274

Ball RD (2001) Bayesian methods for quantitative trait loci mapping based on model selection: approximate analysis using the Bayesian Information Criterion. Genetics 159:1351–1364

Ball RD (2005) Experimental designs for reliable detection of linkage disequilibrium in unstructured random population association studies. Genetics 170:859–873

Barker CL, Donald T, Pauquet J, Ratnaparkhe MB, Bouquet A, Adam-Blondon AF, Thomas MR, Dry I (2005) Genetic and physical mapping of the grapevine powdery mildew resistance gene, Run1, using a bacterial artificial chromosome library. Theor Appl Genet 111:370–377

Beaulieu J, Plourde A, Daoust G, Lamontagne L (1996) Genetic variation in juvenile growth of *Pinus strobus* in replicated Quebec provenance-progeny tests. For Genet 3:103–112

Beavis WD (1994) The power and deceit of QTL experiments: lessons from comparative QTL studies. In: Proc 49th Annu Corn and Sorghum Industry Res Conf, ASTA, Washington, DC, pp 250–266

Bedell JA, Budiman MA, Nunberg A, Citek RW, Robbins D, Jones J, Flick E, Rholfing T, Fries J, Bradford K, McMenamy J, Smith M, Holeman H, Roe BA, Wiley G, Korf IF, Rabinowicz PD, Lakey N, McCombie WR, Jeddeloh JA, Martienssen

RA (2005) Sorghum genome sequencing by methylation filtration. PLoS Biol 3:13

Bell JC, Powell M, Devey ME, Moran GF (2004) DNA profiling, pedigree lineage analysis and monitoring in the Australian breeding program of radiata pine. Silvae Genet 53:130–134

Blada I (1994) Inter-specific hybridization of Swiss Stone Pine *Pinus cembra* L. Silvae Genet 43:14–20

Borzan Z, Papes D (1978) Karyotype analysis in *Pinus*: contribution to the standardization of the karyotype analysis and review of some applied techniques. Silvae Genet 27:144–150

Bowe LM, Coat G, dePamphilis CW (2000) Phylogeny of seed plants based on all three genomic compartments: Extant gymnosperms are monophyletic and Gnetales' closest relatives are conifers. Proc Natl Acad Sci USA 97:4092–4097

Bradshaw HD, Foster GS (1992) Marker-aided selection and propagation systems in trees: advantages of cloning for studying quantitative inheritance. Can J For Res 22:1044–1049

Bramlett DL, Burris LC (1995) Topworking young scions into reproductively-mature loblolly pine. In: Weir RJ, Hatcher AV (eds) Proc 23rd South For Tree Improv Conf, Asheville, NC, pp 234–241

Brendel O, Pot D, Plomion C, Rozenberg P, Guehl JM (2002) Genetic parameters and QTL analysis of ä13C and ring width in maritime pine. Plant Cell Environ 25:945–953

Brettschneider R (1998) RFLP analysis. In: Karp A, Isaac PG, Ingram DS (eds) Molecular Tools for Screening Biodiversity Plants and Animals. Chapman & Hall, London, pp 83–95

Brinker M, van Zyl L, Liu WB, Craig D, Sederoff RR, Clapham DH, von Arnold S (2004) Microarray analyses of gene expression during adventitious root development in *Pinus contorta*. Plant Physiol 135(3):1526-1539

Brown GR, Kadel III EE, Bassoni DL, Kiehne KL, Temesgen B, van Buijtenen JP, Sewell MM, Marshall KA, Neale DB (2001) Anchored reference loci in loblolly pine (*Pinus taeda* L.) for integrating pine genomics. Genetics 159:799–809

Brown GR, Bassoni DL, Gill GP, Fontana JR, Wheeler NC, Megraw RA, Davis MF, Sewell MM, Tuskan GA, Neale DB (2003) Identification of quantitative trait loci influencing wood property traits in loblolly pine. III. QTL verification and candidate gene mapping. Genetics 164:1537–1546

Brown GR, Gill GP, Kuntz RJ, Langley CH, Neale DB (2004) Nucleotide diversity and linkage disequilibrium in loblolly pine. Proc Natl Acad Sci USA 101:15255–15260

Burban C, Petit RJ (2003) Phylogeography of maritime pine inferred with organelle markers having contrasted inheritance. Mol Ecol 12:1487–1495

Burdon RD (1997) Genetic diversity for the future: conservation or creation and capture? In: Burdon RD, Moore JM (eds) "IUFRO '97 Genetics of Tadiata Pine", Proc IUFRO Conf 1–4 Dec and workshop 5 Dec, Rotorua, New Zealand. FRI Bull 203:263–274

Burdon RD (2001) Genetic diversity and disease resistance: some considerations for research, breeding and deployment. Can J For Res 32:596–606

Burdon RD (2002) An introduction to pines. In: Pines of Silvicultural Importance. CABI, Wallingford, UK pp x–xxi

Burdon RD (2004) Breeding goals: Issues of goal-setting and applications. In: Walter C, Carson M (eds) Plantation Forest Biotechnology for the 21st Century, Research Signpost, Kerala, India, pp 101–118

Burdon RD, Bannister MH, Low CB (1992) Genetic Survey of *Pinus radiata* 4: variance structures and heritabilities in juvenile clones. NZ J For Sci 22:187–210

Burdon RD, Firth A, Low CB, Miller MA (1998) Multi-site provenance trials of *Pinus radiata* in New Zealand. In: Forest Genetic Resources 26:3–8, FAO, Rome

Burdon RD, Gea LD (2006) Pursuit of genetic gain and biotic risk management for *Pinus radiata* in New Zealand. In: Proc 13th Australasian Plant Breed Conf, Christchurch, New Zealand pp 75–85, http://www.apbc.org.nz

Burdon RD, Kumar S (2003) Stochastic modelling of the impact of four generations of pollen contamination in unpedigreed gene resources. Silvae Genet 52:1–7

Burdon RD, Kumar S (2004) Forwards versus backwards selection: trade-offs between expected genetic gain and risk avoidance. NZ J For Sci 34:3–21

Burdon RD, Low CB (1992) Genetic survey of *Pinus radiata*. 6: Wood properties: variation, heritabilites, and interrelationship with other traits. NZ J For Sci 22:228–245

Burdon RD, Miller JT (1992) Introduced forest trees of New Zealand: Recognition, role and seed source. 12. Radiata pine (*Pinus radiata* D. Don). NZ For Res Inst Bull 124/12 pp 59

Byun KO, Kim MZS, Shim SY, Hong SH, Sohn SI (1989) Review of pitch- loblolly hybrid pine (*Pinus rigida* and *P. taeda*) breeding researches in Korea and future strategy. Res Rep, For Genet Res Inst, Korea 25:204–211

Cahalan CM (1981) Provenance and clonal variation in growth, branching and phenology in *Picea sitchensis* and *Pinus contorta*. Silvae Genet 30:40–46

Cai Q, Zhang D, Liu ZL, Wang XR (2006) Chromosomal localization of 5S and 18S rDNA in five species of subgenus *Strobus* and their implications for genome evolution of *Pinus*. Ann Bot 97:715–722

Cánovas FM, Dumas-Gaudot E, Recorbet G, Jorrin J, Hans-Peter Mock HP, Rossignol M (2004) Plant proteome analysis. Proteomics 4:285–298

Cantón FR, García-Gutiérrez A, Gallardo F, de Vicente A, Cánovas FM (1993) Molecular characterization of a cDNA clone encoding glutamine synthetase from a gymnosperm: *Pinus sylvestris*. Plant Mol Biol 22:819–828

Cantón FR, Le Provost G, Garcia V, Barré A, Frigerio J-M, Paiva J, Fevereiro P, Avila C, Mouret J-F, de Daruvar A, Cánovas FM, Plomion C (2003) Transcriptome analysis of wood formation in maritime pine In: Espinel S, Barredo Y, Ritter E (eds) Sustainable Forestry, Wood Products and Biotechnology. DFA-AFA Press, Vitoria-Gasteiz, Spain, pp 333–347

Cato S, Mc Millan L, Donaldson L, Richardson T, Echt C, Gardner R (2006a) Wood formation from the base to the crown in

Pinus Radiata: gradients of tracheid wall thickness, wood density, radial growth rate and gene expression. Plant Mol Biol 60:565–581

Cato SA, Pot D, Kumar S, Douglas J, Gardner RC, Wilcox PL (2006b) Balancing selection in a dehydrin gene associated with increased wood density and decreased radial growth in *Pinus radiata* (Abstr). In: Plant & Animal Genome XIV Conf, San Diego

Cato SA, Richardson TE (1996) Inter- and intra-specific polymorphism at chloroplast SSR loci and the inheritance of plastids in *Pinus radiata* D. Don. Theor Appl Genet 93:587–592

Cervera MT, Plomion C, Malpica CA (2000a) Molecular markers and genome mapping in woody plants. In: Jain SM, Minocha SC (eds) Molecular Biology of Woody Plants. Kluwer, Dordrecht, pp 375–394

Cervera MT, Remington D, Frigerio JM, Storme V, Ivens B, Boerjan W, Plomion C (2000b) Improved AFLP analysis of tree species. Can J For Res 30:1608–1616

Chagné D, Brown G, Lalanne C, Madur D, Pot D, Neale DB, Plomion C (2003) Comparative genome and QTL mapping between maritime and loblolly pines. Mol Breed 12:185–195

Chagné D, Chaumeil P, Ramboer A, Collada C, Guevara A, Cervera M-T, Vendramin GG, Garcia V, Frigerio JM, Echt C, Richardson T, Plomion C (2004) Cross species transferability and mapping of genomic and cDNA SSRs in pines. Theor Appl Genet 109:1204–12

Chagné D, Lalanne C, Madur D, Kumar S, Frigerio JM, Krier C, Decroocq S, Savouré A, Bou-Dagher-Kharrat M, Bertocchi E, Brach J, Plomion C (2002) A high density genetic map of Maritime pine based on AFLPs. Ann For Sci 59:627–636

Chambers PGS, Borralho NMG (1999) A simple model to examine the impact of changes in wood traits on the cost of thermomechanical pulping and high-brightness newsprint production with radiata pine. Can J For Res 29:1615–1626

Chang S, Puryear JD, Dias MADL, Funkhouser EA, Newton RJ, Cairney J (1996) Gene expression under water deficit in loblolly pine (*Pinus taeda*): isolation and characterization of cDNA clones. Physiol Planta 97:139–148

Chaumeil P (2006) Plasticité moléculaire de deux écotypes de pin maritime soumis à un stress osmotique. Thèse de Doctorat. Université Henri Poincaré, Nancy I, Nancy

Ciavatta VT, Morillon R, Pullman GS, Chrispeels MJ, Cairney J (2001) An aquaglyceroporin is abundantly expressed early in the development of the suspensor and the embryo proper of loblolly pine.Plant Physiol 127:1556–1567

Claros MG, Cantón FR, Cánovas FM (2004) Construction of a pine BAC genomic library and bioinformatics platform for automated high-throughput sequence analysis (Abstr). 2nd Annual Meeting of the Spanish Forest Functional Genomics Network. Pontevedra, Spain, 29 Nov 2004. http://www.difo.uah.es/forestgenomics/pontevedra_04/Gonzalo%20CLAROS.pdf

Colbert T, Till BJ, Tompa R, Reynolds S, Steine MN, Yeung AT, McCallum CM, Comai L, Henikoff S (2001) High-throughput screening for induced point mutations. Plant Physiol 126:480–484

Conkle MT (1981) Isozyme variation and linkage in six conifer species. In: Proc Isozymes of North American Forest Trees and Insects, USD Forest Service, General Technical Report PSW-48, pp 11–17

Costa P (1999) Réponse moléculaire, physiologique et génétique du pin maritime à une contrainte hydrique. PhD thesis. Univ of Nancy

Costa P, Plomion C (1999) Genetic analysis of needle protein in Maritime pine. 2. Quantitative variation of protein accumulation. Silvae Genet 48:146–150

Costa P, Pot D, Dubos C, Frigerio JM, Pionneau C, Bodenes C, Bertocchi E, Cervera MT, Remington DL, Plomion C (2000) A genetic map of Maritime pine based on AFLP, RAPD and protein markers. Theor Appl Genet 100:39–48

Costa, P, Bahrman, N, Frigerio, J-M, Kremer A, Plomion C (1998) Water-deficit-responsive proteins in maritime pine. Plant Mol Biol 38:587–596

Cornelius J (1994) Heritabilities and additive genetic coefficients of variation in forest trees. Can J For Res 24:372–379

Danjon F (1995) Observed selection effects on height growth, diameter and stem form in maritime pine. Silvae Genet 44(1):10-19

Darvasi A, Soller M (1994) Selective DNA pooling for determination of linkage between a molecular marker and a quantitative trait locus. Genetics 138:1365–1373

Darvasi A, Weinreb A, Minke V, Weller JI, Soller M (1993) Detecting marker-QTL linkage and estimating QTL gene effect and map location using a saturated genetic map. Genetics 134:943–951

Devey ME, Fiddler TA, Liu BH, Knapp SJ, Neale DB (1994) An RFLP linkage map for loblolly pine based on a three-generation outbred pedigree. Theor Appl Genet 88:273–278

Devey ME, Fino-Mix A, Kinloch BB Jr, Neale DB (1995) Random amplified polymorphic DNA markers tightly linked to a gene for resistance to white pine blister rust in sugar pine. Proc Natl Acad Sci USA 92:2066–2070

Devey ME, Bell JC, Smith DN, Neale DB, Moran GF (1996) A genetic linkage map for *Pinus radiata* based on RFLP, RAPD, and microsatellite markers. Theor Appl Genet 6:673–679

Devey ME, Sewell MM, Uren TL, Neale DB (1999) Comparative mapping in loblolly and radiata pine using RFLP and microsatellite markers. Theor Appl Genet 99:656–662

Devey ME, Bell JC, Uren TL, Moran GF (2003) A set of microsatellite markers for fingerprinting and breeding applications in *Pinus radiata*. Genome 45:984–989

Devey ME, Carson SD, Nolan MF, Matheson AC, Te Riini C, Hohepa J (2004a) QTL associations for density and diameter in *Pinus radiata* and the potential for marker-aided selection. Theor Appl Genet 108:516–524

Devey ME, Groom KA, Nolan MF, Bell JC, Dudzinski MJ, Old KM, Matheson AC, Moran GF (2004b) Detection and verification of quantitative trait loci for resistance to Dothistroma needle blight in *Pinus radiata*. Theor Appl Genet 108:1056–1063

Dhakal LP, White TL, Hodge GR (1996) Realized genetic gains from slash pine (*Pinus elliottii*) tree improvement. Silvae Genet 45:190–197

Dilbirligi M, Erayman M, Sandhu D, Sidhu D, Gill KS (2004) Identification of wheat chromosomal regions containing expressed resistance genes. Genetics 166:461–481

Doudrick RL, Heslop-Harrison JS, Nelson CD, Schmidt T, Nance WL, Schwarzacher T (1995) Karyotype of slash pine (*Pinus elliottii* var *elliottii*) using patterns of fluorescence *in situ* hybridization and fluorochrome banding. J Hered 86:289–296

Drewry A (1982) G-banded chromosomes in *Pinus resinosa*. J Hered 73:305–306

Du Toit B, Smith C, Carlson C, Esprey L, Allen R, Little K (1998) Eucalypt and pine plantations in South Africa. In: Nambiar EKS, Cossalter C, Triarks A (eds) Workshop Proceedings. CIFOR, 16-20 Feb 1998, South Africa, pp 23–30

Dubos C, Plomion C (2003) Identification of water-deficit responsive genes in maritime pine (*Pinus pinaster* Ait.) roots. Plant Mol Biol 51(2):249-262

Dubos C, Provost G, Pot D, Salin F, Lalane C, Madur D, Frigerio JM, Plomion C (2003) Identification and characterization of water-stress-responsive genes in hydroponically grown maritime pine (*Pinus pinaster*) seedlings. Tree Physiol 23:169–179

Dvornyk V, Sirviö A, Mikkonen M, Savolainen O (2002) Low nucleotide diversity at the *pal1* locus in the widely distributed *Pinus sylvestris*. Mol Biol Evol 19:179–188

Echt CS, DeVerno LL, Anzidei M, Vendramin GG (1998) Chloroplast microsatellites reveal population genetic diversity in red pine, *Pinus resinosa* Ait. Mol Ecol 7:307–317

Echt CS, Burns R (1999) PCR primer pair sequences and SSRs for various *P. taeda* ESTs. Available on the dendrome Web site at http://dendrome.ucdavis.edu/Gen_res.htm, accessed 20 March 2006

Echt CS, Nelson CD (1997) Linkage mapping and genome length in eastern white pine (*Pinus strobus* L.). Theor Appl Genet 94:1031–1037

Echt CS, Saha S, Deemer D, Nelson CD (2006) Evaluation of pine EST-SSR markers. In: Plant & Animal Genome XIV Conf, San Diego, W-117 (http://www.intl-pag.org/14/abstracts/PAG14_W117.html, accessed 20 March 2006)

Edwards MD, Page NJ (1994) Evaluation of marker-assisted selection through computer simulation. Theor Appl Genet 88:376–382

Edwards MD, Stuber CW, Wendel JF (1987) Molecular-marker-facilitated investigations of quantitative-trait loci in maize. I. Numbers, genomic distribution and types of gene action. Genetics 116:113–125

Egertsdotter U, van Zyl LM, Mackay J, Peter G, Kirst M, Clark C, Whetten R, Sederoff R (2004) Gene expression during formation of earlywood and latewood in loblolly pine: expression profiles of 350 genes. Plant Biol 6:654–663

Elsik CG, Williams CG (2000) Retroelements contribute to the excess low-copy-number DNA in pine. Mol Gen Genet 264:47–55

Elsik CG, Williams CG (2001) Low copy microsatellite recovery form a conifergenome. Theor Appl Genet 103:1189–1195

Emebiri LC, Devey ME, Matheson AC, Slee MU (1997) Linkage of RAPD markers to NESTUR, a stem growth index in radiata pine seedlings. Theor Appl Genet 95:119–124

Emebiri LC, Devey ME, Matheson AC, Slee MU (1998) Age-related changes in the expression of QTLs for growth in radiata pine seedlings. Theor Appl Genet 97:1053–1061

Epperson BK, Allard RW (1989) Spatial autocorrelation analysis of the distribution of genotypes within population of lodgepole pine. Genetics 121:369–377

Falkenhagen E (1991) Provenanace variation in *Pinus radiata* at six sites in South Africa. Silvae Genet 40:41–50

Farjon A (1984) Pines. Drawings and description of the genus *Pinus*. Brill, Leiden

Farjon A (2001) World checklist and bibliography of conifers, 2nd edn. The Royal Botanic Gardens, Kew, UK

Farjon A, Styles BT (1997) *Pinus* (Pinaceae). Monograph 75, Flora Neotropica. New York Botanical Gardens, New York

Farnir F, Grisart B, Coppieters W, Riquet J, Berzi P, Cambisano N, Karim L, Mni M, Moisio S, Simon P, Wagenaar D, Vilkki J, Georges M (2002) Simultaneous mining of linkage and linkage disequilibrium to fine map quantitative trait loci in outbred half-sib pedigrees: revisiting the location of a quantitative trait locus with major effect on milk production on bovine chromosome 14. Genetics 161:275–287

Fernando RL, Grossman M (1989) Marker-assisted selection using best linear unbiased prediction. Genet Sel Evol 21:246–477

Fisher PJ, Gardner RC, Richardson TE (1996) Single locus microsatellites isolated using 5' anchored PCR. Nucleic Acids Res 24:4369–4371

Fisher PJ, Gardner RC, Richardson TE (1998) Characteristics of single and multi-copy microsatellities from *Pinus radiata*. Theor Appl Genet 96:969–979

Flint-Garcia SA, Thornsberry JM, Buckler ES (2003) Structure of linkage disequilibrium in plants. Annu Rev Plant Biol 54:357–374

Folkertsma RT, Spassova MI, Prins M, Stevens MR, Hille J, Goldbach RW (1999) Construction of a bacterial artificial chromosome (BAC) library of *Lycopersicon esculentum* cv. Stevens and its application to physically map the *Sw-5* locus. Mol Breed 5:197–207

Friesen N, Brandes A, Heslop-Harrison JS (2001) Diversity, origin, and distribution of retrotransposons (gypsy and copia) in conifers. Mol Biol Evol 18:1176–1188

Fulton TM, Van der Hoeven R, Eannetta NT, Tanksley SD (2002) Identification, analysis, and utilization of conserved ortholog set markers for comparative genomics in higher plants. Plant Cell 14:1457–1467

García-Gil MR, Mikkonen M, Savolainen O (2003) Nucleotide diversity at two phytochrome loci along a latitudinal cline in *Pinus sylvestris*. Mol Ecol 12:1195–1206

García-Gutiérrez A, Cantón FR, Gallardo F, Sánchez-Jiménez F, Cánovas FM (1995) Expression of ferredoxin-dependent glutamate synthase in dark-grown pine seedlings. Plant Mol Biol 27:115–128

Gaussen H (1960) Les Gymnospermes actuelles et fossils. Fascicule VI. Généralités, genre *Pinus*; Faculté des Sciences, Toulouse

Geada López G, Kamiya K, Harada K (2002) Phylogenetic relationships of *Diploxylon* pines (subgenus *Pinus*) based on plastid sequence data.Int J Plant Sci 163:737–747

Geldermann H (1975) Investigations on inheritance of quantitative characters in animals by gene markers. I. Methods. Theor Appl Genet 46:319–330

Georges M, Nielsen D, Mackinnon M, Mishra A, Okimoto R, Pasquino AT, Sargeant LS, Sorensen A, Steele MR, Zhao X, Womack JE, Hoeschele I (1995) Mapping quantitative trait loci controlling milk production in dairy cattle by exploiting progeny testing. Genetics 139:907–920

Gerber S, Rodolphe F (1994) An estimation of the genome length of maritime pine (*Pinus pinaster* Ait). Theor Appl Genet 88:289–292

Gerber S, Rodolphe F, Bahrman N, Baradat P (1993) Seed-protein variation in maritime pine (*Pinus pinaster* Ait.) revealed by two-dimensional electrophoresis: genetic determinism and construction of a linkage map. Theor Appl Genet 85:521–528

Gernandt DS, Geada Lopez G, Ortiz Garcia S, Liston A (2005) Phylogeny and classification of *Pinus*. Taxon 54:29–42

Gernandt DS, Liston A, Pinero D (2003) Phylogenetics of *Pinus* subsections *Cembroides* and *Nelsoniae* inferred from cpDNA sequences. Syst Bot 28:657–673

Gion JM, Lalanne C, Le Provost G, Ferry-Dumazet H, Paiva J, Chaumeil P, Frigerio JM, Brach J, Barré A, de Daruvar A, Claverol S, Bonneu M, Sommerer N, Negroni L, Plomion C, Gion J-M, Lalanne C, et al. (2005) The proteome of maritime pine wood forming tissue. Proteomics 5:3731–3751

Gómez MA, González-Martínez SC, Collada C, Climent J, Gil L (2003) Complex population genetic structure in the endemic Canary Island pine revealed using chloroplast microsatellite markers. Theor Appl Genet 107:1123–1131

González-Martínez SC, Gerber S, Cervera MT, Martínez-Zapater J, Gil L, Alía R (2002) Seed gene flow and fine-scale structure in a Mediterranean pine (*Pinus pinaster* Ait.) using nuclear microsatellite markers. Theor Appl Genet 104:1290–1297

González-Martínez SC, Robledo-Arnuncio JJ, Collada C, Díaz A, Williams CG, Alía R, Cervera MT (2004) Cross-amplification and sequence variation of microsatellite loci in Eurasian hard pines. Theor Appl Genet 109:103–11

González-Martínez SC, Ersoz E, Brown GR, Wheeler NC, Neale DB (2006a) DNA sequence variation and selection of tag SNPs at candidate genes for drought-stress response in *Pinus taeda*. Genetics 172:1915–1926

González-Martínez SC, Krutovsky KV, Neale DB (2006b) Forest-tree population genomics and adaptive evolution. New Phytol 170:227–238

Govindaraju D, Lewis P, Cullis C (1992) Phylogenetic analysis of pines using ribosomal DNA restriction fragment length polymorphisms. Plant Syst Evol 179:141–153

Grace SL, Hamrick JL, Platt WJ (2004) Estimation of seed dispersal in an old-growth population of Longleaf Pine (*Pinus palustris*) using maternity exclusion analysis. Castanea 69:207–215

Grattapaglia D, Sederoff R (1994) Genetic linkage maps of *Eucalyptus grandis* and *Eucalyptus urophylla* using a pseudo-testcross mapping strategy and RAPD markers. Genetics 137:1121–1137

Greaves BL (1999) The value of tree improvement: A case study in radiata pine grown for structural sawn timber and liner-board. In: Nepveu G (ed) Connection between silviculture and wood quality through modelling approaches and simulation software. Proc IUFRO Workshop, 5-12 Dec 1999, France, pp 448–459

Groover AT, Devey ME, Lee JM, Megraw R, Mitchell-Olds T, Sherman B, Vujcic S, Williams C, Neale DB (1994) Identification of quantitative trait loci influencing wood specific gravity in an outbred pedigree of loblolly pine. Genetics 138:1293–1300

Groover AT, Williams CG, Devey ME, Lee JM, Neale DB (1995) Sex-related differences in meiotic recombination frequency in *Pinus taeda*. J Hered 86:157–158

Grotkopp E, Rejmanek M, Sanderson MJ, Rost TL (2004) Evolution of genome size in pines (*Pinus*) and its life-history correlates: supertree analyses. Evolution 58:1705–0729

Guevara MA, Chagné D, Almeida MH, Byrne M, Collada C, Favre JM, Harvengt L, Jeandroz S, Orazio C, Plomion C, Ramboer A, Rocheta M, Sebastiani F, Soto A, Vendramin GG, Cervera MT (2005a) Isolation and characterization of nuclear microsatellite loci in *P. pinaster*. Mol Ecol Notes 5:57–59

Guevara MA, Soto A, Collada C, Plomion C, Savolainen O, Neale DB, González-Martínez SC, Cerevera MT (2005b) Genomics applied to the study of adaptation in pine species. Invest Agrar Syst Recur For 14:292–306

Gugerli F, Senn J, Anzidei M, Madaghiele A, Buchler U, Sperisen C, Vendramin GG (2001) Chloroplast microsatellites and mitochondrial *nad1* intron 2 sequences indicate congruent phylogenetic relationships of Swiss stone pine (*Pinus cembra*), Siberian stone pine (*P. sibirica*) and Siberian dwarf pine (*P. pumila*). Mol Ecol 10:1489–1497

Gupta PK, Rustgi S, Kulwal PL (2005) Linkage disequilibrium and association studies in higher plants: Present status and future prospects. Plant Mol Biol 57:461–485

Gwaze DP, Harding KJ, Purnell RC, Bridgwater FE (2002) Optimum selection age for wood density in loblolly Pine. Can J For Res 32:1393–1399

Gwaze DP, Zhou Y, Reyes-Valdés MH, Al-Rababah MA, Williams CG (2003) Haplotypic QTL mapping in an outbred pedigree. Genet Res 81:43–50

Hamrick JL, Godt JW (1990) Allozyme diversity in plant species. In: Brown A, Clegg MT, Kahler AL, Weir BS (eds) Plant Population Genetics, Breeding and Genetic Resources. Sinauer, Sunderland, MA, pp 43–63

Hamrick JL, Godt MJW (1996) Effects of life history traits on genetic diversity in plant species. Phil Trans R Soc Lond B Biol Sci 351:1291–1298

Hannrup B, Ekberg I (1998) Age-age correlations for tracheid length and wood density in *Pinus sylvestris*. Can J For Res 28:1373–1379

Hannrup B, Ekberg I, Persson A (2000) Genetic correlations among wood, growth capacity and stem traits in *Pinus sylvestris*. Can J For Res 15:161–170

Harkins DM, Johnson GN, Skaggs PA, Mix AD, Dupper GE, Devey ME, Kinloch Jr BB, Neale DB (1998) Saturation mapping of a major gene for resistance to white pine blister rust in sugar pine. Theor Appl Genet 97:1355–1360

Harry DE, Temesgen B, Neale DB (1998) Codominant PCR-based markers for *Pinus taeda* developed from mapped cDNA clones. Theor Appl Genet 97:327–336

Hartmann HT, Kester DE, Davies FT Jr (1990) Plant Propagation Principles and Practices, 5th edn. Prentice Hall, Englewood Cliffs, NJ

Hayashi E, Kondo T, Terada K, Kuramoto N, Goto Y, Okamura M, Kawasaki H (2001) Linkage map of Japanese black pine based on AFLP and RAPD markers including markers linked to resistance against the pine needle gall midge. Theor Appl Genet 102:871–875

Hayashi E, Kondo T, Terada K, Kuramoto N, Kawasaki S (2004) Identification of AFLP markers linked to a resistance gene against pine needle gall midge in Japanese balck pine. Theor Appl Genet 108:1177–1181

Heath LS, Ramakrishnan N, Sederoff RR, Whetten RW, Chevone BI, Struble CA, Jouenne VY, Chen D, van Zyl LM, Grene R (2002) Studying the functional genomics of stress responses in loblolly pine with the "Expresso microarray experiment management system". Comp Funct Genom 3:226–243

Hirschhorn JN, Daly MJ (2005) Genome-wide association studies for common diseases and complex traits. Nat Rev Genet 6:95–108

Hizume M, Arai M, Tanaka A (1990) Chromosome banding in the genus *Pinus*. III. Fluorescent banding pattern of *P. luchuensis* and its relationships among the Japanese diploxylon pines. Bot Mag Tokyo 103:103–111

Hizume M, Ohgiku A, Tanaka A (1989) Chromosome banding in the genus *Pinus*. II. Interspecific variation of fluorescent banding patterns in *P. densiflora* and *P. thunbergii*. Bot Mag Tokyo 102:25–36

Hizume M, Shibata F, Matsusaki Y, Garajova Z (2002) Chromosome identification and comparative karyotypic analyses of four *Pinus* species. Theor Appl Genet 105:491–497

Hodge GR, Dvorak WS (1999) Genetic parameters and provenance variation of *Pinus tecunumanii* in 78 international trials. For Genet 6:157–180

Hodge GR, Volker PW, Potts BM, Owen JV (1996) A comparison of genetic information from open-pollinated and control-pollinated progeny tests in two Eucalypt species. Theor Appl Genet 92:53–63

Hospital F, Goldringer I, Openshaw S (2000) Efficient marker-based recurrent selection for multiple quantitative trait loci. Genet Res 75:357–368

Hospital F, Moreau L, Lacoudre F, Charcosset A, Gallais A (1997) More on the efficiency of marker-assisted selection. Theor Appl Genet 95:1181–1189

Hurme P, Sillanpää M, Repo T, Arjas E, Savolainen O (2000) Genetic basis of climatic adaptation in Scots pine by Bayesian QTL analysis. Genetics 156:1309–1322

Hyun SK (1976) Inter-specific hybridization in pines with the special reference to *Pinus rigida* and *P. taeda*. Silvae Genet 25:188–191

Ingvarsson PK (2005) Nucleotide polymorphism and linkage disequilibrium within and among natural populations of European Aspen (*Populus tremula* L., Salicaceae). Genetics 169:945–953

Isik F, Li B, Frampton J, Goldfarb B (2004) Efficiency of seedlings and rooted cuttings for testing and selection in *Pinus taeda*. For Sci 50:44–53

Islam-Faridi MN, Chang Y-L, Zhang HB, Kinlaw C, Doudrick RL, Neale DB, Echt CS, Price HJ, Stelly DM (1998) Construction of a pine BAC library. In: Plant & Animal Genome VI Conf, San Diego

Islam-Faridi MN, Mujeeb-Kazi A (1995) Visualization of *Secale cereale* DNA in wheat germplasm by FISH. Theor Appl Genet 90:595–600

Islam-Faridi MN, Nelson CD, Kubisiak TL (2007) Reference karyotype and cytomolecular map for loblolly pine (*Pinus taeda* L.). Genome (in press)

Islam-Faridi MN, Nelson CD, Kubisiak TL, Gullirmo MV, McNamara VH, Price HJ, Stelly DM (2003) Loblolly pine karyotype using FISH and DAPI positive banding. In: Proc 27th Southern For Tree Improv Conf, 25-27 June 2003, Stillwater, OK, pp 184–188

Ivkovic M, Kumar S, Wu H (2006) Adding value to the end-products of radiata pine: A review of breeding for structural timber production. In: Proc 13th Australasian Plant Breed Conf, 18-22 April 2006, Christchurch, NZ, pp 273–278

Jackson RB, Schlesinger WH (2004) Curbing the U.S. carbon deficit. Proc Natl Acad Sci USA 101:15827–15829

Jacobs MD, Gardner RC, Murray BG (2000) Cytological characterization of heterochromatin and rDNA in *Pinus radiata* and *P. taeda*. Plant Syst Evol 223:71–79

Jansen RC (1993) Interval mapping of multiple quantitative trait loci. Genetics 135:205–211

Jewell DC, Islam-Faridi MN (1994) Details of a technique for somatic chromosome preparation and C-banding of maize. In: Freeling M, Walbot V (eds) The Maize Hand Book. Springer, Berlin Heidelberg New York, pp 484–493

Jiang C, Zeng ZB (1995) Multiple trait analysis of genetic mapping for quantitative trait loci. Genetics 140:1111–1127

Jiang JM, Liu ZX, Lu BS, Jiang JM, Liu ZX, Lu BS (1999) Provenance genetic variation analysis of loblolly pine and determination of suitable provenances (areas). For Res Beijing 12:485–492

Johnson GR, Wheeler NC, Strauss SH (2000) Financial feasibility of marker-aided selection in Douglas-fir. Can J For Res 30:1942–1952

Kamm A, Doudrick RL, Heslop-Harrison JS, Schmidt T (1996) The genomic and physical organization of *Ty*1-copia-like sequences as a component of large genomes in *Pinus elliottii* var. *elliottii* and other gymnosperms. Proc Natl Acad Sci USA 93:2708–2713

Kao C-H, Zeng J, Teasdale RD (1999) Multiple interval mapping for quantitative trait loci. Genetics 152:1203–1216

Karhu A, Hurme P, Karjalainen M, Karvonen P, Kärkkäinen K, Neal DB, Savolainen O (1996) Do molecular markers reflect patterns of differentiation in adaptive traits of conifers? Theor Appl Genet 93:215–221

Kärkkäinen K, Savolainen O (1993) The degree of early inbreeding depression determines the selfing rate at the seed stage: model and results from *Pinus sylvestris* (Scots pine). Heredity 71:160–166

Kaya Z, Neale DB (1995) Utility of random amplified polymorphic DNA (RAPD) markers for linkage mapping in Turkish red pine (*Pinus brutia* Ten). Silvae Genet 44:110–116

Kaya Z, Sewell MM, Neale DB (1999) Identification of quantitative trait loci influencing annual height- and diameter-increment growth in loblolly pine (*Pinus taeda* L.). Theor Appl Genet 98:586–592

Kayihan GC, Huber DA, Morse AM, White TL, Davis JM (2005) Genetic dissection of fusiform rust and pitch canker disease traits in loblolly pine. Theor Appl Genet 110:948–958

Kent J, Richardson TE (1997) Fluorescently labelled, multiplexed chloroplast microsatellites for high-throughput paternity analysis in *Pinus radiata*. NZ J For Sci 27:305–312

Khoshoo TN (1961) Chromosome numbers in gymnosperms. Silvae Genet 10:1–9

Kinlaw CS, Neale DB (1997) Complex gene families in pine genomes. Trends Plant Sci 2:356–359

Kinloch BB, Parks GK, Flower CW (1970) White pine blister rust: Simply inherited resistance in sugar pine. Science 167:193–195

Kirst M, Johnson AF, Baucom C, Ulrich E, Hubbard K, Staggs R, Paule C, Retzel E, Whetten R, Sederoff R (2003) Apparent homology of expressed genes from wood forming tissues of loblolly pine (*Pinus taeda* L.) with *Arabidopsis thaliana*. Proc Natl Acad Sci USA 100:7383–7388

Kirst M, Myburg AA, De Leon JP, Kirst ME, Scott J, Sederoff R (2004) Coordinated genetic regulation of growth and lignin revealed by quantitative trait locus analysis of cDNA microarray data in an interspecific backcross of *Eucalyptus*. Plant Physiol 135:2368–2378

Kirst M, Cordeiro CM, Rezende GDSP, Grattapaglia D (2005) Power of microsatellite markers for fingerprinting and parentage analysis in *Eucalyptus grandis* breeding populations. J Hered 96:161–166

Knott S, Neale DN, Sewell MM, Haley C (1997) Multiple marker mapping of quantitative trait loci in an outbred pedigree of loblolly pine. Theor Appl Genet 94:810–820

Komulainen P, Brown GR, Mikkonen M, Karhu A, Garcia-Gil MR, O'Malley DM, Lee B, Neale DB, Savolainen O (2003) Comparing EST-based genetic maps between *Pinus sylvestris* and *Pinus taeda*. Theor Appl Genet 107:667–678

Kondo T, Terada K, Hayashi E, Kuramoto N, Okamura M, Kawasaki H (2000) RAPD markers linked to a gene for resistance to pine needle gall midge in Japanese black pine (*Pinus thunbergii*). Theor Appl Genet 100:391–395

Korol AB, Ronin YI, Kirzhner VM (1995) Interval mapping of quantitative trait loci employing correlated traits complexes. Genetics 140:1137–1147

Korol AB, Ronin YI, Nevo E, Hayes PM (1998) Multi-interval mapping of correlated trait complexes. Heredity 80:273–284

Kossack DS, Kinlaw CS (1999) IFG, a gypsy-like retrotransposon in *Pinus* (Pinaceae), has an extensive history in pines. Plant Mol Biol 39:417–426

Kremer A (1992) Predictions of age-age correlations of total height based on serial correlations between height increments in Maritime pine (*Pinus pinaster* Ait.). Theor Appl Genet 85:152–158

Kriebel HB (1985) DNA sequence components of the *Pinus strobus* nuclear genome. Can J For Res 15:1–4

Kruglyak L (1999) Prospects for whole-genome linkage disequilibrium mapping of common disease genes. Nat Genet 22:139–144

Krupkin AB, Liston A, Strauss SH (1996) Phylogenetic analysis of the hard pines (*Pinus* subgenus *Pinus*, Pinaceae) from chloroplast DNA restriction site analysis. Am J Bot 83:489–498

Krutovsky KV, Neale DB (2005) Nucleotide diversity and linkage disequilibrium in cold hardiness and wood quality related candidate genes in Douglas-fir. Genetics 171:2029–2041

Krutovsky KV, Troggio M, Brown GR, Jermstad KD, Neale DB (2004) Comparative mapping in the Pinaceae. Genetics 168:447–461

Kuang H, Richardson TE, Carson SD, Bongarten B (1999b) Genetic analysis of inbreeding depression in plus tree 850.55 of *Pinus radiata* D. Don. II. Genetics of viability genes. Theor Appl Genet 99:140–146

Kuang H, Richardson TE, Carson SD, Wilcox PL, Bongarten B (1999a) Genetic analysis of inbreeding depression in plus tree 850.055 of *Pinus radiata* D. Don. 1. Genetic map with distorted markers. Theor Appl Genet 98:697–703

Kubisiak TL, Nelson CD, Nance WL, Stine M (1995) RAPD linkage mapping in a longleaf pine × slash pine F_1 family. Theor Appl Genet 90:1119–1127

Kubisiak TL, Nelson CD, Nance WL, Stine M (1996) Comparison of RAPD linkage maps constructed for a single longleaf pine from both haploid and diploid mapping populations. For Genet 3:230–211

Kubisiak TL, Nelson CD, Nowak J, Friend AL (2000) Genetic linkage mapping of genomic regions conferring tolerance to high aluminum in slash pine. J Sust For 10:69–78

Kubisiak TL, Nelson, CD, Stine M (1997) RAPD mapping of genomic regions influencing early height growth in longleaf pine × slash pine F_1 hybrids. In: Proc 24th Southern For Tree Improv Conf, Orlando, FL, 9-12 June 1997, pp 198–206

Kumar S (2004) Genetic parameter estimates for wood stiffness, strength, internal checking and resin bleeding for radiata pine. Can J For Res 34:2601–2610

Kumar S, Garrick DJ (2001) Genetic response to within-family selection using molecular markers in some radiata pine breeding schemes. Can J For Res 31:779–785

Kumar S, Richardson TE (2005) Inferring heritability and relatedness using molecular markers in radiata pine. Mol Breed 15:55–64

Kumar S, Spelman R, Garrick D, Richardson TE, Wilcox PL (2000) Multiple marker mapping of quantitative trait loci on chromosome *three* in an outbred pedigree of radiata pine. Theor Appl Genet 100:926–933

Kumar S, Echt C, Wilcox PL, Richardson TE (2004) Testing for linkage disequilibrium in the New Zealand radiata pine breeding population. Theor Appl Genet 108:292–298

Kumar S, Gerber S, Richardson TE (2006) Pedigree reconstruction using SSR markers in a radiata pine breeding programme. In: Proc 13th Australasian Plant Breed Conf, 17-21 April, Christchurch, New Zealand pp 578–583, http://www.apbc.org.nz

Kutil BL, Williams CG (2001) Triplet-repeat microsatellites shared among hard and soft pines. J Hered 92:327–32

Lambeth C (2000) Realized genetic gains for first generation improved loblolly pine in 45 tests in coastal North Carolina. J Appl For 24:140–144

Lambeth C, Lee BC, O'Malley D, Wheeler N (2001) Polymix breeding with parental analysis of progeny: an alternative to full-sib breeding and testing. Theor Appl Genet 103:930–943

Lamoureux D, Peterson DG, Li W, Fellers JP, Gill BS (2005) The efficacy of Cot-based gene enrichment in wheat (*Triticum aestivum* L.). Genome 48:1120–1126

Lamprecht H (1990) Silviculture in the Tropics: tropical forest ecosystems and their tree species-possibilities and methods for their long-term utilization. GTZ, Eschborn, Germany

Lande R, Thompson R (1990) Efficiency of marker-assisted selection in the improvement of quantitative traits. Genetics 124:743–756

Lander ES, Botstein D (1989) Mapping Mendelian factors underlying quantitative traits using RFLP linkage maps. Genetics 121:185–199

Laurie CC, Chasalow SD, LeDeaux JR, McCarroll R, Bush D, Hauge B, Lai C, Clark D, Rocheford TR, Dudley JW (2004) The genetic architecture of response to long-term artificial selection for oil concentration in the maize kernel. Genetics 168:2141–2155

Le Dantec L, Chagné D, Pot D, Cantin O, Garnier-Géré P, Bedon F, Frigerio JM, Chaumeil P, Léger P, García V, Laigret F, de Daruvar A, Plomion C (2004) Automated SNP detection in expressed sequence tags: statistical considerations and application to maritime pine sequences. Plant Mol Biol 54:461–470

Le Provost G, Paiva J, Pot D, Brach J, Plomion C (2003) Seasonal variation in transcript accumulation in wood forming tissues of maritime pine (*Pinus pinaster* Ait.) with emphasis on a cell wall Glycine Rich Protein. Planta 217:820–830

Ledig FT (1998) Genetic variation in *Pinus*. In: Richardson DM (ed) Ecology and Biogeography of *Pinus*. Cambridge University Press, Cambridge, UK, pp 251–280

Ledig FT, Conkle MT (1983) Gene diversity and genetic structure in a narrow endemic Torrey pine (*Pinus torreyana* Parry ex Carr). Evolution 37:79–85

Leitch IJ, Hanson L, Winfield M, Parker J, Bennett MD (2001) Nuclear DNA C-values complete familial representation in gymnosperms. Ann Bot 88:843–849

Lerceteau EC, Plomion C, Andersson B (2000) AFLP mapping and detection of quantitative trait loci (QTLs) for economically important traits in *Pinus sylvestris*: a preliminary study. Mol Breed 6:451–458

Li C, Yeh FC (2001) Construction of a framework map in *Pinus contorta* subsp. *latifolia* using random amplified polymorphic DNA markers. Genome 44:147–153

Liston A, Robinson WA, Pinero D, Alvarez-Buylla ER (1999) Phylogenetics of *Pinus* (Pinaceae) based on nuclear ribosomal DNA internal transcribed spacer region sequences. Mol Phylogenet Evol 11:95–109

Liu ZL, Zhang D, Hong DY, Wang XR (2003) Chromosomal localization of 5S and 18S-5.8S-25S ribosomal DNA sites in five Asian pines using fluorescence *in situ* hybridization. Theor Appl Genet 106:198–204

Lopez-Upton J, White TL, Huber DA (2000) Species differences in early growth and rust incidence of loblolly and slash pine. For Ecol Mng 132:211–222

Lorenz WW, Dean JFD (2002) SAGE profiling and demonstration of differential gene expression along the axial developmental gradient of lignifying xylem in loblolly pine (*Pinus taeda*). Tree Physiol 22:301–310

Lorenz WW, Sun F, Liang C, Kolychev D, Wang H, Zhao X, Cordonnier-Pratt MM, Pratt LH, Dean JFD (2005) Water stress-responsive genes in loblolly pine (*Pinus taeda*) roots identified by analyses of expressed sequence tag libraries. Tree Physiol 26:1–16

Lorenz WW, Sun F, Liang C, Kolychev D, Wang H, Zhao X, Cordonnier-Pratt MM, Pratt LH, Dean JF (2006) Water stress-responsive genes in loblolly pine (*Pinus taeda*) roots identified by analyses of expressed sequence tag libraries. Tree Physiol 26:1–16

Lowe WJ, Byram TD, Bridgwater FE (1999) Selecting loblolly pine parents for seed orchards to minimise the costs of producing pulp. For Sci 45:213–216

Lubaretz O, Fuchs J, Ahne R, Meister A, Schubert I (1996) Karyotyping of three *Pinaceae* species via fluorescent *in situ* hybridization and computer-aided chromosome analysis. Theor Appl Genet 92:411–416

Luikart G, England P, Tallmon D, Jordan S, Taberlet P (2003) The power and promise of population genomics: from genotyping to genome typing. Nat Rev Genet 4:981–994

Lund MS, Sorensen P, Guldbrantsen B, Sorensen DA (2003) Multitrait fine mapping of quantitative trait loci using combined linkage disequilibrium and linkage analysis. Genetics 163:405–410

Ma XF, Szmidt A, Wang WR (2006) Genetic structure and evolutionary history of a diploid hybrid pine *Pinus densata* inferred from the nucleotide variation at seven gene loci. Mol Biol Evol 23:807–816

MacPherson P, Filion WG (1981) Karyotype analysis and the distribution of constitutive heterochromatin in five species of *Pinus*. J Hered 72:193–198

Margulies M, Egholm M, Altman WE, Attiya S, Bader JS, Bemben LA, Berka J, Braverman MS, Chen YJ, Chen Z, Dewell SB, Du L, Fierro JM, Gomes XV, Godwin BC, He W, Helgesen S, Ho CH, Irzyk GP, Jando SC, Alenquer ML, Jarvie TP, Jirage KB, Kim JB, Knight JR, Lanza JR, Leamon JH, Lefkowitz SM, Lei M, Li J, Lohman KL, Lu H, Makhijani VB, McDade KE, McKenna MP, Myers EW, Nickerson E, Nobile JR, Plant R, Puc BP, Ronan MT, Roth GT, Sarkis GJ, Simons JF, Simpson JW, Srinivasan M, Tartaro KR, Tomasz A, Vogt KA, Volkmer GA, Wang SH, Wang Y, Weiner MP, Yu P, Begley RF, Rothberg JM (2005) Genome sequencing in microfabricated high-density picolitre reactors.Nature 437:376–380

Mariette S, Chagné D, Decroocq S, Vendramin GG, Lalanne C, Madur D, Plomion C (2001) Microsatellite markers for *Pinus pinaster* Ait. Ann For Sci 58:203–206

Markussen T, Fladung M, Achere V, Favre JM, Faivre-Rampant P, Aragones A, Da Silva Perez, Havengt L, Ritter E (2003) Identification of QTLs controlling growth, chemical and physical wood property traits in *Pinus pinaster* (Ait.). Silvae Genet 52:8–15

Marquardt PE, Epperson BK (2004) Spatial and population genetic structure of microsatellites in white pine. Mol Ecol 13:3305–3315

Matheson AC, Lindgren D (1985) Gain from the clonal and the clonal seed orchard options compared for tree breeding programs. Theor Appl Genet 71:242–249

Matziris DI (1995) Provenance variation of *Pinus radiata* grown in Greece. Silvae Genet 44:88–96

McKeand S, Mullin T, Byram T, White TL (2003) Deployment of genetically improved loblolly and slash pines in the South USA. J For 101:32–37

Mergen F (1958) Natural polyploidy in slash pine. For Sci 4:283–295

Miksche JP, Hotta Y (19931973) DNA base composition and repetitious DNA in several conifers. Chromosoma 41:29–36

Millar CI (1998) Early evolution of pines. In: Richardson DM (ed) Ecology and Biogeography of *Pinus*. Cambridge University Press, Cambridge, UK, pp 69–91

Mitton JB (1992) The dynamic mating system of conifers. New For 6:187–216

Moore G, Devos KM, Wang Z, Gale MD (1995) Grasses, line up and form a circle. Curr Biol 5(7):737-739

Moran GF, Bell JC, Hilliker AJ (1983) Greater meiotic recombination in male vs. female gametes in *Pinus radiata*. J Hered 74:62

Morgante M, Salamini F (2003) From plant genomics to breeding practice. Curr Opin Biotechnol 14:214–219

Morgante M, Vendramin GG, Rossi P, Olivieri AM (1993) Selection against inbreds in early life-cycle phases in *Pinus leucodermis*. Heredity 70:622–627

Morin PA, Luikart G, Wayne RK, and the SNP workshop group (2004) SNPs in ecology, evolution and conservation. Trends Ecol Evol 19:208–2169

Morse AM, Nelson CD, Covert SF, Holliday AG, Smith KE, Davis JM (2004) Pine genes regulated by the necrotrophic pathogen Fusarium circinatum. Theor Appl Genet 109:922–932

Morse AM, Nelson CD, Covert SF, Smith KE, Davis JM (2004) Pine genes regulated by the necrotrophic pathogen, *Fusarium circinatum*. Theoretical and Applied Genetics 109:922-932

Morton NE (1991) Parameters of the human genome. Proc Natl Acad Sci USA 88:7474–7476

Mosig MO, Lipkin E, Khutoreskaya G, Tchourzyna E, Soller M, Friedmann A (2001) A whole genome scan for quantitative trait loci affecting milk protein percentage in Israeli-Holstein cattle, by means of selective milk DNA pooling in a Daughter design, using an adjusted false discovery rate criterion. Genetics 157:1683–1698

Moura VPG, Dvorak WS, Hodge GR (1998) Provenance and family variation of *Pinus oocarpa* grown in the Brazilian cerrado. For Ecol Manage 109:315–322

Müller-Starck G (1998) Isozymes. In: Karp A, Isaac PG, Ingram DS (eds) Molecular Tools for Screening Biodiversity: Plants and Animals. Chapman and Hall, London, UK, pp 75–81

Muona O, Harju A (1989) Effective population sizes, genetic variability and mating system in natural stands and seed orchards of *Pinus sylvestris*. Silvae Genet 38:221–228

Muona O, Yazdani R, Rudin R (1987) Genetic change between life stages in *Pinus sylvestris*: allozymes variation in seeds and planted seedlings. Silvae Genet 35:39–42

Murray B (1998) Nuclear DNA amounts in gymnosperms. Ann Bot 82(Suppl A):3-15

Myburg H, Morse AM, Amerson HV, Kubisiak TL, Huber DA, Osborne JA, Garcia SA, Nelson CD, Davis JM, Covert SF, van Zyl LM (2006) Differential gene expression in loblolly pine (*Pinus taeda* L.) challenged with the fusiform rust fungus, *Cronartium quercuum* f. sp. *fusiforme*. Physiological and Molecular Plant Pathology 69:79-101

Myers RM, Maniatis T, Lerman LS (1987) Detection and localization of single base changes by denaturing gradient gel electrophoresis. Methods Enzymol 155:501–527

Neale DB, Savolainen O (2004) Association genetics of complex traits in conifers. Trends Plant Sci 9:325–330

Neale DB, Sederoff RR (1989) Paternal inheritance of chloroplast DNA and maternal inheritance of mitochondrial DNA in loblolly pine. Theor Appl Genet 77:212–216

Neale DB, Williams CG (1991) Restriction fragment length polymorphism mapping in conifers and applications to forest genetics and tree improvement. Can J For Res 21:545–554

Nelson CD, Echt CS (2003) New models for marker-assisted selection in tree breeding. In: McKinley C (ed) Proc 27th South For Tree Improv Conf, Stillwater, OK, p 114

Nelson CD, Echt CS (2004) Marker-directed population improvement. In: Li B, McKeand S (eds) Proc IUFRO Joint Conf of Div 2, For Genet Tree Breed in the Age of Genomics: Progress and Future, Charleston, SC, p 255

Nelson CD, Kubisiak TL, Johnson G, Burdine C, Bridgwater FE (2003) Microsatellite analysis of loblolly pine. In: Plant & Animal Genome XI Conf, San Diego, p 560 (http://www.intl-pag.org/11/abstracts/P5i_P560_XI.html, accessed 20 March 2006)

Nelson CD, Kubisiak, TL, Stine M, Nance WL (1994) A genetic linkage map of longleaf pine (*Pinus palustris* Mill.) based on random amplified polymorphic DNAs. J Hered 85:433–439

Nelson CD, Nance WL, Doudrick RL (1993) A partial genetic linkage map of slash pine (*Pinus elliottii* Engelm. var. *elliottii*) based on random amplified polymorphic DNAs. Theor Appl Genet 8:145–151

Nelson JC (1997) QGENE: software for marker-based genomic analysis and breeding. Mol Breed 3:239–245

Nikles DG (2000) Experience with some Pinus hybrids in Queensland, Australia. In: Dungey HS, Dieters MJ, Nikles DG (Compl) Hybrid Breeding and Genetics of Forest Trees. Proc of QFRI/CRC-SPF Symp, 9-14 April 2000, Noosa, Queensland, Australia. Dept of Pri Indus, Brisbane, pp 14–26

No E, Zhou Y, Loopstra CA (2000) Sequences upstream and downstream of two xylem-specific pine genes influence their expression. Plant Sci 160:77–86

NSF (2005) NSF, USDA and DOE Award $32 Million to Sequence Corn Genome. Press Release 05-197. National Science Foundation. 11-15-2005. http://www.nsf.gov/news/news_summ.jsp?cntn_id=104608

O'Farrell PH (1975) High resolution two-dimensional electrophoresis of proteins J Biol Chem 250:4007–4021

O'Malley D, Grattaplagia D. Chaparro JX, Wilcox PL, Amerson HV, Liu B-H, Whetton R, McKeand S, Kuhlman EG, McCord S, Crane B, Sederoff RR (1996) Molecular markers, forest genetics and tree breeding. In: Gustafson JP, Flavell RB (eds) Genomes of Plants and Animals: 21st Stadler Genet Symp. Plenum, New York

Ohri D, Khoshoo TN (1986) Genome size in gymnosperms. Plant Syst Evol 153:119–131

OlssonT, Lindgren D, Li B (2001) Balancing genetic gain and relatedness in seed orchards. Silvae Genet 50:222–227

Orita M, Iwahana H, Kanazawa H, Hayashi K, Sekiya T (1989) Detection of polymorphisms of human DNA by gel electrophoresis as single-strand conformation polymorphisms. Proc Natl Acad Sci USA 86:2766–2770

Ott J (1991) Analysis of Human Genetic Linkage. Johns Hopkins University Press, Baltimore, MD

Pacheco D, Díaz S, Osuna D, Bautista R, Claros MG, Cánovas FM, Cantón FR (2005) Identification of candidate genes conferring specific properties to distinct types of *Pinus pinaster* Ait. wood. IUFRO Tree Biotechnol, Pretoria, South Africa

Paiva J (2006) Phenotypic and molecular plasticity of wood forming tissues in maritime pine (*Pinus pinaster* Ait.). PhD thesis of Universidad Nova de Lisboa and Bordeaux 1

Paran I, Zamir D (2003) Quantitative traits in plants: beyond the QTL. Trends Genet 19:303–306

Park YS (2002) Implementation of conifer somatic embryogenesis in clonal forestry: technical requirements and deployment considerations. Ann For Sci 59:651–656

Paterson AH, Freeling M, Sasaki T (2005) Grains of knowledge: genomics of model cereals. Genome Res 15:1643–1650

Paul AD, Foster GS, Caldwell T, McRae J (1997) Parameters for height, diameter, and volume in a multilocation clonal study with loblolly pine. For Sci 43:87–98

Pavy N, Laroche J, Bousquet J, MacKay J (2005) Large-scale analysis of secondary xylem ESTs in pine. Plant Mol Biol 57:203224

Perlack RD, Wright LL, Turhollow AF, Graham RL, Stokes BJ, Erbach DC (2005) Biomass as a feedstock for a bioenergy and bioproducts industry: the technical feasibility of a billion-ton annual supply. Report No DOE/GO-102005-2135. US Dept of Energy/US Dept of Agri. http://www1.eere.energy.gov/biomass/pdfs/final_billionton_vision_report2.pdf

Peterson DG (2005) Reduced representation strategies and their application to plant genomes. In: Meksem K, Kahl G (eds) The Handbook of Genome Mapping: Genetic and Physical Mapping. Wiley, Weinheim, pp 307–335

Peterson DG, Chouvarine P, Thummasuwan S, Saha S, Mukherjee D, Carlson JE (2006) Exploring the pine genome using Cot filtration and 454 Life Sciences massively parallel shotgun sequencing. In: Plant & Animal Genomes XIV Conf, San Diego. http://www.intl-pag.org/pag/14/abstracts/PAG14_W376.html

Peterson DG, Nelson CD, Islam-Faridi MN, Main DS, Tomkins JP (2005) Accelerating pine genomics through development and utilization of molecular and cytogenetic resources. In: Plant & Animal Genomes XIII Conf, San Diego. http://www.intl-pag.org/pag/13/abstracts/PAG13_P515.html

Peterson DG, Schulze SR, Sciara EB, Lee SA, Bowers JE, Nagel A, Jiang N, Tibbitts DC, Wessler SR, Paterson AH (2002) Integration of Cot analysis, DNA cloning, and high-throughput sequencing facilitates genome characterization and gene discovery. Genome Res 12:795–807

Petit RJ, Duminil J, Fineschi S, Hampe A, Salvini D, Vendramin GG (2005) Comparative organisation of chloroplast, mitochondrial and nuclear diversity in plant populations. Mol Ecol 14:689–701

Petit RJ, Vendramin GG (2007) Phylogeography of organelle DNA in plants: an introduction. In: Weiss S, Ferrand N (eds) Phylogeography of Southern European Refugia. Kluwer, Amsterdam, (in press)

Pflieger S, Lefebvre V, Causse M (2001) The candidate gene approach in plant genetics: a review. Mol Breed 7:275–291

Plomion C, Durel C-E (1996) Estimation of the average effects of specific alleles detected by the pseudo-testcross QTL mapping strategy. Gen Sel Evol 28:223–235

Plomion C, O'Malley DM (1996) Recombination rate differences for pollen parents and seed parents in pine. Heredity 77:341–350

Plomion C, Bahrman N, Durel CE, O'Malley DM (1995a) Genomic analysis in *Pinus pinaster* (Maritime pine) using RAPD and protein markers. Heredity 74:661–668

Plomion C, O'Malley DM, Durel CE (1995b) Genomic analysis in Maritime pine (*Pinus pinaster*). Comparison of two RAPD maps using selfed and open-pollinated seeds of the same individual. Theor Appl Genet 90:1028–1034

Plomion C, Durel C-E, O'Malley D (1996a) Genetic dissection of height in maritime pine seedlings raised under accelerated growth condition. Theor Appl Genet 93:849–858

Plomion C, Yani A, Marpeau A (1996b) Genetic determinism of ?3-carene in maritime pine using random amplified polymorphic DNA (RAPD) markers. Genome 39:1123–1127

Plomion C, Liu BH, O'Malley DM (1996c) Genetic analysis using trans dominant linked markers in an F_2family. Theor Appl Genet 93:1083–1089

Plomion C, Costa P, Bahrman N, Frigerio JM (1997) Genetic Analysis of needle proteins in maritime pine. 1. Mapping dominant and codominant protein markers assayed on diploid tissue, in a haploid-based genetic map. Silvae Genet 46:161–165

Plomion C, Hurme P, Frigerio J-M, Ridolphi M, Pot D, Pionneau C, Avila C, Gallardo F, David H, Neutlings G, Campbell M, Canovas FM, Savolainen O, Bodénès C, Kremer A (1999) Developing SSCP markers in two *Pinus* species. Mol Breed 5:21–31

Plomion C, Pionneau C, Brach J, Costa P, Baillères H (2000) Compression wood-responsive proteins in developing xylem of maritime pine (*Pinus pinaster* Ait.). Plant Physiol 123:959–969

Plomion C, LeProvost G, Pot D, Vendramin G, Gerber S, Decroocq S, Brach J, Raffin A, Pastuszka P (2001) Pollen contamination in a maritime pine polycross seed orchard and certification of improved seeds using chloroplast microsatellites. Can J For Res 31:1816–1825

Plomion C, Bahrman N, Costa P, Dubos C, Frigério J-M, Gerber S, Gion J-M, Lalanne C, Madur D, Pionneau C (2004) Proteomics for genetics and physiological studies in forest trees: application in maritime pine. In: Kumar S, Fladung M (eds) Molecular Genetics and Breeding of Forest Trees. Haworth, Binghamton, NY, pp 53–80

Pot D (2004) Déterminisme génétique de la qualité du bois chez le pin maritime: du phénotype aux gènes. PhD thesis, University of Rennes I, ENSAR, France

Pot D, Chantre G, Rozenberg P, Rodrigues JC, Jones GL, Pereira H, Hannrup B, Cahalan C, Plomion C (2002) Genetic control of pulp and timber properties in maritime pine (*Pinus pinaster* Ait.). Ann For Sci 59:563–575

Pot D, McMillan L, Echt C, Le Provost G, Garnier-Géré P, Cato S, Plomion C (2005a) Nucleotide variation in genes involved in wood formation in two pine species. New Phytol 167:101–112

Pot D, Rodrigues J-C, Rozenberg P, Chantre G, Tibbits J, Cahalan C, Pichavant F, Plomion C (2005b) QTLs and candidate genes for wood properties in maritime pine (*Pinus pinaster* Ait.). Tree Genet Genom 2:10–26

Price RA, Liston A, Strauss SH (1998) Phylogeny and systematics of *Pinus*. In: Richardson DM (ed) Ecology and Biogeography of *Pinus*. Cambridge University Press, Cambridge, UK, pp 49–68

Provan J, Soranzo N, Wilson NJ, Goldstein DB, Powell W (1999) A low mutation rate for chloroplast microsatellites. Genetics 153:943–947

Quackenbush J, Cho J, Lee D, Liang F, Holt I, Karamycheva S, Parvizi B, Pertea G, Sultana R, White J (2001) The TIGR Gene Indices: analysis of gene transcript sequences in highly sampled eukaryotic species. Nucleic Acids Res 29:159–164

Quencez C, Bastien C (2001) Genetic variation within and between populations of *Pinus sylvestris* L. (Scots pine) for susceptibility to *Melampsora pinitorqua* Rostr. (pine twist rust). Heredity 86:36–44

Rabinowicz PD, Citek R, Budiman MA, Nunberg A, Bedell JA, Lakey N, O'Shaughnessy AL, Nascimento LU, McCombie WR, Martienssen RA (2005) Differential methylation of genes and repeats in land plants. Genome Res 15:1431–1440

Rabinowicz PD, Schutz K, Dedhia N, Yordan C, Parnell LD, Stein L, McCombie WR, Martienssen RA (1999) Differential methylation of genes and retrotransposons facilitates shotgun sequencing of the maize genome. Nat Genet 23:305–308

Rafalski A (1998) Randomly amplified polymorphic DNA (RAPD) analysis. In: Caetano-Anollés G, Gresshoff

PM (eds) DNA Markers Protocols, Applications, and Overviews. Wiley-Liss, New York, pp 75–83

Rake AV, Miksche JP, Hall RB, Hansen KM (1980) DNA reassociation kinetics of four conifers. Can J Genet Cytol 22:69–79

Rat Genome Sequencing Project Consortium (2004) Rat Genome Sequencing Project Consortium, Genome sequence of the Brown Norway rat yields insights into mammalian evolution. Nature 428:493–521

Remington DL, O'Malley DM (2000a) Evaluation of major genetic loci contributing to inbreeding depression for survival and early growth in a selfed family of *Pinus taeda*. Evolution 54:1580–1589

Remington DL, O'Malley DM (2000b) Whole genome characterization of embryonic state inbreeding depression in a selfed loblolly pine family. Genetics 155:337–348

Remington DL, Whetten RW, Liu BH, O'Malley DM (1999) Construction of an AFLP genetic map with nearly complete genome coverage in *Pinus taeda*. Theor Appl Genet 98:1279–1292

Ribeiro MM, Plomion C, Petit RJ, Vendramin GG, Szmidt AE (2001) Variation of chloroplast single-sequence repeats in Portuguese maritime pine (*Pinus pinaster* Ait.). Theor Appl Genet 102:97–103

Richardson BA., Brunsfeld SJ, Klopfenstein NB (2002) DNA from bird-dispersed seed and wind-disseminated pollen provides insights into postglacial colonization of whitebark pine (*Pinus albicaulis*). Mol Ecol 11:215–227

Richardson DM, Rundel PW (1998) Ecology and biogeography of *Pinus*: an introduction. In: Richardson DM (ed) Ecology and Biogeography of *Pinus*. Cambridge University Press, Cambridge, UK, pp 3–46

Ritland K, Zhuang J, Ralph S, Ritland C, Bohlmann J (2006) Comparative conifer genomics and the new "Treenomix" program. In: Plant & Animal Genomes XIV Conf, San Diego, W124

Ritter E, Aragonés A, Markussen T, Acheré V, Espinel S, Fladung M, Wrobel S, Faivre-Rampant P, Jeandroz S, Favre JM (2002) Toward construction of an ultra high density linkage map for *Pinus pinaster*. Ann For Sci 59:637–643

Robledo-Arnuncio JJ, Collada C, Alía R, Gil L (2005) Genetic structure of montane isolates of *Pinus* sylvestris L. in a Mediterranean refugial area. J Biogeogr 32:595–605

Rosvall O, Lindgren D, Mullen TJ (1998) Sustainability, robustness and efficiency of a multi-generation breeding strategy based on within-family clonal selection. Silvae Genet 47:307- 321

Rounsley S, Xiaoying L, Ketchum KA (1998) Large-scale sequencing of plant genomes. Curr Opin Plant Biol 1:136–141

Rozenberg P, Van Loo J, Hannrup B, Grabner M. 2002. Clonal variation of wood density record of cambium reaction to water deficit in *Picea abies* (L.) Karst. Ann For Sci 59:533–540

Rudd S (2003) Expressed sequence tags: alternative or complement to whole genome sequences? Trends Plant Sci 8:321–329

Rudin D, Ekberg I (1978) Linkage studies in *Pinus sylvestris* L. using macrogametophyte allozymes. Silvae Genet 27:1–12

Sanchez AC, Ilag LL, Yang D, Brar DS, Ausubel F, Khush GS, Yano M, Sasaki T, Li Z, Huang N (1999) Genetic and physical mapping of *xa13*, a recessive bacterial blight resistance gene in rice. Theor Appl Genet 98:1022–1028

Sánchez C, Vielva MJ, Vieitez AM, de Mier B, Abarca D, Díaz-Sala C (2005) Identification of genes related to adventitious rooting capacity in pine and chestnut. IUFRO Tree Biotechnol, Pretoria, South Africa

Sathyan P, Newton RJ, Loopstra CA (2005) Genes induced by WDS are differentially expressed in two populations of aleppo pine (*Pinus halepensis*). Tree Genet Genom 1:166–173

Sax K (1923) The association of size differences with seed-coat pattern and pigmentation in *Phaseolus vulgaris*. Genetics 8:552–560

Sax K, Sax HJ (1933) Chromosome numbers and morphology in the conifers. J Arnold Arbor 14:356–375

Saylor LC (1961) A karyotype analysis of selected species of *Pinus*. Silvae Genet 10:77–85

Saylor LC (1964) Karyotype analysis of *Pinus* -group *Lariciones*. Silvae Genet 13:165–170

Saylor LC (1972) Karyotype analysis of *Pinus* -subgenus *Pinus*. Silvae Genet 21:155–163

Saylor LC (1983) Karyotype analysis of the genus *Pinus* – subgenus *Strobus*. Silvae Genet 32:119–121

Schmidt A, Doudrick RL, Heslop-Harrison JS, Schmidt T (2000) The contribution of short repeats of low sequence complexity to large conifer genomes. Theor Appl Genet 101:7–14

Schmidt R (2002) Plant genome evolution: lessons from comparative genomics at the DNA level. Plant Mol Biol 48:21–37

Scotti I, Burelli A, Cattonaro F, Chagné D, Fuller J, Hedley PE, Jansson G, Lalanne C, Madur D, Neale D, Plomion C, Powell W, Troggio M, Morgante M (2005) Analysis of the distribution of marker classes in a genetic linkage map: a case study in Norway spruce (*P. abies* karst). Tree Genet Genom 1:93–102

Scotti-Saintagne C, Bodénès C, Barreneche T, Bertocchi E, Plomion C, Kremer A (2004) Detection of quantitative trait loci controlling bud burst and height growth in *Quercus robur* L. Theor Appl Genet 109:1648–1659

Sewell MM, Bassoni DL, Megraw RA, Wheeler NC, Neale DB (2000) Identification of QTLs influencing wood property traits in loblolly pine (*Pinus taeda* L.). I. Physical wood properties. Theor Appl Genet 101:1273–1281

Sewell MM, Davis MF, Tuskan GA, Wheeler NC, Elam CC, Bassoni DL, Neale DB (2002). Identification of QTLs influencing wood property traits in loblolly pine (*Pinus taeda* L.). II. Chemical wood properties. Theor Appl Genet 104:214–222

Sewell MM, Neale DB (2000) Mapping quantitative traits in forest trees. In: Jain SM, Minocha SC (eds) Molecular Biology of Woody Plants. Kluwer, Dordrecht, pp 407–433

Sewell MM, Sherman BK, Neale DB (1999) A consensus map for loblolly pine (*Pinus taeda* L.). I. Construction and in-

etegration if individual linkage maps from two outbred three-generation pedigrees. Genetics 151:321–330
Shelbourne CJA (1992) Genetic gains from different kinds of breeding population and seed or plant production populations. S Afr For J 160:49–65
Shelbourne CJA (2000) Some insights on hybrids in forest tree improvement. In: Dungey HS, Dieters MJ, Nikles DG (Compl) Hybrid Breeding and Genetics of Forest Trees. Proc QFRI/CRC-SPF Symp, 9-14 April 2000, Noosa, Queensland, Australia. Dept of Pri Indus, Brisbane, pp 53–62
Shelbourne CJA, Apiolaza LA, Jayawickrama KJS, Sorensson CT (1997) Developing breeding objectives for radiata pine in New Zealand. In: Burdon R, Moore JM (eds) "IUFRO '97 Genetics of Radiata Pine", Proc IUFRO Conf 1–4 Dec and workshop 5 Dec, Rotorua, New Zealand. FRI Bull 203:60–168
Shepherd M, Cross M, Dieters MJ, Henry R (2003) Genetic maps for *Pinus elliottii* var. *elliottii* and *P. caribaea* var. *hondurensis* using AFLP and microsatellite markers. Theor App Genet 106:1409–1419
Shepherd M, Cross M, Dieters MJ, Henry R (2002b) Branch architecture QTL for *Pinus elliottii* var. elliottii × *Pinus caribea* var. hondurensis hybrids. Ann For Sci 59:617–625
Shepherd M, Cross M, Maguire TL, Dieters MJ, Williams CG, Henry RJ (2002a) Transpecific microsatellites for hard pines. Theor Appl Genet 104:819–827
Shoulders E (1984) The case for planting longleaf pine. Proc of the Southern Silvicultural Res Conf, 7-8 Nov 1984, Atlanta, GA, pp 255–260
Smith DN, Devey M (1994) Occurrence and inheritance of microsatellites in *Pinus radiata*. Genome 37:977–983
Soranzo N, Alia R, Provan J, Powell W (2000) Patterns of variation at a mitochondrial sequence-tagged-site locus provides new insights into the postglacial history of European *Pinus sylvestris* populations. Mol Ecol 9:1205–1211
Sorensson CT, Shelbourne CJA (2005) Clonal forestry. In: Colley M (ed) Forestry Handbook. NZ Inst of Foresters, pp 92–96
Springer NM, Xu X, Barbazuk WB (2004) Utility of different gene enrichment approaches toward identifying and sequencing the maize gene space. Plant Physiol 136:3023–3033
Stelzer HE, Goldfarb B (1997) Implementing clonal forestry in the southeastern United States: SRIEG satellite workshop summary. Can J For Res 27:442–446
Stirling B, Yang ZK, Gunter LE, Tuskan GA, Bradshaw HD (2003) Comparative sequence analysis between orthologous regions of the *Arabidopsis* and *Populus* genomes reveals substantial synteny and microcollinearity. Can J For Res 33:2245–2251
Strabala TJ (2004) Expressed sequence tag databases from forestry tree species. In: Kumar S, Fladung M (eds) Molecular Genetics and Breeding of Forest Trees. Haworth, Binghamton, NY, pp 19–52
Strauss SH, Doerksen AH (1990) Restriction fragment analysis of pine phylogeny. Evolution 44:1081–1096
Strauss SH, Lande R, Namkoong G (1992) Limitations of molecular marker-aided selection in forest tree breeding. Can J For Res 22:1050–1061
Stromberg LD, Dudley JD, Rufener GK (1994) Comparing conventional early generation selection with molecular marker assisted selection in maize. Crop Sci 34:1221–1225
Stuber CW, Polacco M, Senior ML (1999) Synergy of empirical breeding, marker-assisted selection, and genomics to increase crop yield. Crop Sci 39:1571–1583
Syring J, Willyard A, Cronn R, Liston A (2005) Evolutionary relationships among *Pinus* (Pinaceae) subsections inferred from multiple low-copy nuclear loci. Am J Bot 92:2086–2100
Syvanen AC (2001) Accessing genetic variation. Genotyping single nucleotide polymorphisms. Nat Rev Genet 2:930–942
Szmidt AE, Muona O (1989) Linkage relationships of allozyme loci in *Pinus sylvestris*. Heredity 111:91–97
Tanksley SD, Nelson JC (1996) Advanced backcross QTL analysis: a method for the simultaneous discovery and transfer of valuable QTLs from unadapted germplasm into elite breeding lines. Theor Appl Genet 92:191–203
Telfer EJ, Echt CS, Nelson CD, Wilcox PL (2006) Comparative mapping in *Pinus radiata* and *P. taeda* reveals co-location of wood density-related QTL. In: Plant & Animal Genome XIV Conf, San Diego
Temesgen B, Brown GR, Harry DE, Kinlaw CS, Sewell MM, Neale DB (2001) Genetic mapping of expressed sequence tag polymorphism (ESTP) markers in loblolly pine (*Pinus taeda* L.). Theor Appl Genet 102:664–675
The *Arabidopsis* Genome Initiative (2000) Analysis of the genome sequence of the flowering plant *Arabidopsis thaliana*. Nature 408:796–815
Thiellement H, Plomion C, Zivy M (2001) Proteomics as a tool for plant genetics and breeding. In: Dunn MJ, Pennington S (eds) Proteomics: from protein sequence to function. BIOS, Oxford, pp 289–309
Thoday JM (1961) Location of polygenes. Nature 191:368–370
Thumma BR, Nolan MF, Evans R, Moran GF (2005) Polymorphisms in *Cinnamoyl CoA Reductase* (*CCR*) are associated with variation in microfibril angle in *Eucalyptus* spp. Genetics 171:1257–1265
Travis SE, Ritland K, Whitham TG, Keim P (1998) A genetic linkage map of pinyon pine (*Pinus edulis*) based on amplified fragment length polymorphisms. Theor Appl Genet 97:871–880
Tulsieram LK, Glaubitz JC, Kiss G, Carlson JE (1992) Single tree genetic linkage mapping in conifers using haploid DNA from megagametophytes. Bio/Technology 10:686–690
Tuskan GA, DiFazio SP, Teichmann T (2004) Poplar genomics is getting popular: the impact of the poplar genome project on tree research. Plant Biol (Stuttg) 6:2–4
Van der Burgh J (1973) Hölzer der nierderrheinischen Braunkohlenformation. 2. Hölzer der Braunkohlengruben

‚Maria Theresia' zu Herzogenrath, ‚Zukunft West' zu Eschweiler und ‚Victor' (Zülpich Mitte) zu Zülpich. Nebst einer systematisch-anatomischen Bearbeitung der Gattung *Pinus* L. Rev Palaeobot Palynol 15:73–275

van Tienderen PH, de Haan AA, van der Linden CG, Vosman B (2002) Biodiversity assessment using markers for ecologically important traits. Trends Ecol Evol 17:577–582

van Zyl L, von Arnold S, Bozhkov P, Chen Y, Egertsdotter U, MacKay J, Sederoff R, Weir B, Shen J, Sun Y-H, Whetten R, Zelena L, Clapham D (2002) Heterologous array analysis in Pinaceae: hybridization of *Pinus taeda* cDNA arrays with cDNA from needles and embryogenic cultures of *P. taeda*, *P. sylvestris* or *Picea abies*. Comp Funct Genom 3:306–318

Verhaegen D, Plomion C, Poitel M, Costa P, Kremer A (1998) Quantitative trait disscetion analysis in *Eucalyptus* using RAPD markers: 2. linkage disequilibrium in a factorial design between *E. urophylla* and *E. Grandis*. For Genet 5:61–69

Vos P, Hogers R, Bleeker M, Reijans M, van de Lee T, Hornes M, Friters A, Pot J, Paleman J, Kuiper M, Zabeau M (1995) AFLP: a new technique for DNA fingerprinting. Nucleic Acids Res 23:4407–4414

Wahlenberg WG (1946) Longleaf Pine. Charles Lathrop Pack Forestry Foundation, Washington, DC

Wakamiya I, Price HJ, Messina MG, Newton RJ (1996) Pine genome size diversity and water relations. Physiol Planta 96:13–20

Walker S, Haines R, Dieters M (1996) Beyond 2000. In: Dieters MJ, Matheson AC, Nikles DG, Harwood CE (eds) Clonal Forestry in Queensland. Queensland For Res Inst. Tree Improvement for Sustainable Tropical Forestry. QFRI-IUFRO Conf, Caloundra, Queensland, Australia, 27 Oct-1 Nov 1996, vol 2, pp 351–354

Walter R, Epperson BK (2001) Geographic pattern of genetic variation in *Pinus resinosa*: area of greatest diversity is not the origin of postglacial populations. Mol Ecol 10:103–111

Wang XR, Szmidt AE, Nguyen HN (2000) The phylogenetic position of the endemic flat-needle pine *Pinus krempfii* (Pinaceae) from Vietnam based on PCR-RFLP analysis of chloroplast DNA. Plant Syst Evol 220:21–36

Wang XR, Szmidt AE, Savolainen O (2001) Genetic composition and diploid hybrid speciation of a high mountain pine, *Pinus densata*, native to the Tibetan plateau. Genetics 159:337–346

Wang XT, Tsumara Y, Yoshimaru H, Nagasaka K, Szmidt A (1999) Phylogenetic relationships of Eurasian pines (*Pinus*, Pinaceae) based on chloroplast *rbcL*, *matK*, *rpl20-rps18* spacer, and *trnV* intron sequences. Am J Bot 86:1742–1753

Ware D, Jaiswal P, Ni J, Pan X, Chang K, Clark K, Teytelman L, Schmidt S, Zhao W, Cartinhour S, McCouch S, Stein L (2002) Gramene: a resource for comparative grass genomics. Nucleic Acids Res 30:103–105

Watkinson JI, Sioson AA, Vasquez-Robinet C, Shukla M, Kumar D, Ellis M, Heath LS, Ramakrishnan N, Chevone B, Watson LT, van Zyl L, Egertsdotter U, Sederoff RR, Grene R, et al (2003) Photosynthetic acclimation is reflected in specific patterns of gene expression in drought-stressed loblolly pine. Plant Physiol 133:1702–1716

Wei RP, Lindgren D, Yeh FC (1997) Expected gain and status number following restricted individual and combined-index selection. Genome 40:1–8

Weng C, Kubisiak TL, Nelson CD, Stine M (2002) Mapping quantitative trait loci controlling early growth in a (longleaf pine × slash pine) × slash pine BC1 family. Theor Appl Genet 104:852–859

Weng C, Kubisiak TL, Stine M (1998) SCAR markers in a longleaf pine × slash pine F1 family. For Genet 5:239–247

Whetten R, Sun Y, Zhang Y Sederoff RR (2001) Functional genomics and cell wall biosynthesis in loblolly pine. Plant Mol Biol 47:275–291

Whitelaw CA, Barbazuk WB, Pertea G, Chan AP, Cheung F, Lee Y, van Heeringen S, Karamycheva S, Bennetzen JL, SanMiguel P, Lakey N, Bedford J, Yuan Y, Budiman MA, Resnick A, van Aken S, Utterback T, Riedmuller S, Williams SM, Feldblyum T, Schubert K, Beachy R, Fraser CM, Quackenbush J (2003) Enrichment of gene-coding sequences in maize by genome filtration. Science 302:2118–2120

Wilcox PL (1995) Genetic dissection of fusiform rust resistance. PhD Thesis, Dept of Forestry, North Carolina State University, Raleigh, NC

Wilcox PL (1997) Linkage groups, map length, and recombination in *Pinus radiata*. In: Burdon RD, Moore JM (eds) "IUFRO '97 Genetics of Radiata Pine", Proc IUFRO Conf 1–4 Dec and workshop 5 Dec, Rotorua, New Zealand. FRI Bull 203

Wilcox PL, Burdon RD (2006) Application of association genetics to coniferous forest trees. In: Proc 13th Australasian Plant Breed Conf, Christchurch, New Zealand pp 651–659, http://www.apbc.org.nz

Wilcox PL, Amerson HV, Kuhlman EG, Liu B-H, O'Malley DM, Sederoff RR (1996) Detection of a major gene for resistance to fusiform rust disease in loblolly pine by genomic mapping. Proc Natl Acad Sci USA 93:3859–3864

Wilcox PL, Richardson TE, Carson SD (1997) Nature of quantitative trait variation in *Pinus radiata*: insights from QTL detection experiments. In: Burdon RD, Moore JM (eds) "IUFRO '97 Genetics of Radiata Pine", Proc IUFRO Conf 1–4 Dec and workshop 5 Dec, Rotorua, New Zealand. FRI Bull 203:304–312

Wilcox PL, Richardson TE, Corbet GE, Ball RD, Lee JR, Djorovic A, Carson SD (2001a) Framework Linkage Maps of *Pinus radiata* D. Don based on pseudotestcross markers. For Genet 8:109–117

Wilcox PL, Carson SD, Richardson TE, Ball RD, Horgan GP, Carter P (2001b) Benefit-cost analysis of DNA marker-based selection in progenies of *Pinus radiata* seed orchard parents. Can J For Res 31:2213–2224

Wilcox PL, Echt CS, Cato SA, McMillan LK, Kumar S, Ball RD, Burdon RD, Pot D (2003) Gene assisted selection – a new paradigm for selection in forest tree species? In: Plant & Animal Genomes XI Conf, San Diego

Wilcox PL, Cato S, McMillan L, Power M, Ball RD, Burdon RD, Echt CS (2004) Patterns of linkage disequilibrium in *Pinus radiata*. In: Plant & Animal Genome XII Conf, San Diego, W89. http://www.intl-pag.org/12/abstracts/W22_PAG12_89.html

Wilcox PL, Echt CS, Burdon RD (2007) Gene-assisted selection: applications of association genetics for forest tree breeding. In: Oraguzie N, Rikkerink EHA, Gardiner SA, Nihal De Silva H (eds) Association Mapping in Plants. Springer, Berlin Heidelberg New York, pp 211–247

Williams CG, Neale DB (1992) Conifer wood quality and marker-aided selection: a case study. Can J For Res 22:1009–1017

Williams C, Megraw RA (1993) Juvenile-mature relationships for wood density in *Pinus taeda*. Can J For Res 24:714–722

Williams JGK, Kubelik AR, Livak KJ, Rafalski JA (1990) DNA polymorphisms amplified by arbitrary primers are useful as genetic markers. Nucleic Acids Res 18:6531–6535

Williams CG, Zhou Y, Hall SE (2001) A chromosomal region promoting outcrossing in a conifer. Genetics 159:1283–1289

Wright SI, Gaut BS (2005) Molecular population genetics and the search for adaptive evolution in plants. Mol Biol Evol 22:506–519

Wu H (2002) Study of early selection in tree breeding. 4. Efficiency of marker aided early selection (MAES). Silvae Genet 51:261–269

Wu J, Krutovskii KV, Strauss SH (1999) Nuclear DNA diversity, population differentiation, and phylogenetic relationship in the Californian closed-cone pines based on RAPD and allozyme markers. Genome 42:893–908

Wu R, Ma C-X, Casella G (2002) Joint linkage and linkage disequilibrium mapping of quantitative trait loci in natural populations. Genetics 160:779–792.

Yan HH, Mudge J, Kim DJ, Shoemaker RC, Cook DR, Young ND (2004) Comparative physical mapping reveals features of microsynteny between Glycine max, *Medicago truncatula*, and *Arabidopsis thaliana*. Genome 47:141–155

Yang SH, Loopstra CA (2005) Seasonal variation in gene expression for loblolly pines (*Pinus taeda*) from different geographical regions. Tree Physiol 25:1063–107325(8):

Yazdani R, Nilsson JE, Plomion C, Mathur G (2003) Marker trait association for autumn cold acclimatation and growth rhythm in *Pinus sylvestris*. Scand J For Res 18:29–38

Yazdani R, Yeh FC, Rimsha J (1995) Genomic mapping of *Pinus sylvestris* (L.) using random amplified polymorphic DNA markers. For Genet 2:109–116

Yin TM, Huang MR, Wang MX, Zhu LH, Zhai WX (1997) Construction of molecular linkage map in masson pine using RAPD markers and megagametophytes from a single tree. Acta Bot Sin 39:607–612

Yin TM, Wang XR, Andersson B, Lerceteau-Köhler E (2003) Nearly complete maps of *Pinus sylvestris* L. (Scots pine) constructed by AFLP marker analysis in a full-sib family. Theor Appl Genet 106:1075–1083

Yuan Y, SanMiguel PJ, Bennetzen JL (2003) High-Cot sequence analysis of the maize genome. Plant J 34:249–255

Zeng ZB (1993a) Theoretical basis for separation of multiple linked gene effects in mapping quantitative trait loci. Proc Natl Acad Sci USA 90:10972–10976

Zeng ZB (1993b) Precision mapping of quantitative trait loci. Genetics 136:1457–1468

Zeng ZB, Liu J, Stam LF, Kao CH, Mercer JM, Cathy CL (2000) Genetic architecture of a morphological shape difference between two *drosophila* species. Genetics 154:299–310

Zhang H-B, Wu C (2001) BACs as tools for genome sequencing. Plant Physiol Biochem 39:195–209

Zheng Y, Ennos R, Wang HR, Zheng YQ, Wang HR (1994) Provenance variation and genetic parameters in a trial of *Pinus caribaea* Morelet var. *bahamensis* Barr and Golf. For Genet 1:165–174

Zhou Y, Bui T, Auckland L, Williams CG (2002) Undermethylation as a Source of microsatellites in large plant genomes. Genome 34:91–99

Zhou Y, Gwaze DP, Reyes-Valdes MH, Biu T, Williams CG (2003) No clustering for linkage map based on low-copy and undermethylated microsatellites. Genome 46:809–816

3 Spruce

Jean Bousquet[1], Nathalie Isabel[2], Betty Pelgas[1], Joan Cottrell[3], Dainis Rungis[4,5], and Kermit Ritland[4]

[1] Département des Sciences du bois et de la forêt, Université Laval, Québec G1K 7P4, Canada
[2] Ressources naturelles Canada, Service canadien des forêts, Centre de foresterie des Laurentides, 1055 rue du P.E.P.S., C.P. 10380 succ. Sainte-Foy, Québec G1V 4C7, Canada
[3] Forest Research, UK Forestry Commission, Northern Research Station, Roslin, Midlothian EH25 9SY, UK
[4] Department of Forest Sciences, University of British Columbia, Vancouver, British Columbia V6T1Z4, Canada
e-mail: kermit.ritland@ubc.ca
[5] Current address: LVMI Silava, 111 Rigas st, Salaspils, LV-2169, Latvia

3.1 Introduction

3.1.1 The spruce genus, *Picea*

The genus *Picea* consists of up to 35 species, three quarters of which are Eurasian and one quarter are North American (although reevaluation of the East Asian taxa would reduce this number, c.f. Farjon 1990; Sigurgeirsson and Szmidt 1993). The oldest recognizable spruce fossil dates back to the middle Eocene, around 45 million years ago (LePage 2001). Spruce species generally occur in the subtropical high-altitude, temperate, and boreal regions of the northern hemisphere. In the southern part of their distribution, they mainly occur in mountainous areas, while in the north, spruce occurs throughout the boreal forest, often being the dominant tree species across vast tracts of Scandinavia, Russia, Alaska, and Canada. In the mountains of southwest China and in Japan, spruce is not dominant yet shows the great species diversity, suggesting that eastern Asia is the center of origin for spruce (Wright 1955).

There is lack of agreement among taxonomists regarding the subdivision of the genus *Picea* (Schmidt-Vogt 1977). Based on morphology and crossability studies, early taxonomists divided the genus into three sections: Eupicea (or Morinda), Casicta, and Omorika. Mikkola (1969) recommended recognition of only two sections: Abies and Omorika. Part of the disagreement stems from the different traits used for morphological classification and variable results from crossability studies and because of the too little morphological and anatomical differentiation among spruce taxa (Wright 1955; Mikkola 1969; Weng and Jackson 2000). On the basis of crossability studies, Fowler (1983, 1987) further divided Omorika into two subsections: Omorikoides and Glaucoides, with white spruce, Sitka spruce, and Engelmann spruce assigned to the latter. The lack of agreement among taxonomists suggests that, in comparison to the pines (*Pinus*), the genus *Picea* is relatively monophyletic.

In work yet to be published (Presby-Germano 2003; Bouillé and Bousquet 2006), up to 35 spruce species were sequenced for chloroplast DNA (*rbcL*, *trnTLF*, *trnK*), mitochondrial DNA (*nad1 B/C* and *nad7 1/2* introns), and nuclear DNA (portions of 4CL). Sequence divergence was relatively low, yielding low bootstrap support for many clades in the inferred phylogenies, and phylogenies were significantly incongruent between genomes. This is indicative of recent speciation and/or more or less recent reticulate evolution. Also, mtDNA phylogenies were geographically more structured than cpDNA phylogenies, and incomplete lineage sorting is evident at nuclear loci (Bouillé and Bousquet 2005; Campbell et al. 2005). Further work involving more intensive gene coverage and population sampling is needed to resolve species groups.

In this chapter, we focus on the spruce species that have received the predominant attention from molecular mapping and breeding perspectives. These are the most economically important species: white spruce, Sitka spruce, black spruce, and Norway spruce.

White Spruce

White spruce [*Picea glauca* (Moench) Voss] is widely distributed in Canada, from the Atlantic to the Pacific coast. It is used for lumber, pulp and paper, and ranks as the second-most important conifer species for re-

Genome Mapping and Molecular Breeding in Plants, Volume 7
Forest Trees
C. Kole (Ed.)

forestation in Canada, with over 100 million seedlings planted yearly. Since the 1950s, the Canadian Forest Service and Provincial ministries in Canada (mainly British Columbia, Quebec, and New Brunswick) have invested considerable effort on white spruce breeding. Some programs are now entering the third generation of breeding (Beaulieu 1996; Tosh and Fullarton 2001; Yanchuk 2001).

White spruce is highly diverse genetically for quantitative characters (Furnier et al. 1991; Jaramillo-Correa et al. 2001), and low to moderate genetic control has been shown for a variety of growth and adaptive traits, and for wood characters (Nienstaedt 1985; Kiss and Yeh 1988; Corriveau et al. 1991; Li et al. 1993, 1997; Rweyongeza et al. 2004). These trends indicate that significant gains could be realized from selection and breeding in this species, and that genomics can provide new inputs into these programs.

Sitka Spruce

Sitka spruce [*Picea sitchensis* (Bong.)] has a narrow distribution along the mainland and offshore islands of the Pacific coast of North America, from central Alaska to northern California. In Alaska, Sitka spruce is the most important timber species (Arno and Hammerly 1977), and in 1995, southeast Alaska accounted for 70% of the total estimated North American production of Sitka spruce. Its wood has a high strength-to-weight ratio, making it useful for diverse applications including turbine blades, sailboat masts, oars, piano sounding boards, and guitars (Hosie 1969; Viereck and Little 1972; Harris 1990).

Outside its natural range, Sitka spruce has played an important role in plantation forestry, particularly in northern Europe (Hermann 1987). It is planted extensively in Britain, Ireland, and Denmark and is a minor plantation species in France and Germany (Hermann 1987). David Douglas introduced Sitka spruce to Britain in 1831; its popularity has increased since then, with plantings increasing from 27% in 1925 to 60% in 1985 (Joyce and O'Carroll 2002). Sitka spruce now accounts for almost 70% of the annual conifer planting stock (Malcolm 1997), with plantations covering over 20% of the British woodland (Cannell and Milne 1995). Breeding strategies for Sitka spruce in Britain have also been conducted (Lee 1993), with molecular approaches implemented (see below).

Sitka spruce harbors much diversity in productive traits among provenances in North America (Fletcher 1992) that are used to develop productive plantations in Europe. However, in North America, Sitka spruce has an endemic susceptibility to the white pine weevil [*Pissoides strobi* (Peck)]. Young trees are attacked and cannot maintain growth leaders, resulting in a bushy growth form. Only in the Queen Charlotte Islands, where the weevil is absent (Hall 1994), can Sitka spruce be reestablished. The reestablishment of plantation forestry in the much more fertile region of Vancouver Island (British Columbia) in the face of this pest is the primary goal of breeding for Sitka spruce in British Columbia.

Black Spruce and Ally Red Spruce

Black spruce [*Picea mariana* (Mill.) B.S.P.] is an abundant transcontinental boreal species in North America (Farrar 1995). Its natural range overlaps with white spruce; it extends across the boreal forest from Newfoundland/northern Quebec to Alaska, but it also occurs in some northern US states. Black spruce is economically very important in New Brunswick, Nova Scotia, and Quebec. In contrast, the range of red spruce [*Picea rubens* (Sarg.)] is more restricted and extends from the southern Appalachians in the United States, where it is often restricted to mountain plateaus, to the Maritime Provinces of eastern Canada (Farrar 1995). Red spruce is a minor component of commercial plantations in the Maritime Provinces.

Black spruce harbors large amounts of genetic variation in quantitative characters, which is an indication of the adaptive capacities of its populations (Khalil 1984; Beaulieu et al. 2004). In addition, various patterns of clinal variation have been reported for germination rate, survival rate, phenology, juvenile growth characters, and hardiness (Dietrichson 1969; Morgenstern 1969; Corriveau 1981; Beaulieu et al. 1989; Parker et al. 1994). Red spruce appears to be genetically less variable than the sympatric black and white spruce (Rajora et al. 2000).

Norway Spruce

Norway spruce [*Picea abies* (L.) Karst.] is economically the most important conifer tree species in Europe. Its natural distribution ranges across the Pyrenees, Alps, and Balkans, northwards to southern Germany and Scandinavia and eastwards through the Carpathian Mountains and Poland, to western Russia. It has a long history of cultivation in central Europe (Schmidt-Vogt 1977), including introductions in Belgium, Germany, and central France; more recent introductions have occurred in North America (SE Canada and NE USA), but only a few million seedlings are planted every year due to the sensitivity of Norway spruce to white pine weevil. Breeding programs exist in several European countries, dating from the late 1940s. In the last two decades, Norway spruce has suffered severe forest decline in central Europe; as a result, focus has shifted from breeding to gene conservation and forest health. In part to guide these efforts, a Norway spruce network, EUFORGEN (http://www.bioversityinternational.org/networks/euforgen/networks/conifers/picea_abies/pabies.htm), was created.

3.1.2 Natural Hybridization

In southern British Columbia, white spruce hybridizes with Engelmann spruce (*P. engelmannii* Parry ex Engelm.), and in coastal northern British Columbia, white spruce hybridizes with Sitka spruce (Roche 1969; Sutton et al. 1991). In southern British Columbia, hybridization with Engelmann spruce is quite extensive, so much that the spruce in the region is termed "interior spruce" due to the mixed heritage of trees in this area (Kiss 1989). Black spruce and red spruce share an extensive sympatric zone in eastern Canada and northeastern USA, where they hybridize and introgress naturally (Perron and Bousquet 1997). They represent a recent progenitor-derivative species pair (Perron et al. 2000; Jaramillo-Correa and Bousquet 2003). In this region, like that of interior spruce, they form a species complex: black spruce × red spruce (*P. mariana* × *P. rubens*). As well, in northern Europe, Norway spruce exhibits introgressive hybridization with Siberian spruce (*Picea obovata* Ledeb.) (Farjon 1990; Krutovskii and Bergmann 1995), which is also considered a subspecies of *P. abies* (*P. abies* var. *obovata*) (Schmidt-Vogt 1977).

3.1.3 Molecular Genetic Variation and Mating System

Diversity at neutral nuclear loci is high for Sitka, white, black, and Norway spruce, with heterozygosity values generally exceeding 0.2 to 0.3 for isozymes (Yeh and El Kassaby 1980; Yeh and Arnott 1986; Yeh et al. 1986; Furnier et al. 1991; Krutovskii and Bergmann 1995; Isabel et al. 1995; Jaramillo-Correa et al. 2001), random amplified polymorphic DNAs (RAPDs) (Isabel et al. 1995), and expressed sequence tag polymorphisms (ESTPs) (Perry and Bousquet 1998a, b; Perry et al. 1999; Jaramillo-Correa et al. 2001). As expected, much higher diversity is observed at simple sequence repeats (SSRs) (Hodgetts et al. 2001; Rajora et al. 2001; Scotti et al. 2002).

Genetic diversity in spruce is partitioned mainly within populations: the majority (90 to 99%) occurs within populations for Sitka, white, black, and red spruce (Yeh and El-Kassaby 1980; Hawley and DeHayes 1994; Isabel et al. 1995; Rajora et al. 2000; Jaramillo-Correa et al. 2001; Perry and Bousquet 2001; Gamache et al. 2003). When maternally inherited cytoplasmic markers were surveyed in Norway and black spruce as well as in the Mexican *P. chihuahuana*, population differentiation was much higher (Sperisen et al. 2001; Bastien et al. 2003; Gamache et al. 2003; Jaramillo-Correa et al. 2004, 2006). Genetically distinct ancestral lineages could be identified from cpDNA or mtDNA polymorphisms (Vendramin et al. 2000; Sperisen et al. 2001; Jaramillo-Correa et al. 2004, 2006; Jaramillo-Correa and Bousquet 2005).

In contrast to nuclear loci, quantitative traits demonstrate much more population differentiation in boreal spruce (Li et al. 1997; Jaramillo-Correa et al. 2001; Beaulieu et al. 2004; Lagercrantz and Ryman 1990; Collignon et al. 2002; Acheré et al. 2005).

In terms of detecting the underlying loci responsible for such selective differentiation, a genomic scan of amplified fragment length polymorphism (AFLP), SSR, and ESTP markers indicated a small number of outlier loci that were more differentiated (Acheré et al. 2005). The frequency of outlier loci was higher from a genome-wide scan relying on gene SNPs (Namroud et al. 2006). Further inputs from developing genomic programs will greatly aid in identifying the specific loci underlying these adaptive differentations.

At the nucleotide level, white, black, and Norway spruce have moderate levels of intraspecific diversity (ca. 0.5%) but high levels of haplotype diversity, with *H* in excess of 90% (Bouillé and Bousquet 2005; Bousquet et al., personal communication). This is comparable to that found in pines (Brown et al. 2004; Pot et al. 2005). High nucleotide diversity has also been found for the ITS1, but this might be caused by interlocus divergence (Campbell et al. 2005). We also found, as in pines (Brown et al. 2004; Neale and Savolainen 2004), that linkage disequilibrium appears to decay rapidly within spruce genes. Finally, in an analysis of in silico SNP found in ESTs of white spruce, Pavy et al. (2006) found about 0.3% heterozygosity in clusters containing four or more cDNAs. These are all examples of how data from the current spruce genome projects are helping to identify large-scale patterns of spruce genomic variation and differentiation.

With regard to outcrossing rates, boreal spruces are predominantly outcrossing (Cheliak et al. 1985; Boyle and Morgenstern 1986; Shea 1987; Chaisurisri and El-Kassaby 1994; Xie and Knowles 1994; Cottrell and White 1995; Rajora et al. 2000; Perry et al. 1999). Red spruce and subtropical taxa are exceptions, with lower outcrossing rates (e.g., Ledig et al. 2000; Rajora et al. 2000). Selection against inbreds has been noted in boreal spruces (Isabel et al. 1995) and in subtropical species (Ledig et al. 2002).

3.1.4 Cytogenetics

Chromosome number in somatic cells of spruces is $2n = 24$ (Burley 1965; Fox 1987; Murray 1998), and small supernumerary B chromosomes have been observed in some individuals of Sitka (Moir and Fox 1972; Teoh and Rees 1976; Kean et al. 1982) and white spruce (Nkongolo 1996). DNA content per diploid cell is 19 pg for Sitka (Ingle et al. 1975), 30 pg for Norway, from 17 to 40 pg for white, and from 22 to 32 pg for black spruce, depending on the sample and the method used (Murray 1998). These numbers correspond to estimated genome sizes ranging from 17 to about 40×10^9 bp (Murray 1998). The chromosomal locations of ribosomal RNA genes in Sitka spruce have been determined using fluorescent in situ hybridization. The 5s rDNA was restricted to one chromosome, whereas 18s-5.8s-26s rDNA occurred on chromosome 5 as well as four other chromosomes (Brown and Carlson 1997). The chromosomal location of large tandem repeats on the genome of white and Sitka spruce was similar, emphasizing the little divergence between the two species (Brown et al. 1998).

3.1.5 Economic and Breeding Issues

Economic Importance and Fundamental Breeding Issues

It is clear that the need for wood fiber and wood volume will increase (Brooks 1997). Modern economies must adapt to these changes. But forests are diverse, and besides providing for wood fiber and volume, they also provide for many other values, including biodiversity, recreation, employment for local communities, and the lore of native people's legends. Genetically improved planting material can, in principle, accommodate these multiple objectives. However, phenotypic assessments require that trees be grown from less than one half of their rotation age before selections for traits are made (Zobel and Talbert 1984). In the more northern climate of spruce, the rotation age of spruce is 20 to 40 years (British Columbia) to 60 to 100 years (Sweden), making selective breeding based upon traditional practices very slow. Genomics and molecular breeding can speed the progress of identifying trees adapted to future climates and provide insight into the uniqueness of conifers as the hallmark species of northern climes.

The Spruce Weevil Problem

The spruce shoot weevil (a.k.a. white pine weevil, *Pissoides strobi*), budworms, and certain bark beetle species are some of the most destructive insect pests of spruce forests worldwide (Seybold et al. 2000; Alfaro et al. 2002). Larval feeding of the spruce weevil severely damages or kills the leading shoots of susceptible host trees, resulting in reduced growth and tree deformation (Alfaro et al. 2002). In British Columbia, the weevil affects both Sitka and white spruce, but because it affects Sitka so severely, annual plantings of Sitka spruce have been reduced from 10 million seedlings to fewer than a million (King et al. 1997). The spruce weevil is endemic to North America, but as with other insect pests, it could spread to Sitka and Norway spruce forests in Europe. Genetic technologies might aid in fighting these pests.

The first evidence of substantial weevil resistance in spruce was observed in International Union of Forest Research Organization (IUFRO) provenance trials in British Columbia (Ying 1991). Since then, weevil resistance has been demonstrated to have a significant genetic component, as a large (139-family) 10-year-old test in south central British Columbia with 8 years of accumulated weevil attack gave an individual-tree heritability of >0.4 (King et al. 1997). However, to date, most mechanistic studies of conifer defense have been at the anatomical or chemical level (Trapp and Croteau 2001; Huber et al. 2004), not involving genomics technologies.

3.2 Genetic Mapping

3.2.1 First-Generation Genetic Maps in Spruce Species

We define the first generation linkage maps in forest trees as those constructed with anonymous genetic markers, mainly RAPD (Williams et al. 1990) and AFLP (Vos et al. 1995). These technologies are especially useful for species in which genomic knowledge is limited, such as in conifers, as they do not require a priori knowledge of the genome. While both RAPD and AFLP markers exhibit dominance, this handicap is overcome by the use of appropriate pedigrees and configurations such as the two-way pseudotestcross (Grattapaglia and Sederoff 1994) or by the use of haploid megagametophytes (Isabel et al. 1995), which are unique to conifers.

The first map of white spruce was produced using an array of haploid megagametophytes of a single tree mapping population (Tulsieram et al. 1992). This map consisted of 47 RAPD loci distributed on 12 linkage groups (LGs). It covered 874 cM, ca. one third of the estimated 2,700- to 2,900-cM genome length (Gosselin et al. 2002). The first map for Norway spruce (Binelli and Bucci 1994) produced a RAPD genetic linkage map from a population of 72 megagametophytes also derived from a single tree. It had 152 polymorphic loci assigned to 17 LGs covering a total distance of 3,584 cM, a puzzlingly large distance. A summary of the map characteristics is presented in Table 1.

AFLPs have become the marker of choice for anonymous marker maps owing to the large number of detected polymorphisms and the relatively low cost per locus (Schlötterer 2004). It should be mentioned that the usefulness of these markers has broken down the limits of what could be achieved in forest genetics. They have made possible the search for quantitative trait loci (QTLs) underlying specific traits in large undomesticated species such as trees (see other chapters). However, in spruce, the small megagametophyte does not yield sufficient DNA for AFLP reactions.

Paglia et al. (1998) built another single-tree genetic linkage map of Norway spruce using also a panel of 72 megagametophytes with a total of 447 segregating markers [366 AFLPs, 20 selectively amplified microsatellite polymorphic loci (SAMPLs), and 61 SSRs] covering a genetic length of 2,198 cM, which represents 77% of the estimated genome length. Out of the 447 markers, 413 were assigned to 29 LGs. Several of the SSRs developed were later found to be dominant (Pelgas et al. 2005, 2006) and, consequently, less interesting for comparative mapping.

3.2.2 Second-Generation Mapping

Second-generation maps are those that utilize both codominant markers and markers for which homology among species is clear. The first class of such markers are RFLPs, which have been around since the 1990s; but while their homology among species is high, thus allowing transfer of map information, their assay is relatively laborious. Since then, STS markers and ESTP markers (Perry and Bousquet 1998a, b; Perry et al. 1999), as well as single strand conformation polymorphisms (SSCPs) (e.g., Plomion et al. 1999), have emerged. In spruce, Fournier et al. (2002) evaluated the potential of 50 STS markers of arbitrary genes (Perry and Bousquet 1998a) for constructing maps at both intra- and interspecific levels. Several unrelated individuals of white spruce and black spruce and their corresponding progeny were screened for polymorphisms using SSCP. The use of this method of detection allowed to significantly increase the number of codominant markers for comparative genetic mapping purposes but still remained relatively labor intensive (Fournier et al. 2002).

In parallel, Gosselin et al. (2002) conducted a comparative mapping study based on an array of 100 megagametophytes for each of two individuals using dominant (RAPD) and codominant (ESTP) markers. The analysis for the first individual resulted in 165 loci (152 RAPDs, 3 SCARs, and 10 ESTPs) mapping to 23 LGs and covering 2,059 cM. For the second individual, the analysis resulted in 145 loci (137 RAPDs, 1 SCAR, and 7 ESTPs) mapping to 19 LGs and covering 2,007 cM. Both maps covered close to 90% of the entire genome. The percentage of shared loci between the two individual maps was much higher for the codominant gene-based markers than for the anonymous dominant markers (44% vs. 9%). This difference illustrates the need for increasing the number of codominant markers from expressed genes to anchor different linkage maps for a given species.

Tools for Second-Generation Genetic Maps

Second-generation genetic maps ideally utilize genomic information and involve coding or transcribed

Table 1. Linkage maps obtained for *Picea* species

Species	Number of progeny genotyped	Number and types of loci		Number of linkage groups	Software used	Genome length (cM)	Reference
P. glauca	72 gametophytes	47	RAPD	12	MAPMAKER	874	Tulsieram et al. 1992
P. abies	72 gametophytes	152	RAPD	17	MAPMAKER	3584	Binelli et Bucci 1994
P. abies	72 gametophytes	366	AFLP	29	MAPMAKER	2198	Paglia et al. 1998
		20	SAMPL				
		61	SSR				
P. glauca	1) 92 gametophytes	152	RAPD	23	JOINMAP	2059	Gosselin et al. 2002
		3	SCAR		MAPMAKER		
		10	ESTP		GMENDEL		
	2) 96 gametophytes	137	RAPD	19		2007	
		1	SCAR				
		7	ESTP				
*P. glauca**	1) 110 F1	714	AFLP	12	JOINMAP	2168	Pelgas et al. 2006
	2) 103 F1	38	SSR				
		53	ESTP				
P. glauca *	1) 88 F1	361	AFLP	13	JOINMAP	1623	Rungis et al. (pers. comm.)
	2) 88 F1	46	EST-SSR				
		81	COS				
		6	ESTP				
P. sitchensis	192 F1	134	SSR	Under construction	JOINMAP	–	J.E. Cottrell et al. unpublished
*P. mariana**	1) 118 BC1	1014	AFLP	12	JOINMAP	2319	Pelgas et al. 2005
	2) 85 F1	3	RAPD				
		53	SSR				
		54	ESTP				
*P. abies**	73 F1	661	AFLP	12	JOINMAP	2035	Acheré et al. 2004
		74	SSR				
		18	ESTP				
		2	others				
			Additional markers:				Pelgas et al. 2006
		1	SSR				
		13	ESTP				
*P. abies**	85 F1	422	AFLP, SSR, ESTP, IRAP, SSAP	13	JOINMAP	2821	Scotti et al. 2005

*Composite map

sequences, preferably with genes annotated for function. In the past 3 years, the Canadian Genomics Projects Arborea (Quebec) and Treenomix (British Columbia), funded by Genome Canada, Genome British Columbia, and Genome Quebec, as well as other projects funded by the National Science and Engineering Research Council of Canada and the Canadian Biotechnology Strategy, have accelerated the development of markers and made possible new types of genetic markers especially amenable to high-throughput genotyping and transferability across genetic maps. Here, we briefly review these innovations, which should be applicable to mapping projects with any species, plant or animal.

ESTP Markers

The development of ESTP markers is still time consuming and expensive, but their use is easy and informative because they are codominant and derived from locus-specific primers (Perry and Bousquet 1998a, b; Lefort et al. 1999; Perry et al. 1999). Also, they can be indicative of coding regions (Perry and Bousquet 1998a, b). These multiallelic genetic markers are generally orthologous and have been found to be transferable from one species to another within the genus *Picea* (Perry and Bousquet 1998b; Perry et al. 1999). Consequently, they can be used for the assemblage of conspecific maps useful for interspecific comparative mapping. To accelerate the development of ESTPs, Pelgas et al. (2004) and Lamothe et al. (2006) relied on a DNA pool sequencing strategy to look for common (frequency higher than 10%) polymorphisms. This approach was found to be reliable and speeds up marker recovery for the detection of SNPs and/or indels. In spruce, it has rendered availability of more than 100 codominant markers with amenable genotyping strategies such as AGE (agarose gel), DGGE (denaturing gradient gel electrophoresis), or CAPS (cleaved amplified polymorphic DNA) (Pelgas et al. 2004, 2005, 2006; Lamothe et al. 2006).

Genomic-SSR vs. EST-SSR Markers

The first set of microsatellites (SSR) developed from genomic libraries of Sitka was the seven published by van de Ven and Kanamori (1996). Rajora et al. (2001) also developed SSR markers for spruce. However, due to the large, repetitive nature of conifer genomes, robust, single-copy SSR markers from genomic DNA are difficult to find. Based upon our own (KR) experience with several conifers, only 5 to 10% of positive clones identified in hybridization with genomic libraries eventually lead to useful SSRs. On top of this, SSRs that work within the target species rarely work in related species. For example, attempts to transfer primers developed from genomic libraries of *P. abies*, for use in Sitka, have met with little success (A'Hara and Cottrell 2004); Rungis et al. (2004) tested 101 microsatellites developed from genomic libraries of other spruces for amplification in Sitka spruce and found that only 17 amplified in Sitka.

The transcriptome (messenger RNA) is the next logical source to find single-copy conserved SSR markers. Also, since these SSRs are within the expressed portion of the genome, they are more likely to be associated with a particular gene, compared to genomically derived markers, improving their utility for QTL mapping and marker-assisted selection (MAS). To this end, Scotti et al. (2000) screened a *P. abies* cDNA library for $(AG)_n$ and $(AC)_n$ repeats and found ten and six clones, respectively, containing these repeats. They estimated that SSRs in cDNA clones occur about 20 times less frequently than in random (genomic) clones. They designed six primer pairs of which three generated clear patterns similar to genomic SSRs.

EST databases offer the opportunity to quickly identify SSR repeats of all types. From a 20,275 unigene spruce EST set, Rungis et al. (2004) identified 44 candidate EST-SSR markers. Compared to Scotti et al. (2000), this number of 44 enables a good statistical evaluation of EST-SSR diversity and their transferability to other species. Of these 44, 25 amplified and were polymorphic in white, Sitka, and black spruce; 20 amplified in all 23 spruce species tested; and the remaining 5 amplified in all except one species. The 25 EST-SSRs had about 9% less heterozygosity compared to a set of 17 genomically derived SSRs, which also showed consistent amplification across species (mean $H = 0.65$ vs. 0.72). Also, these EST-SSRs showed fewer null alleles. Additional EST-SSRs have been developed from the white spruce EST collection by Forest Research (Scotland) for their work with Sitka spruce (A'Hara and Cottrell 2004, 2007).

Ecotilling

A single-strand specific nuclease, found in extracts of celery juice, can be used to digest heteroduplex DNA and hence identify heterozygous single nucleotide polymorphism (SNP) sites in PCR products (Comai et al. 2004). This method can be used to map genes with relative simplicity and low cost (Rungis et al. 2005). A particular nucleotide substitution does not need to be identified, and in fact, a priori knowledge of the presence of a SNP is not required, as the entire length of the PCR product is interrogated for the presence of SNPs. This feature enables application of this technique to genomes that are not well characterized, to rapidly place particular genes onto linkage maps, and will be used to map COS markers by the Treenomix project in pine and spruce (see below). While this technique is best suited for mapping markers in a backcross configuration, in the F_2 configuration, where alternative homozygotes cannot be discerned by this technique, data are still informative about linkage (Rungis et al. 2005).

Conserved Orthologous Set (COS) Markers

The advent of large-scale EST databases and genome sequences has enabled the identification of a class of genes that are especially amenable for comparative genetic mapping: "conserved orthologous set" (COS) markers. COS markers are single-copy and slowly evolving across species and are identified by self- and cross-BLASTING EST databases. They were originally developed for species of Asterids (Fulton et al. 2002) but also seem appropriate for conifers, as determined by a recent in silico analysis (Krutovsky et al. 2006)

The Treenomix project has undertaken a large-scale screening of COS markers for the white spruce-loblolly pine comparison, with the aim of mapping 1,000 COS markers in a comparative map (using ecotilling). Using the latest EST databases, annotated COS marker genes were identified by reciprocal BLAST. Primers were designed only for sites exhibiting complete sequence identity within these homologs. As of June 2006, 443 such primer pairs were tested for PCR amplification products, and 46% of these gave single bands in both species; interestingly, 31% gave single bands in these two species plus Douglas fir.

Multiple Pedigree Mapping

Another challenge with genetic and QTL mapping, especially when the number of informative markers (gene-based, multiallelic, and codominant) is limited, is that only polymorphic markers showing Mendelian segregation within a given array of progeny can be positioned. One way to alleviate this problem is to use more than one cross for a given species, with at least one parent in common to increase the probability that a given marker segregates (Beavis and Grant 1991; Tani et al. 2003). For example, the use of a common parent between mapping populations allowed Pelgas et al. (2005, 2006) to increase by around 25% the number of useful anchor markers and also to verify the consistency of macrosynteny and macrocolinearity within and between species.

If parents are not shared between pedigrees, the maps can be joined "by eye" using markers in common between maps or with the program JoinMap (Stam 1993). JoinMap starts with individual pairwise recombination estimates derived from different experiments and linearly combines them into a single estimate using weights proportional to LOD scores. However, if estimates of recombination for a pair of markers differ among pedigrees, a joint maximum likelihood procedure, given in Hu et al. (2004), has more statistical power. Combining multiple pedgrees into a single map analysis is especially appropriate in forestry, where mature progeny trials of relatively small size (10 to 40) are very common, as tree breeders tend to maximize the number of families to ensure conservation of genetic diversity. A recent example of the application of this method is a Douglas-fir AFLP map constructed from eight 40-member full-sib families (Ukrainitz et al. 2007).

3.2.3 Second-Generation Genetic Maps (as summarized in Table 1)

White Spruce Maps

Marker maps for white spruce based upon AFLPs, SSRs, ESTPs, and COS markers have recently been completed by the Canadian genome projects. As part of the Arborea project, a composite linkage map for white spruce representing an assemblage of four individual maps and delineating 12 LGs was obtained (Pelgas et al. 2006). The length of the composite map was 2,168 cM. Anonymous and locus- or gene-specific markers (714 AFLPs, 38 SSRs, and 53 ESTPs) were positioned, among which a total number of 108 anchor markers positioned (mostly SSRs and ESTPs). As part of the Treenomix project, an integrated genetic map of white spruce representing the assemblage of two pedigrees contained a total of 452 markers (361 AFLPs, 6 candidate genes, 76 COS markers, 14 orthologous markers [with loblolly pine], and 46 EST-SSR markers) mapped onto 13 major LGs (>5 markers).

The linkage map length was 1,622.7 cM (Rungis et al. pers. comm.).

Black-Red Spruce Maps

In a major effort to further compare genome structure between phylogenetically distant species in the genus and to enable comparisons with other Pinaceae, four individual linkage maps as well as a reference and a composite linkage map were constructed for the black spruce × red spruce species complex (Pelgas et al. 2005). These authors assessed the usefulness of SSR and ESTP codominant markers in combination with the use of two crosses, with one parent in common, for the assembly of a composite map. This map contained a total of 1,124 positioned markers, including 1,014 AFLPs, 3 RAPDs, 53 SSRs, and 54 ESTPs, assembled into 12 major LGs. The map length was 2,319 cM. Information deriving from a second cross contributed in this case to an increase of 24% in the number of anchor markers in comparison with the information from only a single cross.

Norway Spruce Map

Acheré et al. (2004) reported on the construction of the first saturated linkage map based on an F_1 progeny in Norway spruce. The two parental linkage maps generated were integrated into a consensus one. In total, 755 markers (661 AFLPs, 74 SSRs, 18 ESTPs, the 5S rDNA, and the early cone formation trait) were assigned and positioned onto 12 LGs. The resulting consensus map covered 2,035 cM. Additional markers (SSR and ESTP) were positioned on this map for a comparative mapping survey within the *Picea* genus (see below).

With both the same approach and species, Scotti et al. (2005) built two linkage maps covering a total length of 2,316 and 1,667 cM in the female and the male, respectively, as well as a consensus map. Different classes of markers (sequence families), sampling putatively different regions of the genome, were used for a mapping experiment. Using autocorrelation techniques, the authors found that AFLPs and SSRs were in close proximity to one another, while ESTPs did not seem to form clusters. The latter point should be taken cautiously since the number of ESTPs positioned in this study was relatively low. The final composite map consisted of 422 markers assigned to 13 LGs covering 2,821 cM (Scotti et al. 2005).

3.3 Comparative Mapping

In nonmodel species, comparing markers and genetic maps between related species is a useful strategy to transfer genomic information among species in a synergistic way. The anchoring of maps involves both the evaluation of the degree of synteny (gene content of LGs) and colinearity (gene order of LGs). To fulfill these objectives, numerous anchor markers per LG are necessary, which can recognize unambiguously orthologous chromosomal regions between species. EST-based markers are even more important for comparative mapping across species or genera and for the detection of associations between candidate genes and quantitative traits or known metabolic functions.

3.3.1 Comparative Mapping Between Spruce Species

Composite linkage maps have been produced for three *Picea* species (*P. glauca* and *P. abies* and *P. mariana* × *rubens*; Pelgas et al. 2006). In total, 88 (30 SSRs and 58 ESTPs) homologous markers were shared among these maps, and 12 homologous LGs were identified. A high level of conservation of the gene content (synteny) and macrocolinearity was observed (Fig. 1). This conservation is remarkable given that the split between these three divergent taxa occurred millions of years ago (Bouillé and Bousquet 2005).

One incongruity in synteny in these maps, involving *P. glauca* and *P. abies* and *P. mariana* × *rubens*, appears to be due to an insertional translocation, likely from white spruce (Pelgas et al. 2006). Also, a phylogenetic study of the *knox-I* gene family combined with a genetic mapping approach (Guillet-Claude et al. 2004; Pelgas et al. 2006) revealed a putative segmental duplication in spruces on LG 3, the same LG where the lack of synteny between white spruce and the two others was observed. Hence, this part of the spruce genome appears to be the site of various structural modifications that occurred at different times of the genus history.

Evolution by duplication also appears to be frequent in the genome of the Pinaceae, as a succession of three gene duplications and two chromosomal translocations can explain the chromosomal distribution of the *knox-I* gene family in the genera *Picea* and *Pinus* (Guillet-Claude et al. 2004). Mapping work in progress also indicates that several families of reg-

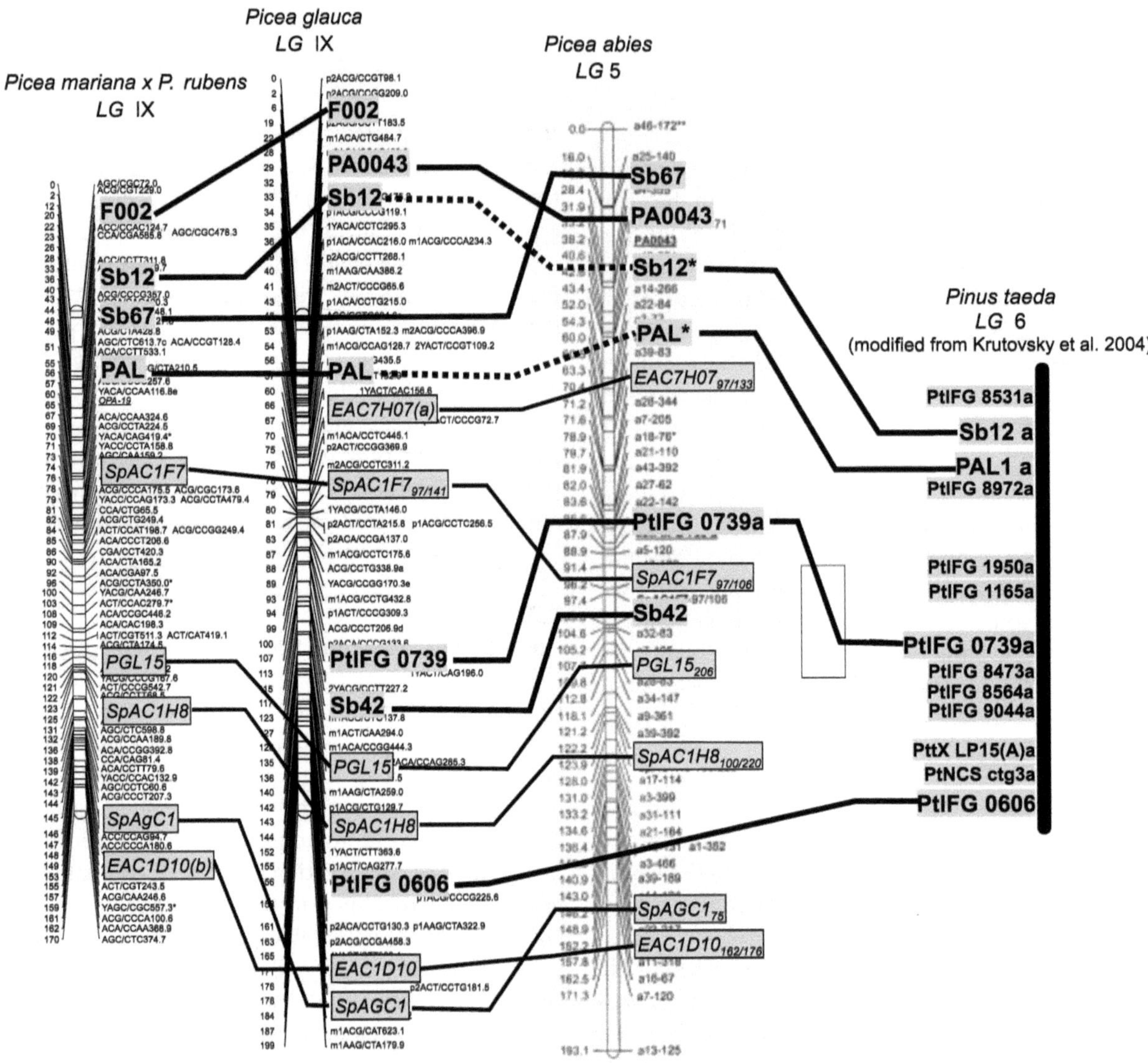

Fig. 1. Synteny between homoeologous composite LGs of *Picea mariana* × *P. rubens* complex, *P. glauca* and *P. abies* species, and *P. taeda*. Markers in *bold* are ESTPs, markers in *italic* are SSRs, and all remaining markers (for spruces) are AFLPs. Orthologous markers are connected by a *solid line*, except when they are connected with homologous anchor markers positioned onto the linkage maps of *P. abies* developed by Scotti et al. (2005; *dotted line*)

ulatory genes are clustered on the same LGs. This indicates that, perhaps, gene duplications are followed by short bursts of rapid evolution involving neofunctionalization (Guillet-Claude et al. 2004).

3.3.2 Comparative Mapping in the Pine Family

Despite their large size (6× human, 100× Arabidopsis), conifer genomes seem to be remarkably conserved. Across the Pinaceae (pine family – pine, spruce, fir, hemlock, Douglas fir, larch), chromosome number is quite conserved ($x = 12$, except $x = 13$ for Douglas fir). Consistent with this conserved chromosomal evolution, Krutovsky et al. (2004) documented synteny and colinearity between the genomes of pine and Douglas fir. This suggests that, as in the grass family, linkages and clusters of gene functions would be preserved among coniferous species.

Because of the large genome size in conifers, the absence of a conifer genome sequence, together with this putative synteny, makes comparative mapping an important tool for integrating information across the

Pinaceae. Among conifers, the largest effort in genetics and genomics has been devoted to loblolly pine, with large EST collections, rich genetic resources, and well-developed genetic and QTL maps (Neale and Wheeler 2004). Species of spruce rank second, largely because of the above-mentioned Canadian genome projects. Loblolly pine and spruce bridge the pine family, and they likely represent the most ancient lineages of the Pinaceae (Frankis 1989; Liston et al. 2003). Besides reinforcing each other, joint genomic patterns in these two genera might enable extension and integration of such genomic knowledge into other species, particularly lodgepole pine (being very proximal to loblolly pine) and Douglas fir (being intermediate between loblolly pine and spruce).

A joint map by Pelgas et al. (2006) suggests that the macrostructure of the Pinaceae genome is well conserved (Fig. 1). Moreover, this study confirmed that a chromosomal fission has presumably occurred in *Pseudotsuga* (Krutovsky et al. 2004; Pelgas et al. 2005). This may have played a central role in generating the difference in haploid chromosome number between *Pseudotsuga* ($n = 13$) and the other Pinaceae ($n= 12$). Some minor rearrangements were also observed that could only be validated by identifying genetic maps or by complete sequencing of specific genomic regions (Pelgas et al. 2006). However, several cases of putative rearrangements were dismissed after DNA sequencing to verify the orthology of the genes compared (Pelgas et al. 2005, 2006). This is an example of how caution should be taken in validating anchor marker gene orthology. Paralogy is rampant in conifers; because of their large repetitive genome, conifers harbor large gene families, giving lots of opportunities for mistaken orthology (Guillet-Claude et al. 2004).

3.4 Molecular Breeding

3.4.1 Marker-Assisted Selection

Achieving genetic gains is a slow and incremental process. The development and implementation of molecular markers predictive of phenotypic performance is expected to greatly improve the accuracy of genetic selection and accelerate the breeding process. In crops, marker-assisted selection (MAS) has been used for traits including rust or other disease resistance (e.g., Gebhardt et al. 2004; Hayden et al. 2004), phenological and adaptative traits (e.g., Saranga et al. 2001; Thornsberry et al. 2001), and crop productivity (e.g., Dirlewanger et al. 2004). Until recently, the use of MAS to shortcut the long breeding cycles in forest trees was more of a concept than a reality. The practical roadblocks include, among many others, a lack of a model system for a conifer, the lack of inbred pedigrees, and the long generation times of conifers.

Ideally, MAS should identify candidate genes for direct selection. Discovery of such genes follows two different strategies. The first involves association analysis of quantitative traits in segregating pedigrees (identifying quantitative trait loci, QTL, c.f. Lander and Botstein 1989). However, large progeny numbers are required to verify marker-trait associations (Beavis 1994). For example, in radiata pine (*Pinus radiata*), the identification and verification of QTLs involved measurements on nearly 4,435 trees from a single full-sib family planted in a commercial forestry plantation (Devey et al. 2004). The alternative strategy is association analysis, which aims to identify DNA polymorphism on genes directly underlying the phenotypic variation. The low level of linkage disequilibrium observed in softwood trees makes them good candidates for such an approach (Neale and Savolainen 2004). Association studies can be regarded as very large linkage studies of unobserved pedigrees that, in theory, permit mapping of a QTL at a higher resolution (reviewed by Cardon and Bell 2001).

Marker-Assisted Selection: Sitka Spruce

With the above concepts in mind, the Conifer Tree Breeding Team in Forest Research (FR) in Scotland has recently initiated a project aimed at using DNA technologies in their Sitka spruce breeding program. The objective is to shorten the breeding cycle and to improve the rate of genetic gain by associating microsatellite and other markers to traits of economic importance, such as wood density. In spring 2005, FR planted three large clonal tests designed to investigate the association between markers and phenotypic variation in Sitka spruce. The experiments involve the same 1,500 progeny from each of three full-sibling crosses. Each progeny (or genotype) is replicated four times at a single site, and the entire experiment is repeated at three climatically contrasting sites. The replicated clones representing each family in the field will be measured for a range of characteristics, such as wood density, stem straightness, or growth rate. It is intended that the top and bottom 150 genotypes for a given trait will be used for marker genotyping. Progress is being made with marker development (A'Hara and Cottrell 2004, 2007), and a genetic map should be available by the time the traits are measurable in the trial.

White Spruce

There are two genome projects in Canada that involve white spruce and their close relatives. The first is centered in Quebec (Arborea) and the second in British Columbia (Treenomix). The goal of the Arborea project is to identify allele-gene combinations that govern naturally occurring phenotypic variation of commercially valuable traits in natural and breeding populations of spruce. These involve investments made by the Canadian Forest Service-Laurentian Forestry Centre (CFS-LFC) for the last 30 years. In British Columbia, the goal of the Treenomix project is to use pedigrees segregating for resistance to the white pine weevil to dissect the chemical ecology of resistance to the pine weevil (and to insects in general). Likewise, the British Columbia Ministry of Forests has made large investments in breeding for weevil resistance in spruce and for general genetic gains. In both projects, large provenance and progeny tests of hundreds of open-pollinated families and full-sib families from crosses are available, involving their respective goals.

For example, in Quebec, mature material from a provenance-progeny test initiated by the CFS in 1976 and replicated on three sites in Quebec will be used (Beaulieu et al. 2006). In British Columbia, a large open-pollinated progeny test ("red rock") will be used in conjunction with the parents of the breeding population, which are also ca. 40 years old. These breeding populations provide genetic material for which "mature" traits, such as wood quality and lifetime fitness, can be measured (King et al. 1997). In both projects, seedling progenies and their replicated lines are being established, which in their young age are adequate to test candidate genes with phenology traits (e.g., timing of bud set).

3.4.2 Somatic Embryogenesis and Genetic Transformation

Since its demonstration in Norway spruce in 1985 (Hakman et al. 1985), somatic embryogenesis (SE) is the most versatile method for vegetative propagation of conifers. SE is divided into five stages: (1) initiation of embryogenic tissue from mature or immature explants, (2) proliferation of embryogenic tissue, (3) embryo maturation, (4) germination, and (5) conversion to plants. The cryopreservation of embryogenic tissue during stage 2 offers an opportunity to select a posteriori elite individuals for clonal plantations (reviewed by Cyr and Klimaszewska 2002). Over the last decade, most of the studies on SE have focused on the proliferation and the maturation processes to get embryos of high quality for both spruce and pine species (reviewed by Stasolla et al. 2002). Recently, new SE protocols have also opened up the opportunity to bring about this technology to a large-scale delivery (Sutton et al. 2004). In addition, SE has become a primary enabling technology for genetic engineering (see below) because embryogenic tissues are amenable to genetic transformation via cocultivation with *Agrobacterium tumefaciens* (Levée et al. 1997).

An impediment to tree improvement is the time required for each progeny to reach sexual maturity. Genetic engineering could circumvent this problem by allowing the transfer of single gene traits into superior genotypes, leading to the integration of desired traits. Accordingly, genetic transformation offers an attractive alternative to breeding, especially for insect and disease resistance for which natural variants with tolerance/resistance phenotype might be nonexistent (reviewed by Pena and Séguin 2001). The first stable transformation in a commercially important conifer species (*P. glauca*) was achieved via particle bombardment of embryogenic tissue (Ellis et al. 1993). This protocol was then successfully adapted for the transformation of other spruce species (Charest et al. 1996). In Sitka spruce, transient expression of the *uidA* gene using expression of the *GUS A* gene in embryogenic cell lines was observed following *Agrobacterium*-mediated transformation, but no stable transformants could be recovered (Drake et al. 1997).

Agrobacterium-mediated transformation is the method of choice for gene transfer in conifers (reviewed by Nehra et al. 2005), primarily owing to the ease of transformation, high efficiency, and clean integration of T-DNA into the host genome. Combined with rapid selection of transgenic embryogenic tissues and highly efficient seedling regeneration via somatic embryo maturation, the production of transgenic spruce has now become routine (Klimaszewska et al. 2001). Nevertheless, this technology is perceived by many as too premature to be commercially implemented.

Many questions and concerns remain regarding genetically modified (GM) trees. In Quebec, the CFS has been conducting some confined trials with GM trees but on a limited basis – only involving the testing of gene function. In British Columbia, testing in any open environment is prohibited. Commercial plantations of GM trees are in fact not permitted in Canada (http://www.inspection.gc.ca/english/ plaveg/bio/pbobbve.shtml.). The bigger questions about the impacts of GM trees revolve around their deployment in operational forestry (Strauss et al. 2001; van Frankenhuyzen and Beardmore 2004). What is the effect of gene flow from GM to native trees of the same species? What are the benefits of GM trees to the local economy? Should we introduce foreign genes into a natural ecosystem?

3.5 Genomics

3.5.1 cDNA Libraries, EST, and SNP Collections

Genome projects in both British Columbia and Quebec have been involved with the construction of cDNA libraries implicating a large variety of tissues. In Quebec, 16 cDNA libraries were constructed, followed by the sequencing and processing of 71,424 spruce ESTs, of which 49,101 were of high quality (at least 100 contiguous nucleotides with a *Phred* score over 20) and retained for the final assembly of 16,578 consensus sequences (average of 797 bp) (Pavy et al. 2005). Half of the clones were sequenced in both directions to accelerate gene discovery, and 45% of the clones were completely sequenced, at least in one direction. A database named SpruceDB is publicly accessible and contains all high-quality ESTs (Pavy et al. 2005). A second-generation publicly accessible database, ForestDB, accommodates ESTs from multiple species (white spruce, loblolly pine, and poplar transcriptomes), and multiple EST assemblies have recently been made available, including SNP tables for each of white spruce contig, encompassing over 12,000 SNPs (Pavy et al. 2006).

In British Columbia, 20 white and Sitka spruce cDNA libraries have been constructed, including normalized and full-length versions, with 184,364 spruce ESTs now submitted to GenBank (average 661 bases per EST). In addition, 6,464 full-length sequences have also been obtained (average length 1,123 bases; not yet submitted), with two complete reads per clone to ensure quality. Using an additional 29,493 ESTs from Genbank, assembly resulted in 26,336 clusters of size at least two, and 52,316 singletons. However, these values are dependent on minimum overlap (set at 40) and minimum percent match (set at 94). There were 12,364 clusters with four or more members, of which 34,658 SNPs were identified. A database of these SNPs and pine SNPs is at http://treenomix2.forestry.ubc.ca/public/spruce_pine/index.htm.

3.5.2 Microarrays

Likewise, these two genome projects have been involved with the design, manufacture, and use of cDNA microarrays. In British Columbia, three successively larger cDNA chips, from 9.7 K to 16.7 K to 21.8 K cDNA clones, have been designed and spotted. The first 9.7 K microarrays were used to investigate the response of Sitka spruce shoot tips to the spruce budworm, and this work represented the first large-scale study of insect-induced transcripts in a gymnosperm (Ralph et al. 2006). In Quebec, a spruce unigene of set 11 K cDNA clones was resequenced and used to fabricate cDNA chips, with the objective of investigating regulatory genes and genes related to wood formation. The microarrays are also being used by various collaborators; customized cDNA microarrays can be obtained, and further collaborations are encouraged.

3.5.3 Proteomics

The complete set of proteins in a particular tissue or in an organism is termed the proteome. In the first proteomic study of conifer insect defense, Lippert et al. (2007) as part of the Treenomix project, studied changes in the proteome of Sitka spruce bark tissue in response to feeding by white pine weevils or mechanical wounding. Two-dimensional polyacrylamide gels coupled with high-throughput tandem mass spectrometry were used to detect and identify induced changes in protein abundance and protein modification. The spruce EST and full-length cDNA sequences mentioned above were essential to identifying proteins from mass spectrometry data; 72% of proteins were identified. Significant changes of protein levels were observed as early as 2 h following the onset of insect feeding. Among the insect-induced proteins are a series of related small heat shock proteins, other stress response proteins, proteins involved in secondary metabolism, oxidoreductases, and a protein to which no homology can be found in other plant species. These proteins and their promoters are good candidate genes for SNP markers to use in molecular breeding.

3.5.4 Transgenic Spruce Lines

To enable functional studies of the biological role of candidate regulatory genes, a high-throughput DNA transformation platform has been established in Quebec in collaboration with CFS using the white spruce line PG-653 (A. Séguin et al. personal communication). A number of gene constructs were tested for each of 37 different conifer genes by focusing on gain-of-function experiments, resulting in the production of 350 uniform and stable transgenic spruce cell lines, further analyzed by quantitative PCR (qPCR). Various unusual phenotypes could be recovered, and gene expression profiling of wild type and transgenic plants is underway (J. Mackay, personal communication).

3.5.5 Genome Composition and Bacterial Artificial Chromosome (BAC) Libraries

Genome Composition

It has long been suspected that the large genome size of spruce and other conifers is due to repetitive DNA. To characterize the types of repeats, Jurman et al. (1999) sequenced 120 clones of Norway spruce and found a large proportion of the clones to be represented by long interspersed repeats (copia- and gypsy-like LTR-retrotransposons), while SINE-like elements were rare; SSRs were not very abundant. Miniature inverted-repeat transposable elements and long interspersed nuclear elements were not represented. Further information about genome structure of noncoding regions will be obtained by sequencing of BAC clones (discussed below).

Pavy et al. (2005) analyzed ca. 15,000 white spruce transcripts and found that ca. 84% of the transcripts had homology to published gymnosperm sequences but only 68% had homologs with rice or *Arabidopsis* (e-value < 1e-10). This is not surprising given the distinct evolution of gene families in gymnosperms (Guillet-Claude et al. 2004). About 70% of transcripts had a significant match to proteins in the Uniref100 database. Gene ontology (GO) terms representing 16 molecular functions could be ascribed to about 40% of these transcripts, with catalytic activity ranking first (Pavy et al. 2005). For transcription factors, a hidden Markov model (HMM) search against *Arabidopsis* and PFAM identified ca. 400 white spruce sequences that represented putative spruce transcription factors (Pavy et al. 2005). The distribution of transcripts with SNPs among GO classes largely reflected the relative abundance of GO classes (Pavy et al. 2006). Algorithms were developed to identify ORFs from contig assemblies and, hence, estimate rates of synonymous and nonsynonymous SNPs. From 3,374 distinct ORFs with reliable SNPs ($P_{Polybayes}$ over 0.95), about 40% of SNPs were nonsynonymous and 60% synonymous, and the ratio of nonsynonymous to synonymous SNPs per site was found to be 0.17 (Pavy et al. 2006), surprisingly the same as that reported for 240 ORFs in *Arabidopsis* (Zhang et al. 2002).

A 3 × Unpooled Spruce BAC Library

Due to the large genome size of conifers, BAC libraries have been extremely limited for conifers. However, a library for *Pinus pinaster* is reportedly being expanded to 3× coverage (Plomion et al. this volume). More dramatic are the plan for a 10× arrayed library for *P. taeda* (loblolly pine), consisting of 1.7 to 2 million clones (Peterson 2005), which would make it the largest library ever constructed. When coverage reaches 3×, isolation of individual genes is planned (Plomion et al. this volume).

In spruce, a 3× unarrayed library is currently being constructed for the Treenomix project by Bio S&T (Montreal), and halfway through (June 2006), insert sizes are averaging ca. 150 kb. In the Treenomix project, about ten randomly selected BAC clones will be shotgun-sequenced to obtain a picture of genome structure, another ten clones enriched for coding genes will be sequenced to get a picture of upstream regulatory sequence, and a final ten sequenced for targeted genes also sequenced in loblolly pine sequenced clones to get a picture of microsynteny and local rearrangements. To target genes, BACs are stored as pools, about 1,000 BACS per pool, arrayed in 96-well plates; each plate is pooled for rows and columns and screened by PCR to identify wells containing the target gene; colonies from the target well are grown and individually rearrayed for a second PCR screen (see Isidore et al. 2005).

3.5.6 High-Throughput SNP Genotyping

Tests of the Illumina High-Throughput Genotyping Technology

In the framework of the Arborea and Treenomix projects, proof-of-principle experiments have been conducted to test the Illumina SNP bead-array genotyping technology, which is based on a highly multiplexed (either 96, 384, 768, or 1536 SNPs) primer-extension approach (Fan et al. 2003). The technology involves an allele-specific extension step using allele-specific oligos (ASOs), followed by ligation to a locus-specific oligo (LSO) carrying, for each SNP, a unique address sequence. Amplification is done with two labelled universal primers (Cy3 and Cy5) complementing each ASO, and a third universal primer matching the LSO. The dye-labelled DNA products are hybridized onto a plate array and a bead array reader is used for scanning fluorescence signals. With this assay, homozygous DNA should be detected by a signal in either green (Cy3) or red (Cy5) channel and heterozygous DNA in both channels.

In the Arborea project, SNPs were identified in silico from contig assemblies (e.g., Pavy et al. 2006) and from resequencing genomic DNA. A 768-SNP array has been tested on pedigree and unstructured populations and the success rate was 70%, using the stringent criteria of human SNP projects such as HapMap (that is, over 98.5% call rate success for any given SNP with normal segregation). Across six white spruce natural populations emcompassing 160 individuals, observed and expected heterozgosities (H), were respectively 0.28 and 0.26 for the 534 valid SNPs. These SNPs are currently used for the mapping of candidate genes and in genomic scans involving natural populations with significant differentiation for adaptive characters (Q_{ST}) (Jaramillo-Correa et al. 2001).

In the Treenomix project, this system was tested using 384 SNPs identified in silico from a collection of 25,000 ESTs derived from a single spruce tree (PG29) used the project. Two mapping populations of size 75, both of which had PG29 as a parent, were assayed. Two hundred seventy-three of 384 SNPs (71%) appeared to work, showing 1:1 or 1:2:1 ratios. Two instances of null alleles were found (showing 2:1:1 or 1:0:1 ratios). Given the repetitive nature of conifer genomes, this represents an unexpectedly high success rate. SNP development under the Illumina system is always a two-stage process: development of an initial testing plate of a minimum of 96 SNPs, then a final "production" plate, which can be expensive for low-throughput projects.

Second-Generation SNP Plates

Since the cost per genotype is much lower with the high-throughput systems (ca. 6 cents and likely to drop), a "universal SNP plate" could be developed for spruce in order to facilitate comparative genomic studies and accelerate molecular breeding research. However, the plated SNPs must be characterized a priori at the population level in order to verify their occurrence and abundance in each targeted taxon. A large proportion of SNPs are singletons or at low frequency (Bousquet et al. personal communication). Such SNPs are likely to be useless for most applications. In addition, common SNPs are usually different between *P. glauca*, *P. mariana*, and *P. abies* (Guillet-Claude et al. 2004; Bousquet et al. personal communication). The occurrence of common SNPs could be estimated in EST databanks from contig assemblies containing sequences from a large number of distinct clones (Pavy et al. 2006). Resequencing on genomic DNA appears necessary to ascertain the occurrence and frequency in nontarget species.

As for mapping, unless SNPs are prescreened for high heterozygosity, there will not be sufficient numbers of segregating SNPs in any one cross to warrant the cost of the plate. The above-mentioned Treenomix test provides a preliminary indication of how SNPs detected in EST databases might lead to a good SNP plate for mapping. In a survey of 27 individuals across the white spruce range, mean heterozygosity (h) for the 273 SNPs was 0.24, and h for SNPs where all three parents were heterozygous in the two pedigrees was 0.36. (It should be remembered that these h values are conditioned on the observation of heterozygosity in the tree PG29; true values of h would be considerably lower.) With an average h of 0.36, a SNP panel of 384 highly heterozygous loci would segregate for about 200 loci for either of the two parents of a pedigree. However, it should be remembered that these SNP polymorphisms are largely species specific.

3.6 Summary

In spruce, as in most coniferous tree species, probably the greatest hindrance for effective molecular genetic dissection and understanding of traits important for tree breeding is the lack of appropriate pedigree materials. Most pedigrees are small (10 to 30 progeny), meant to test breeding values of individual trees in traditional breeding programs; also the long generation time of spruce (ca. 20 years) inhibits investment in multigenerational pedigrees, let alone the breeding program itself. At least pedigree material has been developed that now makes it possible to invest major efforts in genomics in order to speed up breeding cycles using molecular tools. BC_1 and F_1 of large progeny size, and clonally replicated, have been developed by the CFS and Forest Research (Scotland) for further QTL mapping experiments.

As well, as in all species, anonymous markers have been of limited value for molecular breeding. With the recent availability of catalogs of gene sequences, new cohorts of genetic markers based on candidate genes thought to be involved in specific pathways or characters, such as wood formation and pest resistance, can and are being developed for spruce. The emergence of new methodologies for discovering and assaying SNPs, and for obtaining anchor loci for comparative maps, promises rapid advances in the near future. Using the newly developed markers, the first composite genetic map for each of both white spruce and black spruce has already been assembled (Pavy et al. 2007). In addition, the first comparisons with other Pinaceae have been made. These results have opened new areas of research for spruces not possible before and helped increase the speed of MAS development and implementation for spruce.

Finally, it would seem that genomics can add a new dimension to tree breeding, providing new targets for breeding via better understanding of the genetics of traits such as disease and pest resistance, cold and drought tolerance, and even CO_2 sequestration. One can imagine that genomics might provide novel methods to monitor the health of existing forest stock and make revised predictions about the longer-term sustainability of conifer forests in our landscape. However, the recent investments in forest genomics invite a naively optimistic vision of these new applications. Our actual level of genomics knowledge is still primitive, and, like any new area, the ideas and hypotheses of interest in genomics will change rapidly, as will the potential applications.

References

Acheré V, Faivre-Rampant P, Jeandroz S, Besnard G, Markussen T, Aragones A, Fladung M, Ritter E, Favre JM (2004) A full saturated linkage map of *Picea abies* including AFLP, SSR, ESTP, 5S rDNA and morphological markers. Theor Appl Genet 108:1602–1613

Acheré V, Favre JM, Besnard G, Jeandroz S (2005) Genomic organization of molecular differentiation in Norway spruce (*Picea abies*). Mol Ecol 14:3191–3201

A'Hara SW, Cottrell JE (2004) A set of microsatellite markers for use in Sitka spruce (*Picea sitchensis*) developed from *Picea glauca* ESTs. Mol Ecol Notes 4:659–663

A'Hara SW, Cottrell JE (2007) Characterization of a suite of 40 EST-derived microsatellite markers for use in a Sitka spruce (*Picea sitchensis* (Bong.) Carr) marker aided selection program. Silvae Genet (in press)

Alfaro RI, Borden JH, King JN, Tomlin ES, McIntosh RL, Bohlmann J (2002) Mechanisms of resistance in conifers against shoot infesting insects. In: Wagner MR, Clancy KM, Lieutier F, Paine TD (eds) Mechanisms and Deployment of Resistance in Trees to Insects. Kluwer, Dordrecht, pp 101–126

Arno SF, Hammerly RP (1977) Northwest Trees. Mountaineers, Seattle, WA

Bastien D, Favre JM, Colligon AM, Sperisen C, Jeandroz S (2003) Characterization of a mosaic minisatellite locus in the mitochondrial DNA of Norway spruce (*Picea abies* (L.) Karst.). Theor Appl Genet 107:574–580

Beavis WD (1994) The power and deceit of QTL experiments: lessons from comparative QTL studies. In: Proc 49th Annual Corn and Sorghum Industry Res Conf, American Seed Trade Assosiaction, Washington, DC, pp 250–266

Beavis WD, Grant D (1991) A linkage map based on information from four F_2 populations of maize (*Zea mays* L.). Theor Appl Genet 82:636–644

Beaulieu J (1996) Breeding program and strategy for white spruce in Quebec. Inf Rep LAU-X-11TE. Nat Res Can, Canadian Forest Service, Sainte-Foy, Quebec

Beaulieu J, Corriveau A, Daoust G (1989) Phenotypic stability and delineation of black spruce breeding zones in Quebec. Inf Rep LAU-X-85E. Canadian Forest Service, Sainte-Foy, Quebec

Beaulieu J, Perron M, Bousquet J (2004) Multivariate patterns of adaptive genetic variation and seed source transfer in black spruce (*Picea mariana*). Can J For Res 34:531–545

Beaulieu J, Watson P, Isabel N, Bousquet J, MacKay J (2006) Association mapping for wood candidate genes in white spruce. In: Plant & Animal Genome XIV Conf, San Diego

Binelli G, Bucci G (1994) A genetic linkage map of *Picea abies* Karst., based on RAPD markers, as a tool in population genetics. Theor Appl Genet 88:283–288

Bouillé M, Bousquet J (2005) Trans-species shared polymorphisms at orthologous nuclear gene loci among distant species in the conifer *Picea* (Pinaceae): implications for the maintenance of genetic diversity in trees. Am J Bot 92:63–73

Bouillé M, Bousquet J (2006) Discordant mtDNA and cpDNA phylogenies of the entire genus *Picea* indicate patterns of allopatric speciation related to geographic dispersal and ancient reticulation. In: Proc IUFRO Population Genetics and Genomics of Forest Trees, Madrid, p 152

Boyle TJB, Morgenstern EK (1986) Estimates of outcrossing rates in six populations of black spruce in central New Brunswick. Silvae Genet 35:102–106

Brooks DJ (1997) The outlook for demand and supply of wood: implications for policy and sustainable management. Commonwealth For Rev 76:31–36

Brown GR, Carlson JE (1997) Molecular cytogenetics of the genes encoding 18s-5.8s-26s rRNA and 5s rRNA in two species of spruce (*Picea*). Theor Appl Genet 95:1–9

Brown GR, Newton CH, Carlson JE (1998) Organization and distribution of a *Sau*3A tandem repeated DNA sequence in *Picea* (Pinaceae) species. Genome 41:560–565

Brown GR, Gill GP, Kuntz RJ, Langley CH, Neale DB (2004) Nucleotide diversity and linkage disequilibrium in loblolly pine. Proc Natl Acad Sci USA 101:15255–15260

Burley J (1965) Karotype analysis of Sitka spruce, *Picea sitchensis* (Bong.) Carr. Silvae Genet 14:127–132

Cannell MGR, Milne R (1995) Carbon pools and sequestration in forest ecosystems in Britain. Forestry 68:361–378

Campbell CS, Wright WA, Cox M, Vining TF, Major CS, Arsenault MP (2005) Nuclear ribosomal DNA internal transcribed spacer 1 (ITS1) in *Picea* (Pinaceae): sequence divergence and structure. Mol Phylogenet Evol 35:165–185

Cardon LR, Bell JI (2001) Association study designs for complex diseases. Nat Rev Genet 2:91–99

Chaisurisri K, El-Kassaby YA (1994) Genetic diversity in a seed production population vs. natural populations of Sitka Spruce. Biodivers Conserv 3:512–523

Charest PJ, Devantier Y, Lachance D (1996) Stable genetic transformation of *Picea mariana* (black spruce) via particle bombardment. In Vitro Cell Dev Biol Plant 32:91–99

Cheliak WM, Pitel JA, Murray G (1985) Population structure and the mating system of white spruce. Can J For Res 15:301–308

Collignon AM, Van de Sype H, Favre JM (2002) Geographical variation in random amplified polymorphic DNA and quantitative traits in Norway spruce. Can J For Res 32:266–282

Comai L, Young K, Till BJ, Reynolds SH, Greene EA, Codomo CA, Enns LC, Johnson JE, Burtner C, Odden AR, Henikoff S (2004) Efficient Discovery of DNA polymorphisms in natural populations by ecotilling. Plant J 37:778–786

Corriveau AG (1981) Variabilité spatiale et temporelle de la croissance juvénile des provenances d'épinette noire. 18th Meeting of the CTIA, Duncan, BC, Canada

Corriveau A, Beaulieu J, Daoust G (1991) Heritability and genetic correlations of wood characters of Upper Ottawa Valley white spruce populations grown in Quebec. For Chron 67:698–705

Cottrell JE, White IMS (1995) The use of isozyme genetic markers to estimate the rate of outcrossing in a Sitka spruce (*Picea sitchensis* (Bong.) Carr.) seed orchard in Scotland. New For 10:111–122

Cyr DR, Klimaszewska K (2002) Conifer somatic embryogenesis: II. Applications. Dendrobiology 48:41–49

Devey ME, Carson SD, Nolan MF, Matheson AC, Te Riini C, Hohepa J (2004) QTL associations for density and diameter in *Pinus radiata* and the potential for marker-aided selection. Theor Appl Genet 108:516–524

Dietrichson J (1969) The geographic variation of spring-frost resistance and growth cessation in Norway spruce (*Picea abies* (L.) Karst.). Medd Norske Skogforsøksvesen 27:91–104

Dirlewanger E, Cosson P, Howad W, Capdeville G, Bosselut N, Claverie M, Vosin R, Poizat C, Lafarque B, Baron O, Laigret F, Kleinhentz M, Arus P, Esmenjaud D (2004) Microsatellite genetic linkage maps of myrobalan plum and an almond-peach hybrid location of root-knot nematode resistance genes. Theor Appl Genet 109:827–838

Drake PMW, John A, Power JB, Davey MR (1997). Expression of the *gusA* gene in embryogenic cell lines of Sitka spruce following *Agrobacterium*-mediated transformation. J Exp Bot 48:151–155

Ellis DD, McCabe DE, McInnis S, Ramachandran R, Russell DR, Wallace KM, Martinell BJ, Roberts DR, Raffa KF, McCown BH (1993) Stable transformation of *Picea glauca* by particle acceleration. Biotechnology (NY) 11:84–89

Fan JB, Oliphant A, Shen R, Kermani BG, Garcia F, et al (2003) Highly parallel SNP genotyping. Cold Spring Harb Symp Quant Biol 68:69–78

Farjon A (1990) Pinaceae: drawings and descriptions of the genera *Abies, Cedrus, Pseudolarix, Keteleeria, Nothotsuga, Tsuga, Cathaya, Pseudotsuga, Larix* and *Picea*. Koeltz, Königstein, Germany

Farrar JL (1995) Trees in Canada. Canadian Forest Service and Fitzhenry & Whiteside, Markham, Ontario, Canada

Fletcher AM (1992) Breeding improved Sitka for the 90's. Forestry Commission Bulletin No 103, Edinburgh, pp 11–24

Fournier D, Perry DJ, Beaulieu J, Bousquet J, Isabel N (2002) Optimizing expressed sequence tag polymorphisms by single-strand conformation polymorphism in spruces. For Genet 9:11–18

Fowler DP (1983) The hybrid black × Sitka spruce, implications to phylogeny of the genus *Picea*. Can J For Res 13:108–115

Fowler DP (1987) The hybrid white × Sitka spruce: species crossability. Can J For Res 17:413–417

Fox DP (1987) The chromosomes of *Picea sitchensis* (Bong.) Carr. and its relatives. Proc R Soc Edinburgh Sect B Biol Sci 93:51–59

Frankis MP (1989) Genetic inter-relationships in Pinaceae. Notes R Bot Gard Edinburgh 45:527–548

Fulton T, van der Hoeven R, Eannetta N, Tanksley S (2002) Identification, analysis and utilization of a conserved ortholog set (COS) markers for comparative genomics in higher plants. Plant Cell 14:1457–1467

Furnier GR, Stine M, Mohn CA, Clyde MA (1991) Geographic patterns of variation in allozymes and height growth in white spruce. Can J For Res 21:707–712

Gamache I, Jaramillo-Correa JP, Payette S, Bousquet J (2003) Diverging patterns of mitochondrial and nuclear DNA diversity in subarctic black spruce: Imprint of a founder effect associated with postglacial colonization. Mol Ecol 12:891–901

Gebhardt C, Ballvora A, Walkemeier B, Oberhagemann P, Schüler K (2004) Assessing genetic potential in germplasm collections of crop plants by marker-trait association: a case study for potatoes with quantitative variation of resistance to late blight and maturity type. Mol Breed 13:93–102

Gosselin I, Zhou Y, Bousquet J, Isabel N (2002) Megagametophyte-derived linkage maps of white spruce (*Picea glauca*) based on RAPD, SCAR and ESTP markers. Theor Appl Genet 104:987–997

Grattapaglia D, Sederoff R (1994) Genetic linkage maps of *Eucalyptus grandis* and *Eucalyptus urophylla* using a pseudo-test-cross mapping strategy and RAPD markers. Genetics 137:1121–1137

Guillet-Claude C, Isabel N, Pelgas B, Bousquet J (2004) The evolutionary implications of *knox-I* gene duplications in conifers: correlated evidence from phylogeny, gene mapping, and analysis of functional divergence. Mol Biol Evol 21:2232–2245

Hakman I, Fowke LC, von Arnold S, Eriksson T (1985) The development of somatic embryos in tissue cultures initiated from immature embryos of *Picea abies* (Norway spruce). Plant Sci 38:53–59

Hall PM (1994) Ministry of forests perspectives on spruce reforestation in British Columbia. FRDA Rep 226. Victoria, BC, Canadian Forest Service, pp 1–6

Harris AS (1990) *Picea sitchensis* (Bong.) Carr. Sitka spruce. In: Burns RM, Honkala BH (tech coords) Silvics of North America, vol 1: Conifers. Agric Handbook 654. USDA Forest Service, Washington, DC, pp 260–267

Hawley GJ, DeHayes DH (1994) Genetic diversity and population structure of red spruce (*Picea rubens*). Can J Bot 72:1778–1786

Hayden MJ, Kuchel H, Chalmers KJ (2004) Sequence tagged microsatellites for the *Xgwm533* locus provide new diagnostic markers to select for the presence of stem rust resistance gene *Sr2* in bread wheat (*Triticum aestivum* L.). Theor Appl Genet 109:1641–1647

Hermann RK (1987) North American tree species in Europe: transplanted species offer good growth potential on suitable sites. J For 85:27–32

Hodgetts RB, Aleksiuk MA, Brown A, Clarke C, Macdonald E, Nadeem S, Khasa D (2001) Development of microsatellite markers for white spruce (*Picea glauca*) and related species. Theor Appl Genet 102:1252–1258

Hosie RC (1969) Native Trees of Canada. 7th ed. Canadian Forestry Service, Department of Fisheries and Forestry, Ottawa, Ontario, Canada

Hu XS, Goodwillie C, Ritland K (2004) Joining genetic linkage maps using a joint likelihood function. Theor Appl Genet 109:996–1004

Huber DPW, Ralph S, Bohlmann J (2004) Genomic hardwiring and phenotypic plasticity of terpenoid-based defenses in conifers. J Chem Ecol 30:2401–2420

Ingle J, Timmis JN, Sinclair J (1975) The relationship between satellite deoxyribonucleic acid, ribosomal ribonucleic acid gene redundancy, and genome size in plants. Plant Physiol 55:496–501

Isabel N, Beaulieu J, Bousquet J (1995) Complete congruence between gene diversity estimates derived from genotypic data at enzyme and RAPD loci in black spruce. Proc Natl Acad Sci USA 92:6369–6373

Isidore E, Scherrer B, Bellec A, Budin K, Faivre-Rampant P, Waugh R, Keller B, Caboche M, Feuilllet C, Chalhoub B (2005) Direct targeting and rapid isolation of BAC clones spanning a defined chromosomal region. Funct Integr Genom 9:97–103

Jaramillo-Correa JP, Bousquet J (2003) New evidence from mitochondrial DNA of a progenitor-derivative species relationship between black spruce and red spruce (Pinaceae). Am J Bot 90:1801–1806

Jaramillo-Correa JP, Bousquet J (2005) Mitochondrial genome recombination in the zone of contact between two hybridizing conifers. Genetics 171:1951–1962

Jaramillo-Correa JP, Beaulieu J, Bousquet J (2001) Contrasting evolutionary forces driving population structure at ESTPs, allozymes, and quantitative traits in white spruce. Mol Ecol 10:2729–2740

Jaramillo-Correa JP, Beaulieu J, Bousquet J (2004) Variation in mtDNA reveals multiple glacial refugia in black spruce (*Picea mariana*), a transcontinental North American conifer. Mol Ecol 13:2735–2748

Jaramillo-Correa JP, Beaulieu J, Ledig FT, Bousquet, J (2006) Decoupled mitochondrial and chloroplast DNA population structure reveals Holocene collapse and population isolation in a threatened Mexican-endemic conifer. Mol Ecol 15:2878–2800

Joyce PM, O'Carroll N (2002) Sitka spruce in Ireland. COFORD, Dublin

Jurman I, Zuccolo I, Vischi M, Morgante M (1999) Analysis of structure and organization of repetitive sequences in the Norway spruce genome (*Picea abies* K.). In: Plant & Animal Genome VII Conf, San Diego

Kean VM, Fox DP, Faulkner R (1982) The accumulation mechanism of the supernumerary (B-) chromosome in *Picea sitchensis* (Bong.) Carr. and the effect of this chromosome on male and female flowering. Silvae Genet 31:126–131

Khalil MAK (1984) All-range black spruce provenance study in Newfoundland: Performance and genotypic stability of provenances. Silvae Genet 33:63–71

King JN, Yanchuk AD, Kiss GK, Alfaro RI (1997) Genetic and phenotypic relationships between weevil (*Pissodes strobi*) resistance and height growth in spruce populations of British Columbia. Can J For Res 27:732–739

Kiss GK (1989) Engelmann × Sitka spruce hybrids in central British Columbia. Can J For Res 19:1190–1193

Kiss G, Yeh FC (1988) Heritability estimates for height for young interior spruce in British Columbia. Can J For Res 18:158–162

Klimaszewska K, Lachance D, Pelletier G, Lelu MA, Seguin A (2001) Regeneration of transgenic *Picea glauca*, *P. mariana*, and *P. abies* after cocultivation of embryogenic tissue with *Agrobacterium tumefaciens*. In Vitro Cell Dev Biol Plant 37:748–755

Krutovskii KV, Bergmann F (1995) Introgressive hybridization and phylogenetic relationships between Norway, *Picea abies* (L.) Karst., and Siberian, *P. obovata* Ledeb., spruce species studied by isozyme loci. Heredity 74:464–480

Krutovsky KV, Troggio M, Brown GR, Jermstad KD, Neale DB (2004) Comparative mapping in the *Pinaceae*. Genetics 168:447–461

Krutovsky KV, Elsik CG, Matvienko M, Kozik A, Neale DB (2006) Conserved ortholog sets in forest trees. Tree Genetics and Genomics 3:61–70

Lagercrantz U, Ryman N (1990) Genetic structure of Noway spruce (*Picea abies*): concordance of morphological and allozymic variation. Evolution 44:38–53

Lamothe M, Meirmans P, Isabel N (2006) A set of polymorphic EST-derived markers for *Picea* species. Mol Ecol Notes 6:237–240

Lander ES, Botstein D (1989) Mapping mendelian factors underlying quantitative traits using RFLP linkage maps. Genetics 121:185–199

Ledig FT, Bermejo-Velazquez B, Hodgskiss PD, Johnson DR, Flores-Lopez C, Jacob-Cervantes V (2000) The mating system and genic diversity in Martinez spruce, an extremely rare endemic of Mexico's Sierra Madre Orientale: an example of facultative selfing and survival in interglacial refugia. Can J For Res 30:1156–1164

Ledig FT, Hodgskiss PD, Jacob-Cervantes V (2002) Genetic diversity, mating system, and conservation of a Mexican subalpine relect, *Picea mexicana* Martinez. Conserv Genet 3:113–122

Lee SJ (1993) Breeding strategy for Sitka spruce in Britain. In: Lee SJ (Ed) Proc Nordic Group for Tree Breed. Forestry Research, Roslin, Scotland, pp 95–109

Lefort F, Echt C, Streiff R, Vendramin GG (1999) Microsatellite sequences: a new generation of molecular markers for forest genetics. For Genet 6:15–20

LePage BA (2001) New species of *Picea* A. Dietrich (Pinaceae) from the middle Eocene of Axel Heiberg Island, Arctic Canada. Bot J Linn Soc 135:137–167

Levée V, Lelu MA, Jouanin L, Cornu D, Pilate G (1997) *Agrobacterium tumefaciens*-mediated transformation of hybrid larch (*Larix kaempferi* T *L. decidua*) and transgenic plant regeneration. Plant Cell Rep 16:680–685

Li P, Beaulieu J, Corriveau A, Bousquet J (1993) Genetic variation in juvenile growth and phenology in a white spruce provenance-progeny test. Silvae Genet 42:52–60

Li P, Beaulieu J, Bousquet J (1997) Genetic structure and patterns of genetic variation among populations in eastern white spruce (*Picea glauca*). Can J For Res 27:189–198

Lippert D, Chowrira S, Ralph SG, Zhuang J, Aeschliman D, Ritland C, Ritland K, Bohlmann J (2007) Conifer defense against insects: Proteome analysis of Sitka spruce (*Picea sitchensis*) bark induced by mechanical wounding or feeding by white pine weevils (*Pissodes strobi*). Proteomics 7:248–270

Liston A, Gernandt DS, Vining TF, Campbell CS, Pinero D (2003) Molecular phylogeny of Pinaceae and *Pinus*. Acta Hort 615:107–114

Malcolm DC (1997) The silviculture of conifers in Great Britain. Forestry 70:293–307

Mikkola L (1969) Observations on interspecific sterility in *Picea*. Ann Bot Fenn 6:285–339

Moir RB, Fox DP (1972) Supernumerary chromosomes in *Picea sitchensis* (Bong.) Carr. Silvae Genet 21:182–186

Morgenstern EK (1969) Genetic variation in seedlings of *Picea mariana* (Mill.) B.S.P.: I. Correlation with ecological factors. Silvae Genet 18:151–161

Murray BG (1998) Nuclear DNA amounts in Gymnosperms. Annals Bot 82(Suppl A):3-15

Neale DB, Sewell MM, Brown GR (2002) Molecular dissection of the quantitative inheritance of wood property traits in loblolly pine Ann For Sci 59:595–605

Neale DB, Savolainen O (2004) Association genetics of complex traits in conifers. Trends Plant Sci 9:325–330

Neale DB, Wheeler NC (2004) Mapping of quantitative trait loci in loblolly pine and Douglas-fir: a summary. For Genet 11:173–178

Nehra NS, Becwar MR, Rottmann WH, Pearson L, Chowdhury K, Chang Shujun, Wilde HD, Kodrzycki RJ, Zhang C, Gause KC, Parks DW, Hinchee MA (2005) Forest Biotechnology: Innovative methods, emerging opportunities. Invited Review. In Vitro Cell Dev Biol Plant 41:701–717

Nienstaedt H (1985) Inheritance and correlations of frost injury, growth, flowering, and cone characteristics in white spruce, *Picea glauca* (Moench) Voss. Can J For Res 15:498–504

Nkongolo KK (1996) Chromosome analysis and DNA homology in three *Picea* species, *P. mariana, P. rubens*, and *P. glauca* (Pinaceae). Plant Syst Evol 203:27–40

Paglia GP, Olivieri AM, Morgante M (1998) Towards second-generation STS (sequence-tagged sites) linkage maps in conifers: a genetic map of Norway spruce (*Picea abies* K.). Mol Gen Genet 258:466–478

Parker WH, Van Niejenhuis A, Charette P (1994) Adaptive variation in *Picea mariana* from northwestern Ontario determined by short-term common environment tests. Can J For Res 24:1653–1661

Pavy N, Paule C, Parsons L, Crow JA, Morency M-J, Cooke J, Johnson JE, Noumen E, Guillet-Claude C, Butterfield Y, Barber S, Yang G, Liu J, Stott J, Kirkpatrick R, Siddiqui A, Holt R, Marra M, Séguin A, Retzel E, Bousquet J, Mackay J (2005) Generation, annotation, analysis, and database integration of 16,500 white spruce EST clusters. BMC Genomics 6(144):19

Pavy N, Parsons L, Paule C, Mackay J, Bousquet J (2006) Automated SNP detection from a large collection of white spruce expressed sequences: contributing factors and approaches for the categorization of SNPs. BMC Genomics 7(174):14

Pavy N, Gosselin I, Gagnon F, Lamothe M, Laroche L, Isabel N, Bousquet J (2007) A SNP-based composite map of regulatory genes in the conifer white spruce and structural relationships with other plants In: Plant & Animal Genome XV Conf, San Diego, CA, USA

Pelgas B, Isabel N, Bousquet J (2004) Efficient screening for expressed sequence tag polymorphisms (ESTP) by DNA pool sequencing and denaturing gradient gel electrophoresis in spruces. Mol Breed 13:263–279

Pelgas B, Bousquet J, Beauseigle S, Isabel N (2005) A composite linkage map from two crosses for the species complex *Picea mariana* [Mill.] B.S.P. × *P. rubens* (Sarg.) and analysis of synteny with other Pinaceae. Theor Appl Genet 111:1466–1488

Pelgas B, Beauseigle S, Acheré V, Jeandroz S, Bousquet J, Isabel N (2006) Comparative genome mapping among *Picea glauca, P. abies* and *P. mariana* × *rubens*, and correspondance with other Pinaceae. Theor Appl Genet 113:1371–1393

Pena L, Séguin A (2001) Recent advances in the genetic transformation of trees. Trends Biotechnol (NY) 19:500–506

Perron M, Bousquet J (1997) Natural hybridization between *Picea mariana* and *P. rubens*. Mol Ecol 6:725–734

Perron M, Perry DJ, Andalo C, Bousquet J (2000) Evidence from sequence-tagged-site markers of a recent progenitor-derivative species pair in conifers. Proc Natl Acad Sci USA 97:11331–11336

Perry DJ, Bousquet J (1998a) Sequence-tagged-site (STS) markers of arbitrary genes: development, characterization and analysis of linkage in black spruce. Genetics 149:1089–1098

Perry DJ, Bousquet J (1998b) Sequence tagged-site (STS) markers of arbitrary genes: the utility of black spruce-derived STS primers in other conifers. Theor Appl Genet 97:735–743

Perry DJ, Bousquet J (2001) Genetic diversity and mating system of post-fire and post-harvest black spruce: an investigation using codominant sequence-tagged site (STS) markers. Can J For Res 31:32–40

Perry DJ, Isabel N, Bousquet J (1999) Sequence-tagged-site (STS) markers of arbitrary genes: the amount and nature of variation revealed in Norway spruce. Heredity 83:239–248

Peterson DG (2005) Reduced representation strategies and their application to plant genomes. In: Meksem K, Kahl G (eds) The Handbook of Plant Genome Mapping: Genetic and Physical Mapping. Wiley-VCH, Weinheim, Germany, pp 307–335

Plomion C, Hurme P, Frigerio JM, Ridolfi M, Pot D, Pionneau C, Avila C, Gallardo F, David H, Neutelings G, Campbell M, Canovas FM, Savolainen O, Bodénès C, Kremer A (1999) Developing SSCP markers in two *Pinus* species. Mol Breed 5:21–31

Pot D, McMillan L, Echt C, Le Provost G, Garnier-Géré P,Cato S, Plomion C (2005) Nucleotide variation in genes involved in wood formation in two pine species. New Phytol 167:101–112

Presby-Germano J (2003) Nucleotide and cytoplasmic genetic variation in *Picea:* DNA markers for evaluating past migration, introgression and evolutionary history. PhD Dissertation, University of New Hampshire, Durham, NH

Rajora OP, Mosseler A, Major JE (2000) Indicators of population viability in red spruce, *Picea rubens*. II. Genetic diversity, population structure, and mating behavior. Can J Bot 78:941–956

Rajora OP, Rahman MH, Dayanandan S, Mosseler A (2001) Isolation, characterisation, inheritance and linkage of microsatellite DNA markers in white spruce (*Picea glauca*) and their usefulness in other spruce species. Mol Gen Genet 264:871–882

Ralph SG, Yueh H, Friedmann M, Aeschliman D, Zeznik JA, Nelson CC, Butterfield YSN, Kirkpatrick R, Liu J, Jones SJM, Marra MA, Douglas CJ, Ritland K, Bohlmann J (2006) Conifer defense against insects: Microarray gene expression profiling of Sitka spruce (*Picea sitchensis*) induced by mechanical wounding or feeding by spruce budworms (*Choristoneura occidentalis*) or white pine weevils (*Pissodes strobi*) reveals large-scale changes of the host transcriptome. Plant, Cell and Environment 29:1545–1570.

Roche L (1969) A genecological study of the genus *Picea* in British Columbia. New Phytol 68:505–554

Rungis D, Berube Y, Zhang J, Ralph S, Ritland CE, Ellis BE, Douglas C, Bohlmann J, Ritland K (2004) Robust simple sequence repeat markers for spruce (*Picea* spp.) from expressed sequence tags. Theor Appl Genet 109:1283–1294

Rungis D, Hamberger B, Berube Y, Wilkin J, Bohlmann J, Ritland K (2005) Efficient genetic mapping of single nucleotide polymorphisms based upon DNA mismatch digestion. Mol Breed 16:261–270

Rweyongeza DM, Yeh FC, Dhir NK (2004) Genetic parameters for seasonal height and height growth curves of white spruce seedlings and their implications to early selection. For Ecol Mng 187:159–172

Saranga Y, Menz M, Jiang CX, Wright RJ, Yakir D, Paterson AH (2001) Genomic dissection of genotype × environment interactions conferring adaptation of cotton to arid conditions. Genome Res 11:1988–1995

Schlötterer C (2004) The evolution of molecular markers-just a matter of fashion? Nat Rev Genet 5:63–69

Schmidt-Vogt H (1977) Die Fichte. Band I [The spruces. Vol 1]. Taxonomie-Verbreitung-Morphologie-Ökologie-Waldgesellscahften. Parey, Hamburg

Scotti I, Magni F, Fink R, Powell W, Binelli G, Hedley PE (2000) Microsatellite repeats are not randomly distributed within Norway spruce (*Picea abies* K.) expressed sequences. Genome 43:41–46

Scotti I, Magni F, Paglia G, Morgante M (2002) Trinucleotide microsatellites in Norway spruce (*Picea abies*): their features and the developement of molecular markers. Theor Appl Genet 106:40–50

Scotti I, Burelli A, Cattonaro F, Chagné F, Fuller J, Hedley PE, Jansson G, Lalanne C, Madur D, Neale D, Plomion C, Powell W, Troggio M, Morgante M (2005) Analysis of the distribution of marker classes in a genetic linkage map: a case study in Norway spruce (*Picea abies* karst). Tree Genetics and Genomics 1:93–102

Seybold SJ, Bohlmann J, Raffa KF (2000) The biosynthesis of coniferophagous bark beetle pheromones and conifer isoprenoids: evolutionary perspective and synthesis. Can Entomol 132:697–753

Shea KL (1987) Effects of population structure and cone production on outcrossing rates in Engelmann spruce and subalpine fir. Evolution 41:124–136

Sigurgeirsson A, Szmidt AE (1993) Phylogenetic and biogeographic implications of chloroplast DNA variation in *Picea*. Nord J Bot 13:233–246

Sperisen C, Buchler U, Gugerli F, Matyas G, Geburek T, Vendramin GG (2001) Tandem repeats in plant mitochondrial genomes: application to the analysis of population differentiation in the conifer Norway spruce. Mol Ecol 10:257–263

Stasolla C, Kong L, Yeung EC, Thorpe TA (2002) Maturation of somatic embryos in conifers: morphogenesis, physiology, biochemistry, and molecular biology. In Vitro Cell Dev Biol Plant 38:93–105

Stam P (1993) Construction of integrated genetic linkage maps by means of a new computer package: JoinMap. Plant J 5:739–744

Strauss SH, Difazio SP, Meilan R (2001) Genetically modified poplars in context. For Chron 77:272–279

Sutton BCS, Flanagan DJ, Gawley JR, Newton CH, Lester DT, El-Kassaby YA (1991) Inheritance of chloroplast and mitochondrial DNA in *Picea* and composition of hybrids from introgression zones. Theor Appl Genet 82:242–248

Sutton BCS, Attree SM, El-Kassaby YA, Grossnickle SC, Polonenko DR (2004) Commercialisation of somatic embryogenesis for plantation forestry. In: Walter C, Carson M (eds) Plantation Forest Biotechnology for the 21st Century, Trivandrum, India

Tani N, Takahashi T, Iwata H, Mukai Y, Ujino-Ihara T, Matsumoto A, Yoshimura K, Yoshimaru H, Murai M, Nagasaka K, Tsumura Y (2003) A consensus linkage map for sugi (*Cryptomeria japonica*) from two pedigrees, based on microsatellites and expressed sequence tags. Genetics 165:1551–1568

Teoh SB, Rees H (1976) Nuclear DNA amounts in populations of *Picea* and *Pinus* species. Heredity 36:123–137

Thornsberry JM, Goodman MM, Doebley J, Kresovich S, Nielsen D, Buckler ES (2001) Dwarf8 polymorphisms associate with variation in flowering time. Nat Genet 28:286–289

Tosh KJ, Fullarton MS (2001) Tree improvement progress by the New Brunswick department of Natural Resources and Energy. In: Simpson JD (ed) Proc 27th Meeting of the CTIA. Natural Resources of Canada, Sault-Ste. Marie, Ontario, Canada

Trapp S, Croteau R (2001) Defensive resin biosynthesis in conifers. Annu Rev Plant Physiol Plant Mol Biol 52:689–724

Tulsieram LK, Glaubitz JC, Kiss G, Carlson JE (1992) Single tree genetic linkage mapping in conifers using haploid DNA from megagametophytes. Biotechnology (NY) 10:686–90

Ujino-Ihara T, Kanamori H, Yamane H, Taguchi Y, Namiki N, Mukai Y, Yoshimura K, Tsumura Y (2005) Comparative analysis of expressed sequence tags of conifers and angiosperms reveals sequences specifically conserved in conifers. Plant Mol Biol 59:895–907

Ukrainetz NK, Ritland K, Mansfield SD (2007) An AFLP linkage map for Douglas-fir based upon small multiple full-sib families. Tree Genetics and Genomics (in press)

van de Ven WTG, Kanamori RJ (1996) Microsatellites as DNA markers in Sitka spruce. Theor Appl Genet 93:613–617

van Frankenhuyzen K, Beardmore T (2004) Current status and environmental impacts of transgenic forest trees. Can J For Res 34:1163–1180

Vendramin GG, Anzidei M, Madaghiele A, Sperisen C, Bucci G (2000) Chloroplast microsatellite analysis reveals the presence of population subdivision in Norway spruce (*Picea abies* K.). Genome 43:68–78

Viereck LA, Little EL Jr (1972) Alaska trees and shrubs. Agric Handbook 410. USDA Forest Service, Washington, DC

Vos P, Hogers R, Bleeker M, Reijans M, van de Lee T, Hornes M, Friters A, Pot J, Paleman J, Kuiper M, Zabeau M (1995) AFLP: a new technique for DNA fingerprinting. Nucleic Acids Res 23:4407–4414

Weng C, Jackson ST (2000) Species differentiation of North American spruce (*Picea*) based on morphological and anatomical characteristics of needles. Can J Bot 86:1742–1753

Williams JG, Kubelik AR, Livak KJ, Rafalski JA, Tingey SV (1990) DNA polymorphisms amplified by arbitrary

primers are useful as genetic markers. Nucleic Acids Res 18:6531–6535
Wright W (1955) Species crossability in spruce in relation to distribution and taxonomy. For Sci 1:319–349
Xie CY, Knowles P (1994) Mating system and effective pollen immigration in a Norway spruce (*Picea abies* (L.) Karst) plantation. Silvae Genet 43:48–52
Yanchuk A (2001) The role and implications of biotechnology in forestry. For Genet Resour 30:18–22
Yeh FC, Arnott JT (1986) Electrophoretic and morphological differentiation of *Picea sitchensis*, *Picea glauca*, and their hybrids. Can J For Res 16:791–798
Yeh FC, El-Kassaby YA (1980) Enzyme variation in natural populations of Sitka spruce (*Picea sitchensis*): 1. Genetic variation patterns among trees from 10 IUFRO provenances. Can J For Res 10:415–422
Yeh FC, Khalil MAK, El-Kassaby YA, Trust DC (1986) Allozyme variation in *Picea mariana* from Newfoundland: genetic diversity, population structure, and analysis of differentiation. Can J For Res 16:713–720
Ying CC (1991) Genetic resistance to the white pine weevil in Sitka spruce. Research Note 106, British Columbia Ministry of Forests, Victoria, BC, Canada
Zhang T, Vision T, Gaut B (2002) Patterns of nucleotide substitution among simultaneously duplicated gene pairs in *Arabidopsis thaliana*. Mol Biol Evol 19:1464–1473
Zobel BJ, Talbert JT (1984) Applied Forest Tree Improvement. Wiley, New York

4 Eucalypts

Alexander A. Myburg[1], Brad M. Potts[2], Cristina M. Marques[3], Matias Kirst[4], Jean-Marc Gion[5], Dario Grattapaglia[6], and Jacqueline Grima-Pettenatti[7]

[1] Department of Genetics, Forestry and Agricultural Biotechnology Institute (FABI), University of Pretoria, Pretoria 0002, South Africa, *e-mail:* zander.myburg@fabi.up.ac.za
[2] School of Plant Science, and CRC for Forestry, University of Tasmania, Private Bag 55, Hobart, Tasmania 7001, Australia
[3] RAIZ-Direcçao de Investigacao Florestal ITQB II Av. Republica, Apartado 127, 2781-901 Oeiras, Portugal
[4] School of Forest Resources and Conservation, 363 Newins-Ziegler Hall, University of Florida, Gainesville, FL 32611-0410, USA
[5] Programme Arbres et Plantations, CIRAD Forêt TA 10/C, Campus de Baillarguet 34398, Montpellier Cedex 5, France
[6] EMBRAPA Genetic Resources and Biotechnology SAIN Parque Rural CP 0272 and Graduate Program in Genomic Sciences and Biotechnology, Universidade Catolica de Brasília-SGAN, 916 modulo B, 70790-160 DF Brasilia, Brazil
[7] UMR UPS/CNRS 5546, Pôle de Biotechnologies Végétales, 24 chemin de Borde Rouge, BP42617, Auzeville, Tolosane 31326 Castanet Tolosan, France

4.1 Introduction

4.1.1 History of the Crop

Origin and Evolution

Eucalyptus tree species, commonly referred to as eucalypts, are among the most planted hardwoods in the world (Doughty 2000). They are generally long-lived, evergreen species belonging to the predominantly southern-hemisphere, angiosperm family Myrtaceae (Ladiges et al. 2003). They are native to Australia and islands to its north (Potts and Pederick 2000; Ladiges et al. 2003), where they occur naturally from sea level to the alpine tree line, from high rainfall to semiarid zones, and from the tropics to latitudes as high as 43° south (Williams and Woinarski 1997). Eucalypts are the dominant or codominant species of virtually all vegetation types in Australia except rainforest, the vegetation of the central arid zone, and higher montane regions (Wiltshire 2004). They are generally sclerophyllous and adapted to low nutrient soils (Eldridge et al. 1993; Florence 1996; Specht 1996) and fire (Pryor 1976; Ashton 2000; Burrows 2002).

The eucalypt lineage is old, possibly extending back to the Late Cretaceous – ca. 70 million years ago (Hill et al. 1999; Ladiges et al. 2003; Crisp et al. 2004). Their ancestors were likely to have been widely dispersed on the supercontinent of Gondwana, as there are macrofossils ascribed to eucalypts of Eocene (55 to 34 Mya) age from northeastern Australia (Rozefelds 1996) and possibly Patagonia (Hill et al. 1999) and of Miocene (27 to 10 Mya) age from New Zealand (Pole et al. 1993) and Australia (Hill et al. 1999). The tectonic isolation of Australia (ca. 32 Mya) led to cooler, drier, and more seasonal climates and consequently a transition from a rainforest-dominated flora to Australia's unique sclerophyll flora (Hill et al. 1999; Ladiges et al. 2003; Crisp et al. 2004; Hill 2004). There is little doubt that the current dominance of the Australian continent by eucalypts is relatively recent and linked with the onset of severe aridity during the Late Miocene (10 to 7 Mya) and the present climatic system of extreme wet-dry glacial cycles that commenced around 2.9 Mya (Crisp et al. 2004). The increasing prevalence of fire played a significant role in the transformation of the Australian biota over this drying period (Kershaw et al. 1994), with eucalypts believed to have expanded from drier, central regions of the continent into more coastal environments climatically suitable for fire-sensitive rainforest taxa (Hill et al. 1999). The arrival of Aborigines on the Australian continent at least 55,000 years ago and the instigation of "fire-stick farming" would have continued this shift (Kershaw et al. 1994; Bowman 1998).

The latest molecular dating (Crisp et al. 2004) argues that divergence of the eucalypt genera and subgenera predated the final development of an ocean between Australia and Antarctica ca. 32 Mya. While molecular dating is contentious (Ladiges and Udovicic 2005), it is suggested that diversification proceeded steadily for at least 30 millions years before Australia was isolated and continued thereafter. However, de-

Genome Mapping and Molecular Breeding in Plants, Volume 7
Forest Trees
C. Kole (Ed.)

spite over 100 eucalypt species having been sequenced for ITS (Steane et al. 2002), there is still insufficient sampling to test for more recent rapid radiation (Crisp et al. 2004).

Domestication of the Eucalypts

Following their discovery by Europeans in the late 18th century, eucalypts were spread rapidly around the world (Zacharin 1978; Jacobs 1981; Eldridge et al. 1993; Doughty 2000). They were introduced early on into countries such as India (c. 1790), France (c. 1804), Chile (1823), Brazil (1825), South Africa (1828), and Portugal (1829) (Doughty 2000; Potts et al. 2004) and rapidly spread as the fast growth and good adaptability of eucalypt species were recognized. Initially the botanical gardens of southern Europe played a major role in their introduction to other parts of the world, including Africa and South America (Zacharin 1978). Later in the 19th century, large quantities of seed were distributed directly from Australia. In other cases, seed was collected from local exotic plantings and, where multispecies plantings occurred, this often contained F_1 hybrids (Potts and Dungey 2004). While these F_1 hybrids may have performed well, subsequent seed collection from them often resulted in plantations that performed poorly in subsequent generations and were extremely variable, e.g., the Río Claro hybrid in Brazil (Campinhos and Ikemori 1977; Brune and Zobel 1981) and the Mysore hybrid in India (Varghese et al. 2000). In many countries where eucalypts have been introduced for a long time and continually reproduced from local seed sources, they have formed landraces adapted to the specific environment of the country (Eldridge et al. 1993).

The history of eucalypt breeding is detailed in Eldridge et al. (1993) and Potts (2004). Some of the earliest breeding was undertaken by French foresters in Morocco in 1954-1955 (Eldridge et al. 1993). The rise of industrial eucalypt plantations in the 1960s saw a more formal approach to genetic improvement with, for example, the commencement of the Florida *E. grandis* breeding program in 1961 (Franklin 1986), *E. globulus* breeding in Portugal in 1965-1966 (Dillner et al. 1971; Potts et al. 2004), and establishment of large provenance tests of *E. camaldulensis* in many countries (Eldridge et al. 1993). Major advances occurred in the 1970s with, for example, the first commercial plantings of selected clones derived from hardwood cuttings in the Congo (Martin and Quillet 1974) followed by Aracruz in Brazil and the establishment in many countries of the first large base population trials of species such as *E. urophylla* (Eldridge et al. 1993) and *E. globulus* (Potts et al. 2004). These trials were established from open-pollinated seed lots collected from rangewide provenance collections and formed the base for deployment and breeding populations in many countries (Eldridge et al. 1993). Many other major international base population trials were established through the 1980s for species such as *E. grandis*, *E. tereticornis*, and *E. viminalis* and more intensive collections of elite provenances identified in earlier collections (Eldridge et al. 1993).

Eucalypts are still at the early stages of domestication compared to crop species, with most breeding programs only one or two generations removed from the wild. However, eucalypts are fast becoming among the most advanced genetic material in forestry, with stock originating from the *E. grandis* program in Florida already in its sixth generation (Potts 2004). Domestication of eucalypts has proceeded faster in countries like Brazil and South Africa that rely on plantations for their eucalypt wood than in Australia, where up until the 1990s wood products of eucalypts were derived almost entirely from native forests. As with other forest tree genera with long generation times, there is the potential for eucalypt domestication programs to benefit tremendously from genomic-era molecular technologies that could significantly speed the process of genetic improvement.

4.1.2 Botany

Taxonomy

In the broad sense, eucalypts encompass species of the genera *Eucalyptus* L'Hérit., *Corymbia* (Hill and Johnson 1995), and *Angophora* Cav. (Ladiges 1997, Table 1). A key feature of the majority of eucalypts is the fusion of either the petals and/or sepals to form an operculum from which the eucalypts derive their name (from the Greek *eu*, "well," and *calyptos*, "covered," Eldridge et al. 1993; Ladiges 1997). The operculum appears to have evolved independently in different eucalypt lineages and has not evolved in *Angophora* (Ladiges 1997). There is some debate as to whether the *Corymbia* and *Angophora* genera (bloodwood taxa) warrant separation from the genus *Eucalyptus* (non-bloodwood taxa) in the strict sense. This split is supported by several independent molecular studies (e.g., Sale et al. 1996; Ladiges and Udovicic 2000, Udovicic and Ladiges 2000; Steane et al. 2002; Whittock et al.

2003) and, following Ladiges et al. (2003), adopted herein. The latest taxonomic revision (Brooker 2000) of the eucalypts recognizes just over 700 species that belong to 13 main evolutionary lineages (Table 1) but still treats the bloodwood eucalypts as subgenera of *Eucalyptus*. Most species belong to the subgenus *Symphyomyrtus*, and it is mainly species from three sections of this subgenus that are used in plantation forestry (Table 1).

Eucalypts encompass an exceptional level of genetic diversity. They range in habit from shrubs and multistemmed mallees to giant trees and include the tallest flowering plants on earth (*Eucalyptus regnans* – 96 m; Boland et al. 1985; Wardell-Johnson et al. 1997; Potts and Pederick 2000; Potts et al. 2003). The major subgenera exhibit different ecological and reproductive characteristics (Pryor 1976; Florence 1996; Ladiges 1997), and closely related species are usually ecologically differentiated (Florence 1996; Williams and Woinarski 1997). Within species, marked genetic differentiation between populations is the norm rather than the exception (Pryor and Johnson 1971, 1981; Potts and Wiltshire 1997; Dutkowski and Potts 1999). Genetic variation between populations in quantitative traits is often continuous and clinal, paralleling environmental gradients associated with changes in, for example, latitude, continentality, or altitude (Pryor 1976; Potts and Wiltshire 1997). Indeed, many recognized species intergrade along such gradients, resulting in complexes of closely related species where no clear morphological discontinuity is apparent (Pryor 1976; Jordan et al. 1993; Holman et al. 2003).

Breeding System

Eucalypt flowers are occasionally solitary (e.g., *E. globulus*) but often occur in clusters of three or

Table 1. Major evolutionary lineages within the eucalypts. The alignment of Pryor and Johnson's (1971) genera and subgenera with Brooker's (2000) subgenera. Pryor and Johnson's classification was informal, but widely used for 30 years. The number of species in each of Brooker's subgenera is indicated and examples of well-known forestry species are given. Most species used in plantation forestry, particularly outside Australia, are from Brooker's sections *Maidenaria* (e.g., *E. globulus, E. nitens, E. viminalis*), *Exsertaria* (e.g., *E. camaldulensis, E. tereticornis*), and *Latoangulatae* (e.g., *E. grandis, E. saligna, E. urophylla*) in the subgenus *Symphyomyrtus* (from Potts 2004)

Pryor & Johnson's subgenera/genera	Brooker's subgenera	No. of species	Examples of well-known forestry species
Angophora (genus)	*Angophora*[a]	7	
Blakella	*Blakella*[a]	15	
Corymbia	*Corymbia*[a]	67	*C. torelliana, C. citridora, C. variegata C. maculata*
Eudesmia	*Eudesmia*	19	
Gaubaea	*Acerosa*	1	
Gaubaea	*Cuboidea*	1	
Idiogenes	*Idiogenes*	1	*E. cloeziana*
Monocalyptus	*Primitiva*	1	
Monocalyptus	*Eucalyptus*	110	*E. regnans, E. delegatensis, E. obliqua, E. marginata, E. fastigata*
Symphyomyrtus	*Cruciformes*	1	*E. guilfoylei*
Symphyomyrtus	*Alveolata*	1	*E. microcorys*
Symphyomyrtus	*Symphyomyrtus*	474	*E. camaldulensis, E. exserta, E. globulus, E. grandis, E. nitens, E. paniculata, E. robusta, E. saligna, E. tereticornis, E. urophylla, E. viminalis*
Telocalyptus	*Minutifructus*[b]	4	*E. deglupta*

[a] The subgenera *Blakella* and *Corymbia* had previously been treated as a separate genus *Corymbia* (Hill and Johnson 1995) and the subgenus *Angophora* treated as a genus (Hill and Johnson 1995), and this approach has been adopted in the text

[b] A recent molecular study suggests that these species belong to subgenus *Symphyomyrtus* (Whittock 2003)

more in umbels or terminal inflorescences. The eucalypt flower is normally bisexual, with numerous stamens that expand outward after operculum shed to form the conspicuous floral display (Pryor 1976). Eucalypts are predominantly animal pollinated, with vectors encompassing a wide variety of insects, birds, marsupials, and a few bat species (House 1997; Hingston et al. 2004; Southerton et al. 2004; Barbour et al. 2005). They have a mixed mating system but are generally preferential outcrossers, with high levels of outcrossing maintained by protandry and various incomplete pre- and postzygotic barriers to self-fertilization (Potts and Wiltshire 1997; Pound et al. 2002a, b, 2003). The postzygotic barriers include intense selection against the products of inbreeding (Hardner and Potts 1995, 1997; Potts and Wiltshire 1997). Outcrossing rates may vary between (Butcher and Williams 2002; McDonald et al. 2003) and within trees even of the same species. For example, outcrossing is often higher in denser stands (Hardner et al. 1996) and at the top of the tree canopy (Patterson et al. 2004b; Hingston and Potts 2005). Biparental inbreeding may also occur in open-pollinated progenies from native stands due to related individuals growing in close spatial proximity (Eldridge et al. 1993; Hardner et al. 1998). The seed of most species has no specialized mechanisms for dispersal and is normally deposited within a distance of twice the tree height (Potts and Wiltshire 1997).

The major eucalypt subgenera do not hybridize (Griffin et al. 1988; c.f. Stokoe et al. 2001), and, while there are significant barriers to hybridization between species within subgenera, these are often weak (Potts et al. 2003; Potts and Dungey 2004). Natural hybridization and intergradation between recognized taxa is common in nature (Griffin et al. 1988; Potts and Wiltshire 1997; Byrne and Macdonald 2000; Butcher et al. 2002), often making delineation of species difficult (Pryor and Johnson 1971). Natural introgression may be cryptic, and only detectable at the molecular level (McKinnon et al. 2001a, 2004), indicating ancient hybridization and gene flow between lineages. At the other extreme, first-generation hybridization is actively occurring between exotic plantation eucalypts and native populations (Barbour et al. 2002, 2003, 2005). Artificial hybrid combinations have been produced, and in general hybrid inviability tends to increase with increasing taxonomic distance between the parents, but there are exceptions (Griffin et al. 1988; Potts and Dungey 2004).

Genome Size and Structure

The size and structure of the genome of eucalypts is reviewed in Poke et al. (2005). Consistent with most myrtaceous genera, eucalypts are diploids with a haploid chromosome number of 11 (Eldridge et al. 1993). Reports of higher numbers of chromosomes have not been verified, e.g., *E. cladocalyx* is $2n = 22$ (R. Wiltshire unpublished data) and not $2n = 24$ (Eldridge et al. 1993). While polypoidy has been artificially induced (Janki-Ammal and Khosla 1969), there are no reports of polyploidy in nature. Grattapaglia and Bradshaw's (1994) estimates of the haploid genome size of several eucalypt species and hybrids based on flow cytometry range from 370 to 700 million base pairs (Mbp). They estimated the average haploid genome size for *Symphyomyrtus* species to be 650 Mbp (Grattapaglia and Bradshaw 1994). While no *Angophora* species were included in their study, the two *Corymbia* species examined had a haploid genome size (370 and 390 Mbp) substantially smaller than the *Eucalyptus* species studied. Hybrids had intermediate DNA content, and there was no evidence of polyploidy. Estimates will vary with laboratory technique, and a new estimate for the size of the *E. globulus* genome, for example, is 644 Mbp (Pinto 2005), which is 20% larger than Grattapaglia and Bradshaw's (1994) estimate of 530 Mbp. Nevertheless, the haploid genome size of *Eucalyptus* species would generally appear to be slightly larger than those of the current plants whose genomes have been completely sequenced (125 Mbp for *Arabidopsis thaliana*, 420 to 466 Mbp for two rice varieties, Fukuoka et al. 1998, and 473 Mbp for *Populus tricarpa*, Poke et al. 2005), and is clearly many-fold smaller than the genome size estimated for *Pinus* species, which have large regions of repetitive DNA (Ahuja 2001). There were no studies of the structure of the eucalypt genome before the advent of accessible DNA marker technology (see below).

As in most angiosperms, the inheritance of both the mitochondrial (Vaillancourt et al. 2004) and chloroplast (Byrne et al. 1993; McKinnon et al. 2001b) genomes appears to be fully maternal. While the structure of the mitochondrial genome has not been studied, the *E. globulus* chloroplast has been completely sequenced and shows very high homology of coding regions, with chloroplast sequences from species such as *Nicotiana tobacum* and *Oenothera elata* but considerable divergence in the noncoding regions (Steane 2005). Broadly transferable microsatellite regions have been identified from the *E. globulus* chloroplast sequence (Steaneet al. 2005). A hypervariable region, J_{la}, also occurs in the intergenic spacer on either side of the junction between the large single-copy region and the inverted repeat A in *Symphyomyrtus* but appears absent from *Monocalyptus* species (Vaillancourt and Jackson 2000).

Molecular Diversity

Historically, allozymes were the main molecular markers used for genetic diversity studies in eucalypts and revealed their mixed mating system, high levels of genetic diversity and heterozygosity, and greater population differentiation in regionally distributed species than in widespread or local species (reviewed in Moran 1992; Potts and Wiltshire 1997). However, over the last decade these markers have been replaced in diversity studies by more informative DNA marker technologies such as random amplified polymorphic DNA (RAPD), restriction fragment length polymorphism (RFLP), amplified fragment length polymorphism (AFLP), and microsatellite analysis (Shepherd and Jones 2005), and more recently, by the analysis of direct sequence variation in functional genes (Poke et al. 2003; McKinnon et al. 2005; Thumma et al. 2005). These markers have revealed that eucalypt gene pools have a hierarchy of spatial genetic structure from clonal patches (e.g., Rossetto et al. 1999; Smith et al. 2003; Jones et al. 2005) and family groups (Skabo et al. 1998) that extend over tens of meters, to localized population differention over hundereds of meters (McGowen et al. 2001), and broad-scale genetic differentiation over many hundreds of kilometers (Byrne et al. 1998; Butcher et al. 2002; Jones et al. 2002; Holman et al. 2003; Astorga et al. 2004; Potts et al. 2004). As expected, maternally inherited, haploid chloroplast markers, which are dispersed only through the seed, exhibit greater population differentiation than nuclear markers (Byrne and Macdonald 2000; Jones et al. 2005).

4.1.3 Economic Importance

Eucalypts are renowned as species with fast growth, straight form, valuable wood properties, wide adaptability to soils and climates, and ease of management through coppicing (Eldridge et al. 1993; Potts 2004). They are now found in more than 90 countries where the various species are grown for products as diverse as sawn timber, mine props, poles, firewood, pulp, charcoal, essential oils, honey, and tannin, as well as for shade, shelter, and soil reclamation (Doughty 2000). They are an important source of fuel and building material in rural communities in countries such as India, China, Ethiopia, Peru, and Vietnam. However, it is the great global demand for short-fiber pulp that has driven the massive expansion of eucalypt plantations throughout the world during the 20th century (Turnbull 1999). Their high fiber count relative to other wood components, coupled with the uniformity of fibers relative to other angiosperm species, has caused high demand for eucalypt pulp for coated and uncoated free-sheet paper, bleach board, sanitary products (fluff pulp), and, secondarily, for top liner on cardboard boxes, as a corrugating medium, and as a filler in long-fiber conifer products such as newsprint and containerboard (Kellison 2001). New technologies are also increasing interest in the use of plantation eucalypts for sawn wood, veneer, and medium-density fiberboard and as extenders in plastic and molded timber (Kellison 2001; Bermúdez Alvite et al. 2002; Shield 2004; Waugh 2004).

While precise global figures are difficult to obtain, a conservative estimate is 9.5 million ha of industrial eucalypt plantations in 1999, with the vast majority of these established since the 1950s, and predicted to reach 11.6 million ha in 2010 (Raga 2001). The majority of plantations consist of only a few eucalypt species and hybrids. The most important are *E. grandis*, *E. globulus*, and *E. camaldulensis*, which together with their hybrids account for about 80% of the plantation area; they are followed by *E. nitens*, *E. saligna*, *E. deglupta*, *E. urophylla*, *E. pilularis*, *Corymbia citriodora*, and *E. teriticornis* (Eldridge et al. 1993; Waugh 2004). Market favorites for pulpwood are *E. grandis and E. urophylla* and their hybrids in tropical and subtropical regions and *E. globulus* in temperate regions. There is considerable interest in introgressing superior *E. globulus* pulp traits into the tropical and subtropical genetic backgrounds.

4.1.4 Classical Breeding Objectives

The main breeding programs in eucalypts worldwide are focused on improving profit from industrial pulpwood plantations (Borralho 2001; Kanowski and Borralho 2004; Raymond and Apiolaza 2004). In such systems, the key objective traits traditionally considered to drive profits are volume production per hectare, wood density, and pulp yield (Borralho et al. 1993; Greaves and Borralho 1996; Greaves et al. 1997; Wei and Borralho 1999), although their economic weights will vary depending upon whether plantations are part of a vertically integrated production system (Borralho et al. 1993) or only for wood-chip export (Apiolaza et al. 2005). Traits such as pest and disease resistance (e.g., Coutinho et al. 1998; Soria and Borralho 1998; Milgate et al. 2005), adaptability (e.g., frost resistance; Cauvin and Potts 1991; Tibbits et al. 1991), drought resistance (Toroet al. 1998), and survival (Chambers and Borralho 1997) are only important as far as they impact on one or more of these objective traits. Secondary wood property traits of interest to pulp producers include the quantity or quality of extractives or lignin in the wood that affect the economic and/or environmental cost of pulping (Poke et al. 2004; Raymond and Apiolaza 2004). Breeding programs directly linked with paper mills are also exploring other wood properties that impact on paper quality such as fiber length, courseness, and wall dimensions (Raymond and Apiolaza 2004), as well as studying genetic effects on paper quality (Cotterill et al. 1999).

There is increasing interest in breeding eucalypts for sawn timber, veneers, and reconstituted wood products worldwide (Kube and Raymond 2005; Raymond 2002; Raymond and Apiolaza 2004). Key wood properties believed to affect sawn timber and composites are given in Table 2 (see also Shield 2004). There is increasing work on identifying breeding objectives, economic weights, and cost-effective selection traits for these production systems (Greaves et al. 2004a, b). However, this is complicated by the range of products, silvicultural regimes and production systems, changing technologies over the longer rotations, and requirements of plantations for multiple products through thinning or changes in product pricing (Shield 2004).

While wood and cellulose production is the key focus of most breeding programs, there are small eucalypt breeding programs aiming to improve the yield of alternative products such as leaf volatile oils (Boland et al. 1991; Doran and Bell 1995; Byrne 1999) and for enhanced ecosystem services by selection of salt-resistant genotypes (Meddings et al. 2001; Dale 2002). Decreasing plant propagation costs by, for example, selection for vegetative propagability (de Assis 2001; Cañas et al. 2004) or increased seed production (McGowen et al. 2004) also occurs as a secondary objective in many breeding or deployment programs (Raymond and Apiolaza 2004).

4.1.5 Classical Breeding Achievements

As with most forest trees, large and cost-effective genetic gains have been achieved in the early stages of eucalypt domestication, simply through species (Jacobs 1981) and provenance selection (Eldridge et al. 1993), followed by individual (family) selection and establishment of clonal or seedling seed orchards or clonal propagation of elite selections for direct deployment (Eldridge et al. 1993; Kanowski and Borralho 2004; Potts 2004). Subsequent population improvement has also demonstrated significant genetic gain through recurrent selection in an open-pollinated breeding population (Reddy and Rockwood 1989), or

Table 2. Key wood properties for a range of product classes (modified from Raymond 2002)

Pulp and paper	Sawn timber	Composites
Basic density	Basic density and gradient	Basic density
Pulp yield/cellulose content	Microfibril angle	Lignin content
Fiber length	Strength and stiffness	Extractives content
Lignin content and composition	Dimensional stability	Cellulose content
	Shrinkage and collapse	
	Tension wood	
	Knot size	
	Incidence of decay, spiral grain, and end splits	

sublines (Griffin 2001; Sanhueza and Griffin 2001), possibly coupled with open- or controlled-pollinated nucleus populations of the most elite selections or specialized breeds (Potts 2004). For species that are easily propagated vegetatively, such as *E. grandis*, clonally propagated breeding populations have further enhanced gains (Snedden and Verryn 2004). Overlapping generation breeding using a "rolling front" strategy has also been adopted by programs in Australia and Portugal (Borralho and Dutkowski 1998; McRae et al. 2004b). Major advances were made in the 1990s, with the addition of wood density and pulp yield with growth into the breeding objectives as the two traits that account for over 70% of the benefits from breeding for pulp production (Borralho 2001). With such an objective, two generations of selection of *E. globulus* for a vertically integrated eucalypt pulp production system were expected to increase income by 1.5% and decrease production costs by 16%, saving US$ 7.2 million per annum for a 250,000-ton pulp mill (Kanowski and Borralho 2004).

Eucalypt hybrids, either F_1s or composites, have long been deployed in eucalypt forest through vegetative propagation and are a significant component of eucalypt plantation forestry, particularly in the tropics and subtropics (de Assis 2000; Vigneron and Bouvet 2000; Potts and Dungey 2004). Such hybrids have been classically selected based on large-scale hybrid production and clonal testing, through either (1) initial screening of seedlings for vegetative propagability or (2) remobilizing mature selections through coppicing, sequential grafting, or in vitro techniques and then introducing them into clonal tests. The main hybrids used in industrial plantations are *E. grandis* × *urophylla*, *E. grandis* × *camaldulensis* and varieties including at least one of *E. saligna*, *E. pellita*, *E. exserta*, and *E. tereticornis*. Such hybrids are planted on a relatively large scale in Brazil and the Congo, although sizeable plantations also occur in China, Indonesia, and South Africa. The deployment of selected clones of *E. urophylla* × *grandis* in Brazil and the Congo has been a major success in overcoming canker and disease susceptibity of *E. grandis* (Campinhos and Ikemori 1989; Vigneron and Bouvet 2000). Indeed in Brazil, a combination of improvements in genetics, silviculture, and propagation has increased the mean annual growth of eucalypt plantations from $10\,m^3\,ha^{-1}\,yr^{-1}$ in the 1960s to more than $40\,m^3\,ha^{-1}\,yr^{-1}$ at present (Binkley and Stape 2004).

4.1.6 Future Perspective: Challenges for Molecular Breeding of *Eucalyptus*

The next decade will see major advances in our understanding of the eucalypt genome and molecular breeding (Grattapaglia 2004; Poke et al. 2005; Thumma et al. 2005). The genome of a eucalypt tree (that of an *E. camaldulensis* clone) is being sequenced at Kazusa DNA Research Institute in Japan (T. Hibino and S. Tabata personal communication; Myburg 2004), and the *Eucalyptus* Genome Network (EUCAGEN, www.eucagen.org) has been initiated to coordinate the generation of further genomic resources for *Eucalyptus*. Furthermore, the *Eucalyptus* research community has been invited to submit a proposal for the sequencing of the *E. grandis* genome by the US Department of Energy (DOE), a proposal that will, if successful, lead to the public release of the *E. grandis* genome sequence (6–8 coverage) before the end of 2008 (J. Tuskan personal communication). These milestones are bound to lead to rapid advances in the development of molecular breeding tools and molecular genetic improvement of eucalypts. The development of genetically modified eucalypts has been slow compared with, for example, *Populus* species (Potts et al. 2003). The development of fully tested genetically modified clones to the stage of large-scale planting is a slow process (Griffin 1996), and regulatory and certification requirements associated with the use of genetically modified trees are likely to limit their operational use in the foreseeable future (Burley and Kanowski 2005). Nevertheless, other molecular technologies offer great potential to contribute to eucalypt breeding through many avenues including quantifying genetic diversity and relationships, breeding systems, gene flow and fingerprinting for quantifying contamination and clone identification, QTL detection, and molecular breeding through marker- or gene-assisted selection. Major advances are still to be made through better definition of breeding and deployment objectives and economic weights (Raymond and Apiolaza 2004), enhanced accuracy of genetic evaluation through more sophisticated trial design (Williams et al. 2002) and analysis (Dutkowski et al. 2002; Costa e Silva et al. 2005), and exploitation of the major advances in quantitative genetic analysis that have occurred over the last decade (e.g., Soria et al. 1998; Fernandez and Toro 2001; Costa e Silva et al. 2004; de Resende and Thompson 2004; McRae et al. 2004a). Genetic gain is a function of selection intensity, and there are now clear opportunities for quantum increases in the size of pedigreed breeding populations through the recent advances that have occurred in controlled pollination (Harbard et al. 1999; Williams et al. 1999; Patterson et al. 2004a; de Assis et al. 2005) and vegetative propagation techniques (e.g., mini- and microcuttings; de Assis 2001), increasingly cheaper techniques for nondestructive assessment of key wood properties (e.g., near-infrared reflectance and raman spectroscopy; Schimleck et al. 1996; Downes et al. 1997; Raymond and Apiolaza 2004), and industrial-scale systems for data management and genetic evaluation (McRae et al. 2004a). The challenge facing molecular breeding is its integration into existing eucalypt breeding programs in order to deliver gains to industrial forests above that which can be achieved by equivalent investment in other already well-integrated technologies and strategies.

This chapter provides an overview of the status of genome mapping and molecular breeding of *Eucalyptus* tree species. Emphasis is placed on genetic and physical mapping of eucalypt genomes and on new approaches to locating genes and regulatory regions that control quantitative traits in *Eucalyptus* plantations. The last section of the chapter provides a brief future perspective on genome research in *Eucalyptus* and its impact on the molecular domestication of this important fiber crop. Readers are also referred to other recent reviews of genome research and molecular breeding in *Eucalyptus* (Moran et al. 2002; Potts 2004; Shepherd and Jones 2005; Grattapaglia 2004; Poke et al. 2005).

4.2 Genetic Linkage Mapping of Eucalypt Genomes

Genetic markers are DNA phenotypes that reflect differentiation among individuals, populations, and species. The availability of large numbers of highly polymorphic and neutral genetic markers whose inheritance and segregation can be followed through generations has allowed the construction of genetic linkage maps for many plant species, including several eucalypt species. Most genetic mapping studies in plants have relied on the availability of inbred lines, near-isogenic lines, or backcross progeny of inbred parents. However, available pedigrees for the majority of outbred angiosperm tree species, such as the eucalypts, generally involve only two parents and their full-sib progeny or one parent and its maternal half-sib progeny. Such outbred mapping pedigrees suffer from several limitations (discussed below) that complicate genetic linkage mapping. Novel mapping strategies, such as the "two-way pseudotestcross design" discussed later in this section, have been developed to overcome these limitations, allowing the generation of single-tree genetic linkage maps for selected individuals of tree species. The high level of genetic diversity, the ability to generate large progeny sets, the ability to clone segregating progeny, the relatively small genome size, and the low proportion of repetitive DNA of *Eucalyptus* facilitated early interest in genetic linkage mapping in this genus (Grattapaglia and Bradshaw 1994). Most eucalypt linkage maps have been constructed from segregating half-sib or full-sib families, with family sizes of up to 200 individuals. To maximize the detection of genetic polymorphism at the DNA level, highly heterozygous parents have been selected, most often from different species. Wide interspecific crosses in *Eucalyptus* have revealed genetic barriers to crossing (Myburg et al. 2004) and have resulted in distorted segregation ratios in mapping progeny, which can lead to biased estimates of recombination and false linkage (Lorieux et al. 1995). The development of codominant markers for *Eucalyptus* (Brondani et al. 1998, 2002) stimulated the first comparative mapping studies, which may lead to more efficient use of genetic mapping information for molecular breeding and evolutionary studies in *Eucalyptus*. In this section, we briefly review practical considerations for the construction of genetic linkage maps in *Eucalyptus* and we discuss potential applications of genetic mapping in molecular breeding of these tree species.

4.2.1 DNA Isolation for Genetic Mapping

The isolation of relatively pure genomic DNA from a large number of individuals is the first step in any genetic mapping project. *Eucalyptus* total genomic DNA has been isolated from leaf tissue mostly using modified versions of the published protocols of Doyle and Doyle (1990), Saghai-Maroof et al. (1984), or Wagner et al. (1987). The presence of secondary metabolic compounds and carbohydrates in eucalypt leaves can interfere with DNA quality, especially if leaf samples have to spend a considerable time in transit to the laboratory. Tissue condition before extraction is therefore usually a critical factor (Ferreira and Grattapaglia 1995). Myburg et al. (2001) proposed a high-throughput 96-well DNA extraction protocol based on the QIAGEN 96-well mouse tail DNA extraction kit, modified for plant DNA extraction chemistry, as well as tissue homogenization using the FastPrep instrument (QBiogene). The convenience and throughput of this method is further improved by collecting *Eucalyptus* leaf discs directly into 2-ml tubes that already contain a small amount of a desiccant such as silica gel crystals. Such samples can be stored at room temperature until DNA isolation and are stable for long periods of time.

4.2.2 Marker Availability

DNA-based marker techniques vary in DNA requirements, cost of development and assay, technical expertise, genetic information, and transferability across taxa. Genetic marker analysis and mapping in eucalypts have progressed at the pace of the development of new marker technologies. Isozyme markers were first used in 1983 to study mating systems in natural and exotic populations of *Eucalyptus* (Moran and Bell 1983). These markers were, however, limited in genome coverage and in the amount of genetic polymorphism that could be assayed. Restriction fragment length polymorphism (RFLP) markers were used first in phylogenetic analyses of eucalypt chloroplast DNA (Steane et al. 1991) and later to construct *Eucalyptus* genetic linkage maps (Byrne et al. 1995; Thamarus et al. 2002). RFLP analysis has not found widespread use in *Eucalyptus* as it is technically demanding, labor intensive, and time consuming and requires the previous development and labeling of probes.

The advent of the polymerase chain reaction (PCR) (Mullis and Faloona 1987) allowed the development of many new markers based on DNA polymorphisms. Random amplified polymorphic DNA (RAPD) markers (Williams et al. 1990) greatly facilitated the rapid construction of genetic linkage maps for several eucalypt species (Grattapaglia and Sederoff 1994; Vaillancourt et al. 1994; Verhaegen and Plomion 1996; Bundock et al. 2000; Gan et al. 2003). As in other plant species, the main limiting feature of RAPD markers in *Eucalyptus* is their dominant behavior, relatively low multiplex ratio, and low repeatability across research laboratories (Jones et al. 1997). Amplified fragment length polymorphism (AFLP) analysis (Voset al. 1995) has proven to be more robust, and the technique samples much larger numbers of loci in high-resolution sequencing gels. This has allowed the construction of high-density, AFLP-based genetic maps of individual *Eucalyptus* trees using only small number of oligonucleotide primers and minute amounts of DNA (Marques et al. 1998). Multiplexed AFLP genotyping using fluorescently labeled primers further increased the throughput and convenience of AFLP analysis in *Eucalyptus* (Myburg et al. 2003). As with RAPD markers, the main disadvantage of AFLP markers has been their dominant behavior, but also the fact that there are commercial restrictions on the use of the technology (Keygene, Wageningen, The Netherlands). Furthermore, AFLP markers are generally not shared among different outbred eucalypt parents, which has made the integration of different AFLP maps difficult, except where shared parents are used in mapping pedigrees (Myburg et al. 2004). The development of transportable, PCR-based and codominant simple sequence repeat (SSR), or microsatellite markers for *Eucalyptus* (Byrne et al. 1996; Brondani et al. 1998; Glaubitz et al. 2001; Steane et al. 2001; Brondani et al. 2002) revolutionized the genetic mapping of eucalypt trees. SSR markers have been added to previously constructed genetic linkage maps (Brondani et al. 1998, 2002; Marques et al. 2002) and mapped in combination with other markers (Bundock et al. 2000; Thamarus et al. 2002). Limitations of the use of SSRs for mapping include the effort required to develop markers and the fact that only one locus is sampled with each primer pair.

More recently, gene-based molecular markers have been integrated into existing maps (Bundock et al. 2000; Gion et al. 2000; Thamarus et al. 2002). These are no doubt the "ultimate" markers for genetic linkage mapping, as potential candidate loci may be identified empirically from known genes that collocate with quantitative trait loci (QTLs). However, the multiplex ratio of these markers remains low, as most genes are mapped one at a time. In contrast, microarray technology holds great potential for genotyping large numbers of anonymous markers in *Eucalyptus* (Lezar et al. 2004) and may form the basis of high-density genetic mapping in the future. The use of short oligonucleotide probes (25 mer) on microarrays may in the future allow direct mapping of genes through the detection of single nucleotide polymorphisms (SNPs) in genes (Kirst 2004).

4.2.3 Map Construction

Relative to most crop plant species, eucalypts are still essentially undomesticated and outbred, and as such highly heterozygous, plant species. Linkage analysis is somewhat complicated by the varying numbers of marker alleles (up to four) that may be present at each locus. Moreover, linkage phases of markers are generally unknown. Despite these difficulties, many pollen and seed parent genetic linkage maps have been constructed in *Eucalyptus* in the last decade (Tables 3 and 4) using informative genetic markers that segregate in appropriate pedigrees. Grattapaglia and Sederoff (1994) implemented the "two-way pseudotestcross model" to construct individual-tree genetic linkage maps of the two parents of an interspecific full-sib cross of *E. grandis* and *E. urophylla*. This model allowed the use of dominant markers and inbred line mapping approaches in this oubred pedigree and resulted in three types of segregating markers: (1) 1:1 testcross markers inherited from the pollen parent, (2) 1:1 testcross markers inherited from the seed parent, and (3) 3:1 intercross markers inherited from both parents. Dominant intercross markers (such as AFLPs) and codominant markers (such as SSRs) have been used to establish homology and investigate synteny between parental maps (Verhaegen and Plomion 1996; Brondani et al. 1998, 2002; Marques et al. 1998, 2002; Bundock et al. 2000; Myburg et al. 2003). Higher-resolution comparative mapping was achieved through a "double pseudobackcross" mapping strategy in F_2 interspecific backcross families of *Eucalyptus* (Myburg et al. 2003). This approach allowed the use of shared testcross and intercross markers to align the maps of the F_1 hybrid with those of the two backcross parents.

Genetic linkage mapping is based on recombination rates that may differ between male and female meiosis. Myburg et al. (2003) did not observe differences in whole-genome AFLP recombination rates of seed and pollen parents in an *E. grandis* × *E. globulus* hybrid pseudobackcross with *E. grandis* as the male and *E. globulus* as the female parent. Thamarus et al. (2002) also did not find significant differences in the female and male recombination rates among consecutive pairs of fully informative loci. This has also been verified with SSR markers by Brondani et al. (2002).

Maps constructed with outbred progenies and a reasonable number of fully informative

Table 3. *Eucalyptus* genetic linkage maps constructed with dominant markers

Species	Population	Number of markers	Number of linkage groups	Observed map length (cM)	Reference
E. grandis	62 F_1	240 RAPD	14	1552	Grattapaglia and Sederoff 1994
E. urophylla	62 F_1	251 RAPD	11	1101	Grattapaglia and Sederoff 1994
E. gunnii × *E. globulus*	72 F_2 and 10 BC	75 RAPD	9	255	Vaillancourt et al. 1994
E. urophylla	93 F_1	269 RAPD	11	1331	Verhaegen and Plomion 1996
E. grandis	93 F_1	236 RAPD	11	1415	Verhaegen and Plomion 1996
E. globulus	91 F_1	268 AFLP	16	967	Marques et al. 1998
E. tereticornis	91 F_1	285 AFLP	14	919	Marques et al. 1998
E. urophylla	82 F_1	245 RAPD	23	1505	Gan et al. 2003
E. tereticornis	82 F_1	264 RAPD	23	1036	Gan et al. 2003
E. grandis	156 BC	438 AFLP	11 (comparative)	1335	Myburg et al. 2003
E. globulus	177 BC	367 AFLP	11 (comparative)	1405	Myburg et al. 2003
E. grandis × *E. globulus* F_1 paternal	156 grandis BC	518 AFLP	11 (comparative)	1448	Myburg et al. 2003
E. grandis × *E. globulus* F_1 maternal	177 globulus BC	577 AFLP	11 (comparative)	1318	Myburg et al. 2003

codominant markers (such as RFLP or SSR) have been integrated in species maps for *E. nitens* (Byrne at al. 1995) and *E. globulus* (Thamarus et al. 2002). In most cases, linkage analysis was performed using MAPMAKER (Lander et al. 1987), GMENDEL (Liu and Knapp 1990), or JoinMap (Stam 1993). Often, subsets of markers were selected for their intensity, ease of scoring, reduced missing data, size, and spacing in order to construct framework maps with increased marker order reliability.

4.2.4 Physical Genome Size vs. Genetic Map Size

As discussed in the previous section, the first estimates of nuclear DNA content for *Eucalyptus* (Grattapaglia and Bradshaw 1994) ranged from 0.77 pg/2C for *E. citriodora* to 1.47 pg/2C for *E. saligna*, corresponding to a haploid genome size range of 370 to 700 Mbp. At the time it was not clear how physical map size (in bp) would relate to genetic map sizes (in cM), but several mapping experiments have since shed light

Table 4. *Eucalyptus* genetic linkage maps constructed with codominant markers

Species	Population	Number of markers	Number of linkage groups	Observed map length	Reference
E. nitens	4 grandparents, 2 parents, 118 F_2 outbred progeny	210 RFLP, 125 RAPD, 4 isozyme	12 (integrated)	1462 cM	Byrne et al. 1995
E. grandis	94 F_1	19 SSR	9	Added to Grattapaglia and Sederoff 1994	Brondani et al. 1998
E. urophylla	94 F_1	17 SSR	6	Added to Grattapaglia and Sederoff 1994	Brondani et al. 1998
E. urophylla	201 F_1	8 genes	5	Added to Verhaegen and Plomion 1996	Gion et al. 1996
E. grandis	201 F_1	4 genes	3	Added to Verhaegen and Plomion 1996	Gion et al. 1996
E. globulus (KI2)	165 F_1	153 RAPD, 16 SSR	11	701 cM	Bundock et al. 2000
E. globulus (G164)	94, 71, 165 F_1	173 RAPD, 21 SSR, GPI-2	13	1013 cM	Bundock et al. 2000
E. globulus	73 F_1	34 SSR	8	Added to Marques et al. 1998	Marques et al. 2002
E. tereticornis	73 F_1	34 SSR	8	Added to Marques et al. 1998	Marques et al. 2002
E. grandis	92 F_1	63 SSR	11	Added to Grattapaglia and Sederoff 1994 maps	Brondani et al. 2002
E. urophylla	92 F_1	53 SSR	10	Added to Grattapaglia and Sederoff 1994	Brondani et al. 2002
E. globulus	148 outbred F_1	204 RFLP, (31 EST and 14 genes) 40 SSR, 5 isozyme	12 (integrated)	1375 cM	Thamarus et al. 2002

on this question. Observed map lengths for *E. grandis* average 1,434 cM (Grattapaglia and Sederoff 1994; Verhaegen and Plomion 1996; Myburg et al. 2003), 1,312 cM for *E. urophylla* (Grattapaglia and Sederoff 1994; Verhaegen and Plomion 1996; Gan et al. 2003), 1,092 cM for *E. globulus* (Marques et al. 1998; Bundock et al. 2000; Thamarus et al. 2002; Myburg et al. 2003), 978 cM for *E. tereticornis* (Marques et al. 1998; Gan et al. 2003) and 1,462 cM for *E. nitens* (Byrne et al. 1995) (Tables 3 and 4). Differences among reported map lengths for the same species might result from differences in the number of framework markers and the stringency level adopted for assigning locus order and grouping. Based on Grattapaglia and Bradshaw's (1994) genome estimates of 640 Mbp/1C for *E. grandis*, 650 Mbp/1C for *E. urophylla*, 530 Mbp/1C for *E. globulus*, and 580 Mbp/1C for *E. tereticornis*, 1 cM map distance would be equivalent to 446 kbp in *E. grandis*, 495 kbp in *E. urophylla*, 485 kbp in *E. globulus*, and 593 kbp in *E. tereticornis*. In *Arabidopsis*, 1 cM is equivalent on average to 230 kbp, only about half the genetic/physical ratio for eucalypts (Verhaegen and Plomion 1996). Myburg et al. (2003) suggested that the 20% genome size difference between *E. grandis* and *E. globulus* could be the result of many small and dispersed chromosomal duplications or deletions, reflecting a pattern of dispersed genome expansion in *Eucalyptus*.

4.2.5 Segregation Distortion

The distorted segregation of markers in genetic maps is generally indicative of the segregation of genomic incompatibilities or genetic effects that result in differential fitness of the gametes or progeny through which marker alleles are transmitted. In general, mapping studies in intraspecific crosses of *Eucalyptus* species (Byrne et al. 1995; Thamarus et al. 2002) have reported lower proportions of distorted markers (6% and 0.8%, respectively) than studies in interspecific crosses. However, Grattapaglia and Sederoff (1994) did not find significant numbers of markers with distorted segregation in interspecific F_1 progeny of *E. grandis* and *E. urophylla*. Bundock et al. (2000) reported significant segregation distortion in intraspecific maps of an *E. globulus* female parent, but no distortion in the *E. globulus* male parent. Verhaegen and Plomion (1996) found that 8% of RAPD markers exhibited segregation distortion in an interspecific cross of *E. urophylla* and *E. grandis*. They also observed the clustering of distorted loci, as previously reported in *E. nitens* (Byrne et al. 1995), and subsequently in *E. globulus* (Bundock et al. 2000), *E. urophylla*, and *E. tereticornis* (Gan et al. 2003). Such clustering of loci may indicate a biological basis for the observed distortion (reviewed in Bundock et al. 2000). Marques et al. (1998) reported that 15% of AFLP bands displayed skewed segregation ratios in a cross of *E. tereticornis* and *E. grandis*. Myburg et al. (2004) observed an average of 28% transmission ratio distortion of AFLP markers in the parental maps of a wide interspecific backcross pedigree of *E. grandis* and *E. globulus*. The distorted markers were located in distinct regions of the parental maps. Relatively high levels of distortion were observed in the backcross parents, which was unexpected and suggested that there may be genetic variability within *E. grandis* and *E. globulus* for genetic factors that affect hybrid fitness (Myburg et al. 2003). No overall suppression of recombination was detected in the F_1 hybrid, relative to the *E. grandis* and *E. globulus* backcross parents, suggesting that the observed segregation distortion was not related to genomic incompatibilities that may reduce recombination between *E. grandis* and *E. globulus* homologs. Gan et al. (2003) also reported a high proportion of distorted loci (33%) for *E. tereticornis* in an interspecific cross with *E. urophylla*, which was higher than what had been identified in previous studies (Marques et al. 1998). It is evident from these studies that there is great potential to use high-resolution genome mapping to elucidate the genetic basis of intraspecific and interspecific crossing barriers in *Eucalyptus*. The increased use of codominant markers in genetic mapping studies will further allow the differentiation of gametic and zygotic mechanisms of distortion (Myburg et al. 2003).

4.2.6 Comparative Mapping

Comparative mapping studies rely on a set of common transferable markers that segregate in multiple species or pedigrees of interest. Comparative genetic mapping is a powerful approach to study genetic differentiation, genomic structure, genome evolution, and reproductive isolation in diverging species. In *Eucalyptus*, the verification of synteny and collinearity will add enormous value to the identification and isolation of genes and the validation of gene and QTL positions in multiple pedigrees. One of the immediate objectives would be to align current genetic linkage maps of species in the subgenus *Symphyomyrtus* and particularly those in the sections *Maidenaria*, *Exsertaria*, and *Latoangulatae*, which contain most of the commercially planted eucalypt species (Table 1).

Verhaegen and Plomion (1996) used 25 RAPD intercross markers to identify homeology between LGs (LGs) in *E. urophylla* and *E. grandis*, both members of the section *Latoangulatae*. Despite the limitations of RAPD markers regarding locus specificity, the authors successfully used some of the RAPD markers that had been previously mapped by Grattapaglia and Sederoff (1994) to identify four homologous LGs in *E. urophylla* and three in *E. grandis*. Marques et al. (1998) used 19 AFLP intercross markers to suggest homeology between LGs in *E. globulus* (section *Maidenaria*) and *E. tereticornis* (section *Exsertaria*). Brondani et al. (1998) developed 20 SSR markers for *Eucalyptus* and verified synteny in six *E. grandis* and *E. urophylla* LGs. With the same set of SSR, homeology was extended between these species and two *E. globulus* maps (Bundock et al. 2000) and, later, an *E. tereticornis* and another *E. globulus* map (Marques et al. 2002). Marques et al. (2002) reported synteny of SSR loci and QTLs for vegetative propagation in four eucalypt species. Brondani et al. (2002) contributed an additional 50 SSR markers for comparative mapping in *Eucalyptus*. These markers were used to establish collinearity and synteny among 10 LGs of *E. grandis* and *E. urophylla*. In their double-pseudotestcross mapping pedigree, Myburg et al. (2003) were able to align maps of all 11 LGs of *E. grandis*, *E. globulus*, and their F_1 hybrid, and the authors reported very high collinearity between the genomes of these two species, which represent one of the widest crosses among commercially planted eucalypt species.

Although it is possible that smaller genome rearrangements exist, all of the mapping studies above signify high genome collinearity within the subgenus *Symphyomyrtus*. The morphological similarity and genetic compatibility in sexual crosses among eucalypts from the same subgenus may also suggest high levels of DNA sequence conservation. This is supported by results of SSR primer transferability studies (Byrne et al. 1996; Brondani et al. 2002; Marques et al. 2002), which suggest that prospects for transferring microsatellite mapping information across eucalypt species are excellent. The prospects seem good for obtaining a subgenuswide reference map with SSR and gene-based markers in the foreseeable future.

4.2.7
Future Perspective: Integration and Application of Genetic Linkage Maps

Integrated genetic linkage maps (defined below) will provide a framework for the application of molecular breeding, as well as for addressing questions of population and conservation genetics of eucalypt species. The construction of integrated linkage maps in outbred eucalypt pedigrees relies on the use of codominant markers that are informative (polymorphic) in both parental maps of the pedigree, allowing complete integration of the two parental maps. Ideally, such codominant markers will also be informative in other related and unrelated crosses, allowing integration, or at least alignment of multiple genetic linkage maps. If successful, *Eucalyptus* researchers will in the future be able to use genetic information from multiple mapping pedigrees to guide population (association) genetic studies and, ultimately, perform molecular breeding of eucalypt species and hybrids. The construction of a genuswide reference map for *Eucalyptus*, or a series of connected linkage maps that represent the genus, is now both a commercial and an academic research target. A reasonable number (>400) of codominant markers such as microsatellites, gene and EST markers are currently available, allowing the alignment of existing linkage maps in the three commercially important sections of the subgenus *Symphyomyrtus*. The *Genolyptus* project in Brazil is developing and mapping more than 1,200 new SSR markers identified from different sources, including a large EST database and a growing number of BAC-end sequences (Grattapaglia et al. 2004). Some of these markers will allow the anchoring of genomic clone (BAC) contigs to existing genetic linkage maps, supporting the construction of localized physical maps focused on genomic regions in which a significant concentration of QTLs for important traits have been detected. The identification of genes in these regions will assist functional genomics studies, providing insights on the mechanisms of gene expression and regulation. Besides helping to characterize QTLs, the integration of known genes into linkage maps will also be useful for gross comparative mapping of *Eucalyptus* with model plant species (e.g., *Arabidopsis*) and different tree genera such as *Populus* and *Pinus*. Information exchange and comparison among laboratories and independent experiments will speed candidate gene mapping, QTL verification, and the implementation of marker-assisted selection (MAS) in eucalypt breeding.

4.3 QTL Mapping in *Eucalyptus*

4.3.1 Historical Perspective on QTL Analysis

The foundation for quantitative trait locus (QTL) analysis was established almost a century ago. Thomas H. Morgan (1910) discovered the principles of linkage mapping and demonstrated that genes were linked and could be placed into groups that were equal in number to that of the haploid chromosomes in *Drosophila*. The genetic distance between genes was later defined by Haldane (1919) based on recombination frequencies. Finally, the connection between segregating genetic elements and quantitative traits was proposed by Karl Sax in 1923, who observed that "inherent size differences are apparently dependent on Mendelizing factors," but that in "most cases many factors are involved and simple ratios are not obtained." More sophisticated approaches were later introduced that used the genotypic information from two adjacent loci to more accurately estimate the effect and position of QTLs in marker intervals (Paterson et al. 1988; Lander and Botstein 1989) and control for the effect of other markers associated with the trait (Jansen and Stam 1994; Zeng 1994). Another limitation of the early days of QTL analysis was the lack of broad marker coverage of mapped genomes. Isozymes and restriction fragment length polymorphisms, or RFLPs (Botstein et al. 1980), partially resolved this limitation. High-density maps were later developed using several PCR-based methods (Mullis and Faloona 1987; Williams et al. 1990; Welsh and McClelland 1990; Vos et al. 1995) and, more recently, high-throughout single nucleotide polymorphism (SNP) genotyping methods (Chen and Sullivan 2003). This section provides a brief overview of the application of QTL mapping in *Eucalyptus*, a summary of the traits that have been targeted for QTL mapping, and a discussion of new integrative genomics approaches for the molecular dissection of quantitative traits in *Eucalyptus*.

4.3.2 QTL Mapping in *Eucalyptus*, Limitations and Advantages

Success in QTL identification relies on crossing two individuals with alternative alleles at loci affecting a trait of interest and following the segregation of the parental genomes in the progeny. Multiple genetic loci should be genotyped, with ample genome coverage, so that all of the genomic regions that may affect the trait are evaluated. The first genomewide QTL analysis carried out in plants identified few QTLs with large effect (Paterson et al. 1988, 1991; Stuber et al. 1992), suggesting that loci linked to traits of interest could be rapidly incorporated into traditional breeding programs, through MAS. The early success in crop plants was, however, not expected to be repeated in forest tree species, and the predictions were that QTL identification was going to be limited (Strauss et al. 1992).

The primary limitation of QTL analysis in forest tree genera such as *Eucalyptus* is related to the types of crosses that are the basis for QTL identification methods in agricultural crops (e.g., backcrosses and F_2intercrosses). Traditional backcrosses and F_2s rely on the creation of inbred lines, which cannot be readily generated for forest tree species due to high genetic load and inbreeding depression. Consequently, a typical full-sib tree family may have up to four alleles segregating at any given locus, often in unknown linkage phase. In conifers, the problem can be partially overcome by the analysis of maternally inherited haploid tissue from the seed megagametophyte (Adams and Joly 1980). This alternative is not possible in *Eucalyptus* and other woody angiosperms. Nonetheless, novel mapping designs, such as the two-way pseudotestcross strategy (Grattapaglia et al. 1994), have allowed genetic mapping and QTL analysis to be pursued in highly heterozygous forest tree species using models and software developed for inbred lines.

Another limitation was associated with the long-lived nature of forest tree species. Perennial plants undergo significant morphological changes throughout development, and different sets of genes may contribute to phenotypic variation from the juvenile to mature phase. Developmental changes in wood quality and growth are very significant in conifers (Zobel and Sprague 1998). These changes are also observed in most *Eucalyptus* species, although they are less pronounced. Changes in the set of genes that control a quantitative trait may also be due to environmental variation, as perennial plants have to adapt to changing conditions throughout multiple growing seasons. Seasonal and year-to-year variation in sources of biotic and abiotic stress imply that different physiological mechanisms may need to be activated for plant survival. Some genetic loci may be vital for survival and have significant pleiotropic effects on growth in one year while being essentially "unnecessary" the next, when the stress is no longer existent. Changes in the major genetic loci regulating growth and wood density were documented by Verhaegen et al. (1997), over a period of 18 months, in an F_1 cross of *E. grandis* and *E. urophylla*. None of the QTLs detected at one time (18, 24, and 36 months of age) could be repeated throughout the entire experiment, suggesting that different loci are contributing to phenotypic variation during different stages of development. Weng et al. (2002) showed in *Pinus* that the variance explained by major QTLs decreases over time, suggesting the increased complexity of quantitative traits with aging of the tree. The contribution of different sets of genes to quantitative variation during development may lead to lower power of detecting QTLs, as a phenotype measured at rotation age essentially represents the cumulative effect of many distinct genes.

QTL mapping trials in forest trees generally sample more environmental variation per site than equivalent experiments in crop plants that can often be grown in high density, or sometimes even in greenhouses. Where replicated trails are possible (clonal propagation of mapping progeny) QTLs may be detected in one site while being absent in other sites. Numerous studies in crop plants have identified common QTLs in several environments and QTLs that are specific to certain sites, suggesting genotype by environment interactions (Paterson et al. 1991; Teulat et al. 2001; Hittalmani et al. 2003; Leon et al. 2003). Studies that evaluate the conservation of QTL in different environments have not been published in forest tree species, to our knowledge. A partnership of industry and governmental research institutions in Brazil, the Genolyptus project, has initiated the planting of several clonally propagated populations in a broad range of locations and environments. This study will shed light on the stability of QTLs in different environments in *Eucalyptus* (D. Grattapaglia personal communication). Because QTL experiments in forest tree species typically require large field plantations that are carried out in field sites that are highly heterogeneous, it is likely that confounding effect from the environment (spatial variation) can diminish the power of QTL detection. Environmental variation can be accounted for by analysing spatial variation in field sites (e.g., Dutkowski et al. 2002) and by assessing phenotypes in multiple sites.

These and other limitations suggested early on that QTL identification of traits of interest for commercial forestry would be largely unsuccessful and possibly would be limited to the analysis of traits of chemical or morphological nature (Strauss et al. 1992). However, early studies, carried out in *Eucalyptus*, *Populus*, and *Pinus*, showed that QTLs for growth and wood-quality traits could be readily identified (Groover et al. 1994; Bradshaw and Stettler 1995; Grattapaglia et al. 1995). Nevertheless, QTL studies carried out in *Eucalyptus* and other forest tree species have typically identified fewer QTLs and explained a smaller proportion of the phenotypic variation, compared to agronomic crops. QTL scans for a variety of traits in crop species have been able to identify an average of four major QTLs jointly explaining ca. 46% of the phenotypic variance (Kearsey and Farquhar 1998). The limited power to detect QTLs in forest tree mapping experiments relative to agricultural species may be due to high environmental variation in tree plantations, smaller populations tested, and developmental variation.

Part of the success in the identification of QTLs is forest tree species is likely due to the wide genetic variation in the material that is commonly used in tree breeding programs. Also, the essentially undomesticated character of forest tree species and large population sizes suggest that the majority of commercially important alleles have not been lost or fixed yet by selection. Although there have been major reductions in the gene pool during seed collections destined for breeding programs in other parts of the world, the gene pool remains essentially intact in natural populations in Australia and adjacent islands. The large population sizes and wide pollen dispersion that are characteristic of outcrossing, undomesticated tree species ensure that new alleles are maintained in the population and not lost by genetic drift. Therefore, QTL studies in eucalypt species typically detect QTLs simply because there is large variation. Efforts based on interspecific hybrids of eucalypts, in particular F_2 interspecific progeny (e.g., Myburg 2001), have the added advantage of exploring variation in alleles that underlie differentiation among species, frequently because of the unique characteristics and environmental adaptation of eucalypt species.

4.3.3 QTLs Identified in *Eucalyptus*

QTL mapping in forest trees has been applied to the identification of genetic loci associated with variation in biomass productivity (height, diameter, volume), stem form, wood properties (wood density and composition, fiber traits, bark composition), vegetative propagation, biotic/abiotic stress response, development, foliar chemistry, and inbreeding depression. QTL analysis of transcript levels, measured by cDNA microarray analysis of thousands of genes, has also been reported recently (Kirst et al. 2005). A brief description of the most relevant QTL studies carried out in *Eucalyptus* follows.

Biomass Productivity

Traits associated with biomass productivity (growth) have been the most studied by QTL analysis in *Eucalyptus* (Grattapaglia et al. 1996; Byrne et al. 1997a; Verhaegen et al. 1997; Thamarus et al. 2004; Kirst et al. 2004), and tree species in general, because of their commercial importance and ease of phenotyping. A direct comparison among studies is complicated by the fact that they were generally carried out using different pedigrees, markers, cross designs (e.g.,half-sib, full-sib F_1, and backcrosses), species, biomass productivity estimators (i.e., height, diameter, volume, and growth rate), ages and QTL detection methods (e.g., single-marker vs. composite interval mapping). Grattapaglia and colleagues (1996) first identified QTLs for circumference at breast height in a 6.5-year-old half-sib population of *E. grandis*. A set of 300 individuals was genotyped with 77 RAPD markers, and 3 QTLs explaining 13.7% of the phenotypic variation were identified. Byrne et al. (1997a) identified three QTLs for seedling height (explaining between 10.3 and 14.7% of phenotypic variation) in two three-generation pedigrees of *E. nitens*. A first indication of the level of conservation of QTLs during the lifetime of *Eucalyptus* was established by Verhaegen et al. (1997) in the analysis of an F_1 hybrid of *E. grandis* and *E. urophylla*. A progeny set of 200 F_1 individuals was measured at 18, 26, and 38 months for "vigor" (a combination of height, circumference at breast height, and volume), and QTLs were identified for each age. None of the QTLs were detected consistently throughout the entire experiment, but the majority (68%) were detected in two consecutive ages. The phenotypic variation explained ranged from 5 to 14%. More recently, three QTLs for height and diameter growth were identified in a pseudobackcross of *E. grandis* and *E. globulus* (Kirst et al. 2004). The study was part of an effort to integrate gene expression data and QTL analysis to identify genes controlling quantitative variation (described below).

Wood Quality

Significant effort has been dedicated to the phenotyping and mapping of QTLs for wood physical and chemical composition traits in *Eucalyptus*. The first wood-quality-trait QTLs were identified by Grattapaglia and colleagues (1996) in a half-sib population of *E. grandis*. Five QTLs that controlled almost half of the genetic variation for wood-specific gravity were identified. These initial studies were limited by the technology for phenotyping wood physical and chemical property traits, which carried high cost and labor requirements. Recently, novel methods for wood-quality phenotyping have been developed and applied in forestry. These include near-infrared spectrometry (Schimleck et al. 1996), SilviScan (x-ray densitometry combined with automated scanning x-ray diffraction and image analysis), mass spectrometry (Evans and Ilic 2001), computer tomography X-ray densitometry (CT scan), and pyrolysis molecular beam mass spectrometry (pyMBMS; Tuskan et al. 1999). Some methods are based on predictions based on measurements of associated traits and can be less precise compared to direct measurement methods. Lack of precision can, however, be compensated in part by the size of the populations that can be surveyed by these methods.

Myburg (2001) demonstrated the application of indirect, high-throughput phenotyping of *Eucalyptus* wood-quality traits by NIR for QTL mapping in a pseudobackcross of *E. grandis* and *E. globulus*. Approximately 300 individuals that had been previously genotyped with AFLP markers were analyzed by NIR, and predictions were made for pulp yield, alkali consumption, basic density, fiber length and coarseness, and several wood chemical properties (lignin, cellulose, and extractives). Basic density was also measured directly. A comparison between QTLs identified by direct and indirect measurements yielded a few common and several distinct QTLs between methods. More recently, Thamarus and colleagues (2004) used novel high-throughput and traditional methods to quantify wood density, fiber length, pulp yield, and microfiber angle (MFA) in two full-sib families of *E. globulus* that shared a common parent. Pulp yield and cellulose content were determined by NIR, and MFA was quantified by SilviScan. QTLs for all traits could be detected in both populations (with the exception of fiber length), including three QTLs in common genetic regions on both crosses for wood density, one for pulp yield, and one for microfibril angle. The proportion of phenotypic variation explained by the QTLs identified in both crosses ranged from 3.2 to 15.8%.

Abiotic Stress

Eucalyptus species have adapted to a broad range of environmental conditions. *Eucalyptus* species are naturally found in tropical and temperate regions, in high altitude and coastal plains, and in the central Australian desert. The different species exhibit considerable variation for different sources of abiotic stress.

Eucalyptus species display considerable variation for salt tolerance, and QTLs for this trait were identified by Dale et al. (2000). Six F_1 hybrid populations from three salt-tolerant *E. camaldulensis* and two highly productive, salt-sensitive *E. grandis* genotypes were created and established in field plantations. In addition, five ramets from each of 192 genotypes from one of the populations were established in a greenhouse, and growth and leaf chloride content were phenotyped after a 2-week treatment. QTL analysis of the mapping population identified three QTLs in a RAPD-derived genetic map using multiple-interval mapping. Three QTLs were contributed by the salt-tolerant parent (*E. camaldulensis*) and three by the less tolerant parent (*E. grandis*). Individual QTL effects ranged from 3 to 5% relative to the population mean. The other populations were planted in field trials, but no significant correlation was found between the level of chloride in leaves in the greenhouse population and the same family grown in the field.

Frost tolerance has been another major target for QTL identification (Byrne et al. 1997b; Fullard and Moran 2003), mostly for the purpose of introgressing frost-tolerance alleles from species adapted to temperate regions and high altitude into the fast-growing, commercial species from topical and subtropical areas. Byrne and colleagues (1997b) identified two QTLs explaining 8 and 11% of the variation for frost tolerance in a cross of *E. nitens* involving parents selected from regions of high and low frost incidence. Fullard and Moran (2003) identified QTLs for frost tolerance in a cross involving *E. globulus* (tolerant) and a frost-susceptible hybrid of *E. urophylla* × *E. grandis*.

Biotic Stress

Inheritance of disease resistance in forest species is in many cases explained by Mendelian factors. Major genes that control tolerance to pathogens in forest tree species were primarily identified in loblolly and sugar pines (Devey et al. 1995; Wilcox et al. 1996; Harkins et al. 1998) and in poplar hybrids (Newcombe et al. 1996; Stirling et al. 2001). Examples of quantitative inheritance have also been identified (Newcombe and Bradshaw 1996; Kubisiak et al. 1997), but the proportion of the phenotypic variation explained by the QTL is typically much larger than for traits such as growth and wood quality.

Very few studies have attempted to identify QTLs for disease resistance in *Eucalyptus* (Shepherd et al. 1995; Freeman et al. 2003; Junghans et al. 2003). However, very compelling evidence for the major gene model observed in pines and poplar has also been demonstrated in *Eucalyptus* (Junghans et al. 2003). Analysis of the progeny from ten full-sib families of *E. grandis* by Junghans and colleagues (2003) suggested the presence of a major locus for rust resistance. Bulked segregant analysis (BSA) of one large full-sib population identified a RAPD marker that cosegregates with the resistance locus and may be useful for MAS or introgression of the resistance locus into other genetic backgrounds.

Vegetative Propagation

The ease of propagating certain *Eucalyptus* species/genotypes by rooted cuttings is a major advantage for commercial clonal forestry. With vegetative propagation, the genetic components that contribute to a superior genotype, including dominance and epistatic effects, are captured and potentially multiplied into field plantations. There is substantial variation in rooting ability within and among *Eucalyptus* species. Vegetative propagation traits also have reasonable heritabilities, typically higher than growth traits. Grattapaglia et al. (1995) first described the identification of QTLs for fresh weight of micropropagated shoot, stump sprouting, and rooting ability of cuttings in an F_1 hybrid population of *E. grandis* and *E. urophylla*. For all traits that were quantified, a large proportion (>60%) of the genetic variation could be explained by the detected QTLs. Marques et al. (1999) carried out a similar analysis in a population of 315 genotypes from a pseudotestcross of *E. tereticornis* and *E. globulus*. *E. tereticornis* typically has higher rooting ability, but its pulping quality is inferior to that of *E. globulus*. A broad variety of traits were evaluated, and QTLs could be identified for mortality of cuttings, adventitious rooting, sprouting ability, and other propagation-related traits. The results from these two studies were compared by regenotyping the two populations with a set of common SSR markers to establish synteny between LGs and align them for comparison of QTL location (Marques et al. 2002). Surprisingly, some QTLs could be detected in homeologous LGs, which could be due to variation in orthologous genes. An ancient allele for vegetative propagation ability could have been fixed in a common ancestor to one of the species in each cross, which would explain why it segregated in the two F_1 populations. However, the phylogenetic classification of *E. grandis* and *E. urophylla* into a different section of *Symphyomyrtus* (*Latoangulatae*) than *E. tereticornis* and *E. globulus* (*Exsertaria* and *Maidenaria*, respectively) does not support this hypothesis. Alternatively, considering the broad support interval of QTLs, there is a high probability that two QTLs overlap by chance.

Developmental Traits

QTLs have been reported for a number of developmental traits in *Eucalyptus*, including flowering precocity (Missiaggia and Grattapaglia 2005) and leaf physical and chemical characteristics (Byrne et al. 1997a; Shepherd et al. 1999).

QTL Analysis of Gene Expression Levels

QTL analysis in *Eucalyptus* has focused mostly on traits of importance to the forestry industry. However, molecular phenotypes such as transcript (mRNA) and protein expression levels may also be analyzed using traditional QTL detection methods. Transcript abundance variation has been shown to be genetically controlled and heritable (Dumas et al. 2000; Karp et al. 2000; Brem et al. 2002; Wayne and McIntyre 2002; Schadt et al. 2003; Yvert et al. 2003). A genomic description of transcriptional regulation has recently emerged in our work in *Eucalyptus* (Kirst et al. 2004, 2005) following similar demonstrations in yeast and mice (Brem et al. 2002; Schadt et al. 2003; Yvert et al. 2003). Genetic mapping and transcript level information were integrated to define genomic regions involved in the regulation of gene expression by expression QTL (eQTL) analysis. Transcript level variation was measured using a cDNA microarray platform for 2,605 genes in a segregating population of 91 individuals from a pseudobackcross of an F_1 hybrid of *E. grandis* and *E. globulus* (Myburg et al 2003), which was backcrossed to a different *E. grandis* parent (Kirst et al. 2005). QTL analysis of gene expression variation in the *E. grandis* backcross family identified genomic regions that harbor regulatory sequences controlling, in *cis-* or *trans-*, the expression of 1,067 genes, or 41% of the genes represented on the microarray. Of the 1,067 genes for which eQTLs were detected, 811 were located in the paternal (F_1 hybrid parent) map and 451 in the maternal (*E. grandis* backcross parent) map. The eQTLs for 195 genes mapped to both parental maps, the majority to nonhomologous LGs, suggesting *trans*-regulation by different loci in the two genetic backgrounds. For 821 genes a single eQTL was identified that explained up to 70% of the transcript level variation (Kirst et al. 2005).

The description of the genetic architecture of transcript variation in *Eucalyptus* allowed for inference about the relationship among expressed genes. In several instances, the transcript abundance of genes that were part of the same biochemical pathway (such as the lignin biosynthetic pathway) was shown to be regulated by a single genetic locus (i.e., a shared eQTL, Kirst et al. 2004). Some genes were regulated by multiple genetic loci and shared several eQTLs. Many eQTL hotspots have been shown, in our studies and others (Brem et al. 2002; Yvert et al. 2003; Kirst et al. 2004), to include genes associated with the same metabolic and regulatory pathways, suggesting coordinated regulation of pathway genes by specific regulatory loci. The correlation of gene expression patterns in segregating progeny can also extend our knowledge about genes involved in these pathways. The cDNAs representing previously uncharacterized or hypothetical genes, whose transcript levels are strongly correlated with those of genes with known function, may be associated with the same pathway or biological process. Similarly, new functions can tentatively be assigned to previously characterized genes that had not been described in the context revealing pleiotropic action of these genes. A major limitation in this type of study in *Eucalyptus* is the lack of a completed genome sequence, because without it the relative locations of large numbers of genes and their eQTLs cannot be determined. This information is required to assess whether the genetic control of gene expression variation is in *cis-* or *trans-*, for each gene. The completion of the current *E. camaldulensis* genome sequencing project will provide the first opportunity to dissect the genetic control of gene expression levels in such a fashion. In the meantime, master eQTLs that control gene expression of important biochemical pathways may have great value for molecular breeding in *Eucalyptus*.

4.3.4 Future Perspective: from QTL to Gene

QTL analysis identifies broad genomic regions linked to one or more polymorphisms that cause variation in a phenotype. These broad regions, or QTL support intervals, typically span over 10 to 20 cM or more in most studies carried out in *Eucalyptus* and other forest tree species. Previous estimates of the length of *Eucalyptus* genetic maps have ranged between 1300 and 1,500 cM (Grattapaglia and Sederoff 1994; Byrne et al. 1995; Verhaegen and Plomion 1996; Myburg et al. 2003), with some exceptions (Marques et al. 1998). Therefore, for the haploid genome size of different *Eucalyptus* species (370 to 700 megabase pairs, Grattapaglia and Bradshaw 1994), the average physical distance of 1 cM is expected to range between 300 and 600 kbp. An average QTL would span 3 to 12 Mbp. There are currently no accurate estimates of the number of genes in the *Eucalyptus* genome. Based on the number of genes identified in the sequenced genome of other angiosperms (*Arabidopsis thaliana* ~ 26,000 genes, *Populus trichocarpa* ~ 40,000), an estimate of 30,000 to 50,000 genes seems reasonable. Therefore, a typical QTL interval of 10 to 20 cM may contain anywhere from 200 to 1,000 genes, depending on the precision of the QTL estimate and the gene density in the QTL interval.

Identifying the gene that underlies a QTL remains a major challenge in any organism. The standard strategy has been to fine-scale map the QTL by saturating the region of interest with a large number of makers and evaluating recombination in large progeny sets, to define the location of the QTL at a very high resolution (<1 cM). Chromosome walking can then be undertaken to identify genes in the region. If the resolution is sufficiently high, it can precisely identify the gene that controls the quantitative variation. The large populations that are required and the labor-intensive effort have limited the application of positional cloning to a few examples in crop plants (reviewed in Morgante and Salamini 2003). Positional cloning has not been attempted for any of the QTLs identified in *Eucalyptus*. However, efforts have been made to clone a gene associated with disease-resistance loci in poplar, but the effort has been hindered by the presence of a low-recombination region in the vicinity of the resistance locus. The cost of identifying markers closely linked to quantitative loci and identifying recombinants represents a substantial challenge in forest genetics.

As discussed above, an integrative genomics approach that combines QTL mapping with gene expression analysis may be an alternative to traditional methods of identifying genes underlying QTLs. Genetic analysis of complex traits has normally been carried out by correlating genotypic and phenotypic variation in segregating populations and identifying molecular markers associated with a quantitative trait through QTL analysis and association studies. Complex traits could also be analyzed from the perspective of transcript variation, as it represents an intermediate stage between genotype and phenotype. In contrast to anonymous markers, the transcript levels of thousands of specific genes can be monitored in a segregating population using genomic tools such as microarray analysis that assess genomewide variation in gene expression.

Characterizing the transcriptome of a segregating family can provide valuable information for the dissection of complex traits. Several approaches that integrate genotypic and transcript level variation for the identification of candidate genes have been suggested, but the most powerful one involves the characterization of large progeny sets. Schadt et al. (2003) identified candidate genes by detecting eQTLs that colocalized with QTLs for obesity in mice. Using a similar strategy we have identified candidate genes underlying QTLs for growth and wood density by contrasting the transcript levels for individuals that inherited alternative genotypes for significant phenotypic QTLs in *Eucalyptus* (Kirst et al. 2004). We have also been able to identify specific genetic loci that regulate metabolic pathways and confirmed that variation regulating the expression of metabolically related genes correlates with synthesis of the pathway products (Kirst et al. 2004). For lignin biosynthesis, our previous studies identified genes encoding enzymes of the phenylpropanoid, shikimate, and methionine pathways (all involved in lignin biosynthesis) that have a common eQTL, which overlaps with QTLs for growth in two LGs. Genetic mapping of these genes indicates *trans*- regulation of the pathway genes. These preliminary results indicate that the integration of different genetic information streams (genetic maps, transcript levels, and traditional phenotypic traits) collected from a segregating population can be highly effective in the dissection of the genetic networks that control variation in complex traits such as lignin content and growth. Since most QTL mapping studies in *Eucalyptus* have been performed in interspecific pedigrees, this integrative genomics approach holds the added promise of elucidating how eucalypt species have become differentiated in the genetic control of gene expression patterns and how this affects observed differences in commercially important traits.

4.4 Gene Mapping in *Eucalyptus*

4.4.1 From Anonymous Markers to Candidate Genes

The mapping and the characterization of genes involved in the control of complex traits constitute a new stage in our understanding of genome organization and function in forest trees. New strategies exist to go from the phenotype of a complex trait to the gene(s) contributing all, or part of, its variability. One such strategy would be to screen mutants for a given phenotype (morphological, biochemical, or molecular), followed by fine-scale genetic mapping of the mutant and identification of the mutated gene. Forward genetics approaches such as these have, unfortunately, not been feasible in forest trees, mainly due to the long generation times and outbred nature of forest tree species and the difficulty of screening for the adult phenotypes that we are interested in. Alternatively, complex traits can be broken down into elementary components, each with Mendelian inheritance (i.e., quantitative trait loci, or QTLs), particularly at the physiological or biochemical levels (Paterson et al. 1988; Damerval et al. 1994), as discussed in the previous section. These studies can be followed up with the colocalization of QTLs and candidate genes on genetic maps in order to identify positional candidate genes that can be targeted for further confirmatory studies such as association genetic analysis in tree populations (e.g., Thumma et al. 2005), or transgenic studies.

QTL-candidate gene colocalization studies are presently limited by the low resolution of QTL mapping studies performed so far in *Eucalyptus*. As discussed in the previous section, the confidence interval of a QTL is often between 10 and 20 cM (Mangin et al. 1994). In *Eucalyptus*, such an interval could span up to 12 Mbp (Grattapaglia and Bradshaw 1994), i.e., potentially hundreds of genes. The low resolution of QTL characterization is one of the major factors limiting their usefulness in MAS and gene cloning. Nevertheless, of the possible strategies to identify genes underlying QTLs, the candidate gene approach is certainly the simplest. As QTL information is gathered in increasing numbers of unrelated pedigrees, the association of some candidate genes and QTLs may gain enough support to warrant their classification as positional candidates for the traits involved.

Functional candidate genes affecting the expression of a trait of interest can be selected a priori on the basis of the known biochemical and metabolic pathways affecting this trait. The effect of a candidate gene can be estimated using a traditional QTL analysis in which a polymorphism inside the candidate gene is used as an additional genetic marker. Such analyses, of course, cannot exclude the possibility of another closely linked candidate gene in the same QTL interval and has to be followed up by populationwide association studies. The colocalization of QTLs and genes of known function has already been reported in several plant species (Goldman et al. 1994; Causse et al. 1995; Byrne et al. 1997a; Prioul et al. 1997; de Vienne et al. 1999; Pelleschi et al. 1999). In forest trees like *Eucalyptus* species, such strategies are now being developed to more efficiently identify positional candidate genes. These approaches are generally based on the mapping of expressed sequences, i.e., known or unknown genes,with an a priori knowledge of metabolic pathways involved in the trait of interest. The main interest in direct candidate gene mapping is to get around the requirement of linkage disequilibrium between the gene and flanking markers, directly targeting its functional variability. In this section, we briefly summarize the genes and traits of interest for candidate-gene mapping in *Eucalyptus* and we give an overview of the methods used to detect polymorphism and map candidate genes in this genus.

4.4.2 Traits and Genes of Interest in *Eucalyptus*

In *Eucalyptus*, growth, wood quality, disease resistance, and vegetative-propagation-related traits are considered the most important commercial traits for breeding programs. Several genetic factors with major effect on the variation of these traits have been detected by classical QTL analysis (see previous section). However, the identification of candidate genes underlying these QTLs is usually not reported. Indeed, the mapping of known candidate genes is more or less advanced according to the trait of interest.

Growth

Several studies have reported the mapping of major QTLs for growth in controlled *Eucalyptus* crosses (Grattapaglia et al. 1996; Byrne et al. 1997a; Verhaegen et al. 1997). Even if the QTL detection indicated functional variability of some genes for growth variation, the identity of these genes remains undetermined. Growth-related traits are a good example of how a candidate gene approach is still difficult. The complexity of growth phenotypes (due to the many metabolic pathways and the relatively large environmental effects that may be involved) makes it very difficult to a priori select putative candidate genes to map. Furthermore, major growth QTLs detected in interspecific pedigrees are often related to hybrid abnormality or viability effects, which are generally not useful for breeding purposes.

Vegetative Propagation

Vegetative propagation traits are important for the development and deployment of clonal varieties of *Eucalyptus*. Major QTLs for propagation traits have been detected on *Eucalyptus* genetic maps (Marques et al. 1999, 2002). Gion et al. (2000) mapped two candidate genes regulated by auxin, which are thought to be involved in vegetative propagation traits. These genes were cloned and sequenced from *E. globulus* roots during the symbiosis between *E. globulus* and *Pisolythus* (Carnero Diaz et al. 1996; Nehls et al. 1998). However, the study of colocalization between these genes and propagation-related traits did not reveal any effect of gene variability on phenotypic variation in an interspecific cross of *E. urophylla* and *E. grandis* (Gion 2001).

Wood Quality

Wood quality is today the most important trait for which gene discovery is in progress. Many genomic studies have reported the analysis of genes expressed during wood formation and, more particularly, during xylogenesis (Hertzberg et al. 2001; Israelsson et al. 2003; Yang et al. 2003; Egertsdotter et al. 2004; Yang et al. 2004a, b; Paux et al. 2004, 2005; Foucart et al. 2006). Some important metabolic pathways producing the chemical components of wood, like lignin or cellulose, are well known. Indeed, the first example of a wood property for which genomic data were available for association studies was that of lignification genes. Several structural and regulatory genes involved in lignin biosynthesis are known in *Eucalyptus*, including those encoding components of the common phenylpropanoid pathway (PAL, C3H, C4H, COMT, CCoAOMT, and 4CL) and those of the monolignol-specific pathway like CCR and CAD as well lignin regulatory factors such as MYB transcription factors (Goicoechea et al. 2005). All of these genes seem to be good candidates for QTL colocalization studies with wood-quality and lignin-content QTLs.

Other Traits and Genes

Some other genes that are not directly linked to the targeted traits of breeding programs, like architectural and floral development traits, have also been localized in *Eucalyptus* (Thamarus et al. 2002). Resistance to rust has been studied in *E. grandis* from Brazil (Junghans et al. 2003), foliar oil composition in *E. grandis* (Shepherd et al. 1999), and frost tolerance in *E. nitens* (Byrne et al. 1997b). More than the knowledge of metabolic pathway involved, the choice of mapped genes depends to a great extent on the availability of gene sequences in the public databases.

4.4.3 *Eucalyptus* Species and Populations Used for Gene Mapping

Currently, the genomes of only five of the major *Eucalyptus* species used for plantation forestry have been mapped with relatively high map coverage: *E. globulus*, *E. grandis*, *E. urophylla*, *E. tereticornis*, and *E. nitens* (Tables 3 and 4). Different works have reported a high level of genetic variability in these species for phenotypic data (Potts and Jordan 1994; Chambers et al. 1997; MacDonald et al. 1997) and for molecular data (Byrne et al. 1994, 1996; Martins-Corder and Lopez 1997; Brondani et al. 1998). In spite of the high levels of reported intraspecific variability, other results suggest high conservation at the genome and sequence level between different *Symphyomyrtus* species (Byrne et al. 1996; Gion et al. 2000, 2005), allowing the use of gene sequences from one species to target and map the homologous gene in other species. The relative ease of making interspecific crosses between *Symphyomyrtus* species also suggests high levels of genome conservation, despite differences in genome size of up to 20% (Grattapaglia and Bradshaw 1994). Interspecific hybridization has been exploited in *Eucalyptus* breeding programs to combine desirable traits from different species. This explains why most mapping populations in *Eucalyptus* have been based on interspecific crosses (Tables 3 and 4).

The species generally used for gene sequencing correspond globally to the same *Eucalyptus* species used for genetic mapping. However, the genomes of some species that have been used for gene sequencing like *E. gunnii* and *E. camaldulensis* have not been mapped yet. Fortunately, these species also belong to the *Symphyomyrtus* subgenus (*Maidenaria* and *Exsertaria* sections) and they are closely related to the mapping species. The high degree of marker transferability among *Eucalyptus* species constitutes a unique opportunity for comparative gene and genome mapping in these tree species.

4.4.4 Genomic Resources for Gene Mapping in *Eucalyptus*

High-throughput functional genomics efforts such as EST projects are an important source of sequence data to develop new molecular markers for gene mapping (Gupta and Rustgi 2004). The interest in these sequences for genetic mapping depends on their nature and variability.

Eucalyptus Gene Sequences Available for Marker Development

During the last decade, the number of *Eucalyptus* gene sequences in public databases has increased dramatically (Fig. 1) but trails far behind that available for other trees such as pines and poplar. The first *Eucalyptus* gene sequence deposited into a public database was an mRNA sequence for a lignin gene (CAD) in *E. gunnii* (Feuillet et al. 1993). For the next decade, the number of eucalypt sequences in the database was fewer than one thousand, including several different types of sequences (nuclear, chloroplast, and mitochondrial sequences). The most prevalent sequences were ribosomal sequences used for phylogenetic studies, which represented 39% of available sequences compared to 25% for EST sequences. After 2003, the sequence number increased 40-fold, principally due to EST development from four *Eucalyptus* species: *E. grandis*, *E. tereticornis*, *E. globulus*, and *E. gunnii* (Kirst et al. 2004; Paux et al. 2004; Foucart et al. 2006, P. Sivadon personnal communication). These sequences, which currently represent 92% of the total number of *Eucalyptus* entries, constitute an interesting library for candidate-gene mapping. Around 15,000 entries corresponding to ESTs of *Eucalyptus* are freely available in the EMBL nucleotide database. Private *Eucalyptus* functional genomics projects are also in progress in countries such as Japan (Sato et al. 2005) and Brazil (Grattapaglia 2004), and very large EST databases have been generated by private consortia such as Arborgen. The registration of these sequences in the international databases would constitute a major asset for the international scientific community.

The 15,000 *Eucalyptus* EST and mRNA sequences in EMBL were developed from different types of tissues: flower and carpel (8% of the sequences), seedling (8%), leaf (6%), differentiating xylem (77%), root (<1%), and ectomycorrhiza (<1%). Although xylem ESTs are clearly overrepresented, the multiple origins of around 3,000 ESTs constitute an interesting collection of expressed gene sequences for the mapping of different types of genes affecting traits of interest. The majority of these sequences will be useful for gene mapping because most (99%) are longer than 150 bp, allowing the design of specific primers to amplify genomic DNA in other genotypes and/or species.

Various methods have been used to generate EST databases for *Eucalyptus*. For example, Paux et al. (2004) generated a xylem subtractive library using the suppression subtractive hybridization (SSH) technique (Diatchenko et al. 1996) in order to obtain a cDNA library enriched in xylem-specific sequences. In total, 224 unique sequences (unigenes) were obtained with an average length of 382 bp. The functional classification of these EST sequences, according to the MIPS standard (Paux et al. 2004), revealed a high proportion of sequences classified either as "no hits" (44%) or as "proteins of unknown function" (17%). For those sequences where putative function could be assigned (39%), several functional categories were represented, reflecting the complexity of secondary xylem in woody angiosperms. Some of these ESTs developed from *E. gunnii* were tested for their transferability across the *Eucalyptus* genus (Gion et al. 2005). Primers were designed according to the *E. gunnii* sequences and PCR amplification performed on genomic DNA from 30 different species (3 genotypes/species). The transferability decreases with in-

Fig. 1. Number of *Eucalyptus* sequences in the EMBL database from 1992 to 2006 (May). The relative proportion of sequence types is indicated from 1992 to 2002 and from 2003 to 2006; EST (□), chloroplast or mitochondrial genes (▦), full-length mRNA and complete gene (▧), SSR (▨), ribosomal gene (□), cleaved amplified polymorphic sequence (■), and unassigned DNA sequences (▩)

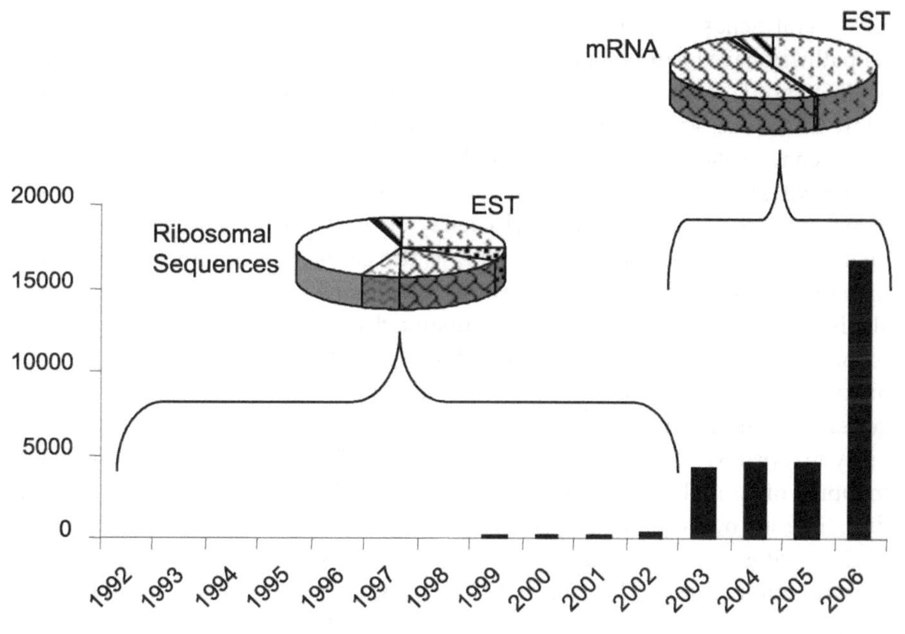

creasing phylogenetic distance from *E. gunnii*, from 94% for the subgenus *Symphyomyrtus* sections *Maidenaria*, *Latoangulatae*, and *Exsertaria*, to 68% for the subgenus *Monocalyptus* and only 48% for the genus *Corymbia*. This work revealed the high potential of *Eucalyptus* ESTs for comparative mapping and association studies between candidate genes and quantitative traits at the subgenus or genus level.

In the near future, we can expect that the number of available *Eucalyptus* sequences in public databases will continue to increase. The idea of a cooperative effort for the development of *Eucalyptus* genomic resources coordinated by the *Eucalyptus* Genome Network (EUCAGEN) could be a major asset for gene-mapping efforts in *Eucalyptus*. It would then be necessary to develop new technologies that would allow the high-throughput mapping of thousands of ESTs in *Eucalyptus*.

4.4.5 Gene Mapping in *Eucalyptus*: Technologies and Perspectives

The first generation of *Eucalyptus* genetic maps was mostly based on dominant, anonymous markers that allowed rapid saturation of LGs in order to achieve high map coverage (Sect. 4.2, Table 3). Several genetic maps were also constructed for *Symphyomyrtus* species using different types of codominant molecular markers (Table 4). These genetic maps have been used to detect the presence of major effect QTLs for complex traits in *Eucalyptus* (discussed in Sect. 4.3.3). More recently, with the availability of *Eucalyptus* genomic resources, it has become feasible to add gene loci to these maps, and several studies have reported the location of important target genes on these maps.

Genes and Marker Types Used for Gene Mapping

Byrne et al. (1995) reported the first gene mapping results corresponding to four isozyme loci mapped on *E. nitens* genetic maps. The three enzyme systems used were malate dehydrogenase (MDH1 and MDH2), shikimic acid dehydrogenase (SDH), and phosphogluconate dehydrogenase (PDH). The polymorphism revealed in *E. nitens* intraspecific F_1 progeny allowed the mapping of the *MDH1*, *MDH2*, *SDH*, and *PDH* gene loci on four different LGs, 6, 1, 12, and 5, respectively. More recently, these same and additional isozyme loci were mapped in the *E. globulus* genome (Thamarus et al. 2002). However, the most commonly used markers for gene mapping in *Eucalyptus* have been PCR-based markers. The recent availability of *Eucalyptus* sequences has allowed the design of specific or degenerate primers for the amplification of the orthologous gene or fragment in the species of interest. Even where the design of *Eucalyptus*-specific primers has not been feasible, degenerate primers based on sequences from other species have been used with success. As an example, the PAL gene was mapped in *E. urophylla* genetic maps by Gion et al. (2000) using degenerated primers developed from ten different accession numbers.

Several genes involved in the lignin biosynthesis pathway have been mapped in *E. grandis*, *E. urophylla*, and *E. globulus* (Gion et al. 2000; Myburg 2001; Thamarus et al. 2002). Some of these genes, such as phenylalanine ammonia-lyase (PAL), Caffeate O-methyltransferase (COMT), 4 coumarate Coenzyme A ligase (4CL), and Caffeoyl coenzyme A O-methyltransferase (CCoAOMT), belong to the common phenylpropanoid pathway. Two others, cinnamyl alcohol dehydrogenase (CAD) and cynnamoyl CoA reductase (CCR), belong to the monolignol-specific pathway. In *E. grandis* and *E. urophylla*, all of these genes were significantly linked (LOD > 11) to previously mapped RAPD markers (Gion et al. 2000). The mapping of lignin biosynthesis genes gives us for the first time an overview of the genome organization in *Eucalyptus* of a set of genes involved in the same biosynthetic pathway. The fact that these genes are located in five distinct genomic regions is a favorable situation for QTL-candidate gene colocalization efforts. It also increases the probability of finding transgressive genotypes with favorable alleles for multiple lignin genes.

In *E. globulus*, 31 cambium-specific ESTs were mapped by Thamarus et al. (2002) and found to be located on 10 different LGs. With an average of 3 ESTs per LG, this EST set will be useful for comparative mapping in other *Eucalyptus* species. Carocha et al. (2004) have started the mapping of 224 expressed candidate genes and 83 additional functional candidate genes in *E. globulus*. A success rate of 90% was achieved with the amplification of targeted gene fragments in *E. globulus* using primers designed on *E. gunnii* sequences. The mapping of the same set of expressional candidate genes is in progress in *E. urophylla* and *E. grandis* in the context of a European collaboration.

Four flowering genes have been mapped in *E. globulus* genetic maps (Moran et al. 2002). *ELF1*, *EAP*, *AGE1*, and *AGE2* were mapped on LG 3, 4, 3, and 9, respectively. *ELF1* was recently also mapped on LG 4 of *E. urophylla* (C. Boudet and J.-M. Gion unpublished data). Junghans et al. (2003) reported the mapping of a monogenic resistance locus for the *Puccinia psidii* resistance gene 1 (*Ppr1*) relative to previously mapped RAPD markers.

Methods Used to Detect Polymorphism in *Eucalyptus*

The genomic resources obtained by EST sequencing, or complete cDNA characterization, constitutes an important source of nonanonymous genetic markers in *Eucalyptus*, provided that polymorphism can be detected in these sequences. For the genetic mapping of such sequence tagged sites (STS), several techniques can be used. The simplest methods include the generation of PCR-RFLP, or cleaved amplified polymorphic sequence (CAPS) markers, which is based on the digestion of amplified gene fragments with specific restriction enzymes. The digestion products are then observed on agarose gels after ethidium bromide staining (Tragoonrung et al. 1992). Other techniques, like thermal gradient gel electrophoresis (TGGE, Riesner et al. 1992) or denaturing gradient gel electrophoresis (DGGE), are based on the comparison of the stability of amplified DNA fragments under specific thermal or denaturing conditions. These techniques seem a priori simple, but they require highly controlled conditions. Finally, the most commonly used method in *Eucalyptus* has been the single-strand conformation polymorphism (SSCP) technique, which is based on the specific secondary structure of single-strand DNA under nondenaturing conditions (Orita et al. 1989).

4.4.6 Future Perspective: Comparative Gene Mapping and Candidate-Gene Analysis in *Eucalyptus*

In addition to its use for QTL characterization, gene mapping is a useful approach to obtain a set of codominant markers for comparative mapping studies in *Eucalyptus*. For the moment, the number of common markers used for genetic mapping of *Eucalyptus* species is not enough to realize detailed comparative mapping of different eucalypt species. Fewer than ten gene-based markers (PGD, MDH, ELF1, 4CL, COMT, CCR, PAL, CAD, and CCoAOMT) have been mapped in more than one genetic linkage map, allowing the identification of homologies between the maps of different species. To date, only a small number of LGs of *E. globulus*, *E. urophylla*, *E. grandis*, and *E. nitens* could be identified as homologs (Fig. 2). The addition of more ESTs on *Eucalyptus* genetic maps will allow higher-resolution comparative mapping and investigation of genome evolution in this genus.

A major goal of gene mapping in *Eucalyptus* has been to determine the proportion of the variation in quantitative traits that can be explained by the segregation of allelic forms of candidate genes in view of using these genes in molecular breeding. This has been achieved by using candidate genes as molecular

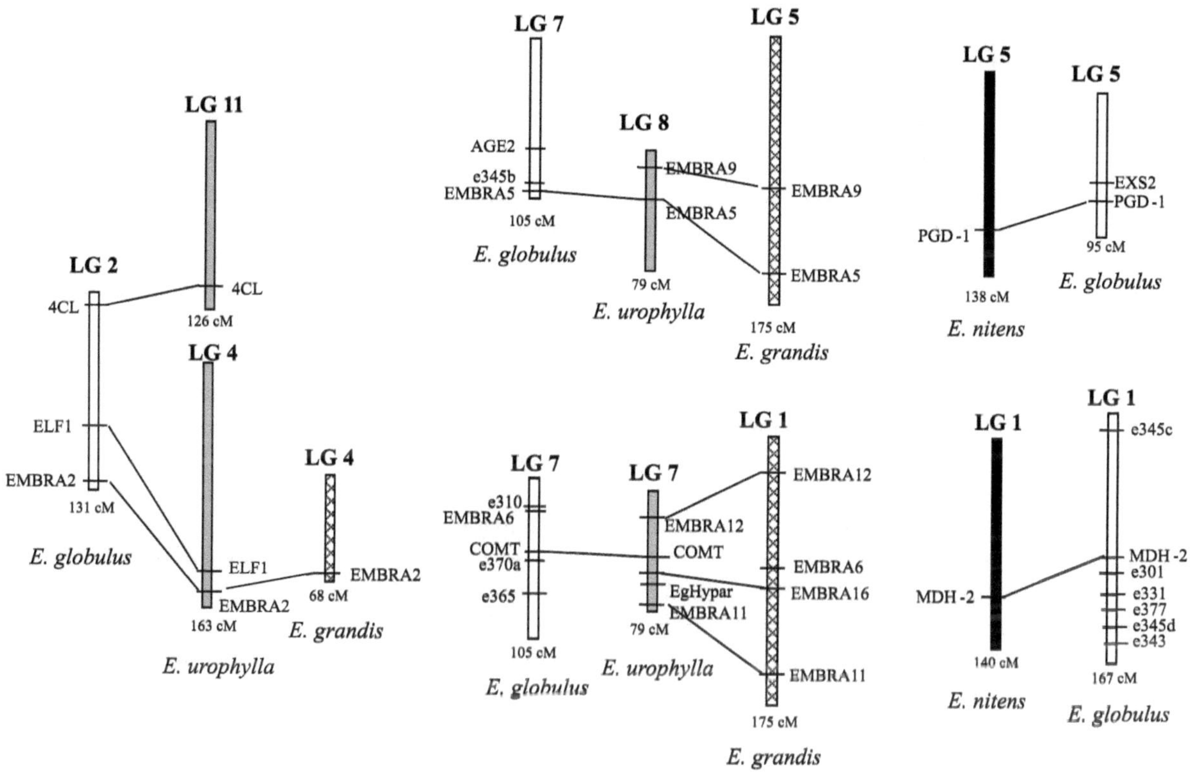

Fig. 2. Comparative mapping of LGs of *E. globulus* (□), *E. urophylla* (□), *E. grandis* (⊠), and *E. nitens* (■) based on Thamarus et al. (2002), Gion et al. (2000), Brondani et al. (1998), and Byrne et al. (1995), respectively. Only codominant markers are represented on the LGs. A total of five LGs are identified as homologs

markers in a normal QTL analysis and so determine the colocalization of genes and QTLs. Some results on gene and QTL colocalization have been published in *Eucalyptus*. In *E. globulus*, the CCR gene locus was mapped in a region of colocating with pulp yield and cellulose QTLs on LG 10 (Thamarus et al. 2004). Another colocalization was found between the *ECS1* gene, a *Eucalyptus* homolog of cellulose synthase, and a fiber length QTL on LG 8. This gene encodes a key enzyme in the deposition of cellulose microfibrils in plant cell walls. A hydroxymethyltransferase gene was also found to colocate with a wood density QTL on LG 1, while a p-glycoprotein was colocated with a pulp yield QTL on LG 4 (Thamarus et al. 2004). The association of candidate-gene loci with QTLs is suggestive, but not indicative, of functional variability. Even tight marker-QTL linkage may represent large genomic regions that may harbor the actual gene or regulatory sequence underlying the QTL. All of these observed colocalizations need to be validated by association studies in more complex populations or by independent QTL studies in other pedigrees. For example, the colocalization between the CCR gene and a QTL for lignin content in *E. globulus* has also been observed in LG 6 of *E. urophylla* (Gion 2001). In this case, the CCR gene explained 15% of the variation in lignin content observed in an interspecific *E. urophylla* × *E. grandis* family. These results in two different species suggest that the CCR gene may be a good candidate gene for molecular breeding for wood quality in *Eucalyptus*.

4.5 Physical Mapping and Map-Based Cloning in *Eucalyptus*

The construction of resources for physical mapping of whole genomes has been a key component of map-based cloning and genome-sequencing efforts in a number of species. Currently, the preferred approach for physical mapping of complex genomes involves the construction of a library of large-insert bacterial artificial chromosome (BAC) clones (Shizuya et al. 1992), followed by their assembly into a structured set using restriction enzyme fingerprinting technologies (Marra et al. 1997; Schein et al. 2004). The rationale behind physical map construction for any genome is the reliance on random breakage of the genome structure and the subsequent ordering of these genome pieces minimizing gaps and aiming for a most likely order. A physical map can therefore be defined as an ordered set of clones comprising a near-contiguous path across the target genome. Such clone-based maps are used to validate sequence assembly order, supply long-range linking information for assembled sequences, anchor sequences to genetic maps, supply contigs for positional cloning efforts, and provide templates for closing gaps in full genome sequencing. Fingerprint maps are also a critical resource for subsequent functional genomics studies because they provide a redundant and ordered sampling of the genome with clones. In other words, once assembled, physical maps are a key resource in the dissemination of clones by known location for use in several disciplines of biology. This section provides a brief overview of physical mapping methodologies and their application in genome mapping in *Eucalyptus*. The potential use of these technologies for map-based cloning and future molecular breeding in *Eucalyptus* is discussed.

4.5.1 Physical Mapping Methodologies and Their Use in *Eucalyptus*

With the technological advances in the ability to clone and fingerprint larger DNA fragments, physical maps have been constructed for a growing number of species. Large-insert BAC fingerprint maps have been developed for humans (McPherson et al. 2001), some animal species including model organisms *Drosophila* (Hoskins et al. 2000), mouse (Gregory et al. 2002), rat (Krzywinski et al. 2004), and economically important species such as chicken (Wallis et al. 2004) and cow (http://www.bcgsc.ca/lab/mapping/bovine). Physical maps have also been published for some plant species including rice (Tao et al. 2001; Chen et al. 2002), sorghum (Klein et al. 2000), *Arabidopsis* (Chang et al. 2001), and soybean (Wu et al. 2004), while work is well advanced for other species such as poplar (http://www.bcgsc.ca/gc/poplar), maize (Coe et al. 2002), wheat (http://wheat.pw.usda.gov), and grape (M. Morgante personal communication).

An excellent and thorough review of the different fingerprinting techniques, detailing the pros and cons of each and the science and statistical issues behind physical map assembly, was recently published (Meyers et al. 2004). Several factors play an important role in the assembly of contigs or a full physical map, including the genome size, the degree of repetitiveness of the genome, the composition and age of repetitive elements, and the fingerprinting method used. The fingerprinting method for building physical maps involves the restriction enzyme digestion of each BAC clone into fragments, which are then separated by electrophoresis and detected. Overlapping clones derived from the same genomic region produce patterns of shared restriction fragments, seen as shared bands on a gel. The proportion of shared bands is indicative of the degree of overlap. This is typically evaluated by the "Sulston cutoff score," i.e., the probability that the number of bands matched between any two clones is coincidental (Sulston et al. 1988). This statistical evaluation and assembly is typically carried out using the only physical map assembly software available today, FPC - Fingerprinted Contigs (Soderlund et al. 2000). The overlap across numerous clones is then used to order the clones into contigs.

Different methods exist to generate clone fingerprints that vary by the number of restriction enzymes used and the detection method. Most published physical maps to date have been constructed using single enzyme digestion, agarose gel electrophoresis, and conventional DNA stains. Variations of this basic method have been published, taking advantage of multiple enzyme digestion, multicolor fluorescent detection technology, and high-resolution capillary electrophoresis (Ding et al. 1999, 2001; Luo et al. 2003). These different methods provide more or less complex fingerprints, i.e., variable numbers of bands that, in turn, have a significant impact on the extent of the overlap among clones and eventually on the quality of the physical map and number of gaps left. Meyers et al. (2004) showed that multiple-enzyme multicolor fluorescent methods that result in more complex fingerprints are superior to single-enzyme, agarose resolution methods. This is intuitive as the observed bands serve as anchors along the BAC clone. The larger the number of anchors available, the greater the confidence in declaring an overlap between clones. Conversely, with larger numbers of bands, a highly significant cutoff score is attained even at lower overlap percentages between clones. This is true, however, only up to a certain number of bands per clone, after which false overlaps will likely occur. Recently, the first physical map constructed using fluorescent fingerprinting technology and capillary electrophoresis was published for *Pennicillium* (Xu et al. 2005). In that work it was shown that the selection of fingerprinting enzymes was crucial to quality map construction and that the fingerprints labeled with one or two colors, resulting in 40 to 70 bands per clone, were assembled into much better quality maps than those labeled with three or four colors.

Physical genome maps are generally built with libraries that offer between 10- and 15-fold redundant coverage, i.e., on average, each region of the map is represented by 10 to 15 clones. Coverage redundancy is critical to achieving map contiguity and is used to provide evidence that individual clones are not cloning artifacts but high-fidelity representations of the underlying genome. Physical maps assembled with higher information content fingerprinting methods will require a lower coverage redundancy as they allow detection of smaller overlaps with high confidence. Furthermore, the way that the BAC clones were produced also determines the final quality of the physical map in terms of gaps. As pointed out by Meyers et al. (2004), two kinds of gaps can occur: (1) assembly gaps resulting from the false-negative rate that is determined with the choice of the cutoff and correspond to our inability to detect existing overlaps between clones; and (2) physical gaps result from regions that are not covered in the clone collection. BAC clones produced by restriction enzyme digestion will have larger gaps and deeper coverage in some regions of the genome than a set of clones produced in a completely random way, such as through mechanical shearing. The use of different libraries produced by digesting genomic DNA with different enzymes will reduce physical gaps that might result from biased restriction site distribution.

In a similar way to genetic map construction, physical map assembly involves initially an automatic step carried out at higher statistical stringency to avoid false positives. Following this automatic assembly, a manual analysis at lower stringency is carried out when contigs are joined to build longer scaffolds. At this stage, linkage and ordering information of mapped microsatellites together with knowledge of the BAC clones to which they were mapped is very valuable for ordering contigs relative to one another. After physical maps are constructed, redundancy is unnecessary for a complete representation of the genome, and most map-driven sequencing efforts use the map to select a minimal tiling path, i.e., a minimal set of overlapping clones that together provides complete coverage across a genomic region. For the human genome, for example, after assembling a physical map with 415,000 BAC clones at a deep coverage of 15×, a set of 32,855 validated clones were selected to cover 99% of the whole genome (Krzywinski et al. 2004).

If on one side genetic mapping of *Eucalyptus* has evolved quite rapidly in the last years, particularly so with the advent of RAPD, AFLP, and, more recently, microsatellite markers, no published report has come out yet on the construction of BAC library resources or physical maps for species of the genus. A complete physical map for *Eucalyptus* will certainly represent a great experimental resource for years to come. With this concept in mind this task was started in the context of the *Genolyptus* project in Brazil (Grattapaglia 2004). A *Eucalyptus grandis* BAC library with over 70% of the inserts averaging 150 kbp has been built (S. Brommonschenkel personal communication). Genomic DNA for library contruction was isolated from a pure *E. grandis* tree of good growth and adaptability in Brazil, and originally from Australian Coffs Harbor provenance. A first set of 20,160 partially digested *Hind*III clones was arrayed, covering an estimated 4× of the 630 Mbp of the *E. grandis* genome. Verification of genome coverage was carried out by PCR using a set of previously mapped single-locus microsatellite markers and STS markers for single-copy genes using a simple bidimensional pooling strategy. Such single-locus markers identified on average two to five BAC clones and provided a satisfactory initial experimental confirmation of the genome coverage of this library. An analysis of cpDNA contamination using chloroplast-specific microsatellites showed it to be less than 1%.

Fingerprinting of the 20,160 *E. grandis* BAC clones is currently under way using the method of Luo et al. (2003). The clones are digested with four rare cutting enzymes and a frequent cutter. Each of the rare cutters leaves a different single-stranded overhang that is then filled in with a distinct, labeled ddNTP using the SNPshot labeling kit (Applied Biosystems). A fifth fluorochrome is used for the internal size standard, and the digested products are resolved on a 5-dye ABI 3100 automated sequencer. To obtain reproducible fingerprints, the DNA minipreps have to yield sufficient and consistent amounts of BAC DNA. As no PCR is involved in the process, the quality of the profile of restriction fragments visualized as fluorescent peaks, i.e., the fingerprint, is fully dependent on the quantity and quality of BAC DNA.

Using the BAC resource, a directed search for the full genomic sequences of a number of candidate genes involved in wood chemical composition has been undertaken (Grattapaglia et al. 2004). Primers were obtained from the literature and from consensus sequences in ORFs derived from *Eucalyptus* EST clusters. A hierarchical pooling strategy was employed to rapidly arrive at the BAC clone containing the gene of interest. BAC DNA midipreps were performed by pooling the 96 clones from a single plate. The 210 pooled DNA samples were, in turn, pooled in 35 superpools of six 96-well plates, i.e., six DNA midiprep samples. Step one of the BAC library screening was carried out with 35 PCR reactions plus the genomic DNA control. The six pools were then screened for a single positive superpool followed by colony-PCR screening of the positive 96-well plate to arrive at the target BAC, thus totaling 140 PCRs to get to a gene. This strategy has allowed us to land on BACs for genes that code for the following proteins: bXylan, UDP-Glucose, Tubulin, Xyloglucan endotransdycosilase, MYB, Pectate lyase, CCR, CCoAOMT, CAD, PAL, COMT, F5H, C4H, and 4CL. Although more than one BAC clone was found for all the genes but 4CL, supporting the 4× coverage, a single BAC was selected for future shotgun sequencing or primer walking. New primer sets pointing out of the amplified segment were designed to provide the full genomic sequence by subsequent walks. Alternatively, shotgun sequencing and assembly of a single BAC clone is being considered by selecting the shortest BAC clone identified in the library that contains the gene of interest. With the full genomic sequence in hand it should be possible to identify regulatory regions and carry out a detailed analysis of polymorphism in a set of individuals by resequencing specific upstream regions in an association mapping approach.

Parallel to clone fingerprinting, both ends of the 20,160 clones have been sequenced. These BAC end sequences could be aligned to the 4× shotgun sequencing now in progress for *E. camaldulensis* at Kazusa DNA Research Institute in Japan (and to the *E. grandis* genome sequence possibly produced by DOE) to provide long-range linking information for assembled sequences and templates for closing gaps in the draft genome sequence. Even though the BAC library is from *E. grandis* and the shotgun sequence from *E. camaldulensis*, the extensive genome homology found across species of the same subgenus *Symphyomyrtus* should allow using the BAC end sequences as useful connectors for closing gaps in interesting regions of the genome. The BAC ends will also be a prolific source of microsatellite and STS markers to help anchor the BAC clones to a linkage map. In a sample sequencing analysis of the *E. grandis* genome based on 7,395 random shotgun clones, Lourenço (2004) used the software TROLL (Castelo et al. 2002) to identify a total of 319 microsatellites for which primers could be designed. Based on this study, it is expected that the almost 40,000 phred-20 BAC end sequences should provide at least 1000 nonredundant microsatellites (R. Brondani personal communication).

In a reverse approach, a set of ~120 framework mapped microsatellite markers has been used to screen the 4× BAC library with the objective of anchoring specific BACs to a reference linkage map. Microsatellites were selected based on (1) robustness of amplification and allele interpretation; (2) absence of locus duplication sometimes seen for microsatellites in *Eucalyptus*; (3) high information content that allowed the marker to be positioned on three independently constructed linkage maps; (4) their relative map position and ordering, attempting to cover the largest extension of the genome from the recombination standpoint, at regular intervals; (5) fluorochrome labeling of the marker so as to allow multiplexed analysis. The same superpool-pool screening strategy used to fish out gene-containing BAC clones is being used to map microsatellites to BACs with the added advantage of simultaneous screening of several microsatellites labeled with different colors in the same PCR. Due to the 4× redundancy, screening of the BAC clones results in the discovery of more than one positive BAC clone. As not all BAC clones will necessarily fall into contigs during the future physical map assembly, all the BAC clones that contain anchoring microsatellites are sought.

Because most *Eucalyptus* maps have been constructed using nontransferable markers (RAPD or AFLP) or with increasingly less used RFLP probes, current numbering of *Eucalyptus* LGs is not consistent across maps. With the increased use of microsatellites this problem could be mitigated. However, even in the existing maps that include microsatellites (Brondani et al. 2002; Thamarus et al. 2002), LG numbering is attributed arbitrarily. BAC clones containing mapped microsatellite markers, once mapped to chromosomes using FISH, will ultimately allow assigning of the correct number to the LGs of *Eucalyptus* maps according to the chromosome number defined by its pachytene length or centromere position.

E. grandis was the obvious choice for a first BAC library and physical map resource as it constitutes the genetic base of most tropically planted elite germplasm. However, due to the special wood properties of the species and the great interest of breeders, a second BAC library from *E. globulus* has been constructed (S. Brommonschenkel personal communication). Accumulated evidence indicates that genomic homology and locus ordering between *E. grandis* and *E. globulus* are very high (Marques et al. 2002). It is therefore not in our immediate plans to build a physical map of *E. globulus* but rather to use the physical/genetic mapping information derived from *E. grandis* to identify and explore specific genomic regions in *E. globulus* by using its BAC library. Based on the map information derived from *E. grandis*, it will be possible, for example, to clone the full homolog gene from *E. globulus* and thus compare in detail both the coding and regulatory regions that could be responsible for differential patterns of gene expression and resulting phenotypic variation.

4.5.2 Map-Based Cloning in *Eucalyptus*

Map-based or positional cloning relies on the identification of closely linked markers to the target trait and then uses chromosome walking or landing to identify, isolate, and characterize the gene(s) responsible for the trait (Wicking and Williamson 1991). Map-based cloning requires genetic segregation for the trait of interest and genetic mapping of the trait close to discrete molecular markers. During chromosome walking, BAC clones are screened with a marker (PCR-based or RFLP) that functions as a starting point. New markers are then developed complementary to sequences of the same BAC clone that are adjacent to the starting point, and these are then used to identify additional BAC clones overlapping the one selected as the starting point. The procedure is used repetitively, working away from the starting point. This is a time-consuming approach hindered by the large amounts of DNA that often have to be crossed and by the high frequency of repetitive DNA. Based on the advances of genetic mapping and high-throughput molecular marker technologies, Tanksley et al. (1995) proposed and demonstrated a new paradigm for positional cloning of genes in complex plant genomes. Chromosome landing involves the isolation of one or more DNA markers at a physical distance from the targeted gene that is less than the average insert size of the genomic library being used for clone isolation. The DNA markers are then used to screen the library and isolate (or "land" on) the clone containing the gene, without any need for chromosome walking and its associated problems. Chromosome landing has been the main strategy used for map-based cloning of genes in plant genomes.

Finding and isolating a gene by map-based cloning in a complex genome requires an integrated and complex set of powerful genetic and genomic tools. Genetic maps based on recombination frequencies among markers are the starting point. However, they provide only megabase-level resolution. In *Eucalyptus*, 1% recombination between two markers or between a marker and a gene should correspond on average to 500,000 base pairs of DNA, which can contain several tens of genes (Grattapaglia and Bradshaw 1994). Furthermore, most QTL-mapping experiments in *Eucalyptus* to date have only achieved resolutions of between 10 and 30 cM, still very far from what is needed for efficient positional cloning. The availability of a physical map that is aligned and anchored to a reference genetic map constructed with transferable microsatellites would greatly facilitate positional cloning in *Eucalyptus*. However, even without a complete physical map, by simply screening the BAC library, in a chromosome landing approach, localized physical maps can be constructed representing specific regions of the genome delimited by molecular markers flanking the QTL of interest. This would require much higher resolution genetic mapping than is currently performed in *Eucalyptus* (i.e., mapping populations of more than 1,000 individuals).

Map-based cloning strategies are particularly appropriate when the biochemical basis of the target trait is unclear, thus precluding the use of a candidate-gene approach. Over the last few years a number of genes responsible for QTLs have been cloned in plants, although still restricted to rice, maize, tomato, and *Arabidopsis* (see Morgante and Salamini 2003 for a review). With the exception of the maize *tb1* gene cloned by transposon tagging, the others were all identified by positional cloning. These genes code for transcription factors, proteins involved in metabolism or active in signal transduction pathways. Two of the three genes identified as responsible for QTLs for flowering times in rice were homologs of *Arabidopsis* genes affecting flowering time. Comparative genome analysis of a *Eucalyptus* genome draft to the *Arabidopsis* genome could significantly accelerate the prediction of the molecular basis of common traits between these two species.

Limited Mendelian genetic analysis has been performed in *Eucalyptus* so far, mostly due to the fact that very few simply inherited traits have been identified in eucalypt species that could be considered as interesting and potential targets for positional cloning. Exceptions could be the recently mapped major QTL for *Puccinia psidii* rust resistance in *E. grandis* (Junghans et al. 2003) and the early-flowering QTL also mapped in *E. grandis* (Missiaggia and Grattapaglia 2005). Most economically relevant traits mapped in *Eucalyptus* are, in fact, multifactorial and controlled by unknown genes that can be genetically mapped as QTLs but not easily identified. Nevertheless, traits related to wood quality typically display high heritability. Major-effect QTLs found and validated for traits such as wood basic density could be targeted for positional cloning. In human genetics the availability of microsatellite maps was a key step toward the subsequent positional cloning of numerous human disease genes (Kong et al. 2002) as it allowed comparing and consolidating segregation data from different affected families. In *Eucalyptus* the situation is similar. With the availability and widespread use of a reference microsatellite map that allows detailed comparative mapping and validation of major QTLs across multiple pedigrees, the perspectives of map-based cloning are tangible. An interesting example could be a QTL for wood density mapped in *E. globulus* by Thamarus et al. (2004). As pointed out in that report, this QTL could be the same as that reported in *E. urophylla/E. grandis* (Verhaegen et al 1997; Gion 2001) since the LGs appear homologous on the basis of common microsatellite markers.

To consider the possibility of positional cloning in *Eucalyptus*, high-resolution mapping experiments are required first, to attempt narrowing down the most probable location of a target QTL to a specific BAC contig. High-resolution mapping involves (1) choosing appropriate segregating pedigrees, (2) generating extended progeny sets of thousands of plants, (3) carrying out accurate phenotyping, and (4) genotyping with high-throughput markers that allow saturation of the target genomic regions followed by the conversion of cosegregating markers to sequence tagged markers. At this level of resolution, crop plants benefit from the availability of near-isogenic lines (NILs) differing specifically for the QTL region and on the analysis of thousands of progenies from their cross. This has been the common way by which genes have been map-based cloned in crop plants and *Arabidopsis*. A more generally applicable alternative has been bulk segregant analysis (BSA), which involves screening phenotypically contrasting bulks for regional mapping (Michelmore et al. 1991).

In genetically heterogeneous *Eucalyptus*, the development of NILs does not seem to be a feasible goal in the near future. BSA, on the other hand, has been successfully used for high-resolution mapping of the *Ppr1* locus that confers resistance to *P. psiidi* rust (Junghans et al. 2003). For complex wood and growth traits, well-planned and large (>2,000) interspecific outbred F_2 populations from species with contrasting phenotypes have to be generated. Phenotypic differences between species may be due to fixed alleles within the species so that populations typically segregate abundantly both genetically and phenotypically. F_2 backcross populations involving *E. grandis* and *E. globulus* have been successfully used for QTL (Myburg 2001) and expression QTL mapping (Kirst et al. 2004, 2005). Within the genetic limitations of *Eucalyptus*, these would be the best populations for map-based cloning of genes responsible for QTLs. Given the improved methods for controlled pollination (Harbard et al. 1999) ,generating large progeny sizes can be routinely done in *Eucalyptus*. Backcross and intercross F_2 families with over 2,000 trees were generated and planted in the experimental network of the *Genolyptus* project in Brazil to be potentially used to clone validated QTLs.

Precise phenotypic assessment of large segregating progeny sets is an essential aspect in the high-resolution mapping step. This should be achievable in *Eucalyptus* for simply inherited binary traits such as disease resistance or flowering mutations. For example, high-resolution mapping allowed landing on single BAC clones containing target genes in heterozygous fruit trees with small genomes such as *Citrus* (Deng et al. 2001) and plum (Claverie et al. 2004). Both reports, however, aimed at discrete disease-resistance phenotypes. Continuously inherited traits such as wood properties represent a challenge. Cloning each one of the segregating progeny individuals would maximize heritability for the traits and provide a way to phenotype plants at a much higher level of accuracy. Cloning would also allow detailed phenotyping by destructive sampling when whole tree measurements are needed. Improved statistical mapping methods will also play a crucial role when attempting to narrow down the most probable location of the QTL and identify a BAC contig comprising the desired gene(s).

High-resolution mapping requires complementary genotyping technologies to microsatellites. Microsatellites are key to arrive at and carry out inter-pedigree validation of genomic segments that contain QTLs. However, the number of microsatellites available in a particular genomic region is still insufficient to allow fine mapping at below centiMorgan resolution. Maybe in the near future, with the availability of a shotgun draft of *Eucalyptus*, it will be possible to develop thousands of microsatellites or at least have the possibility of mining new microsatellites on demand to target specific regions of interest. At this time, markers such as AFLP or variations thereof that intensively scan the genome for single-nucleotide and indel polymorphism would be the choice (Myburg et al. 2001; Lezar et al. 2004). Alternatively, end sequences of the BAC clones that constitute a contig can be used to derive both microsatellite and STS markers. The success of high-resolution mapping, however, is also very much influenced by the rate of recombination in the target region. An instructive example was reported in poplar where high-resolution mapping was carried out around a gene that confers resistance to poplar leaf rust, a clearly defined phenotype with single-gene inheritance. A high-resolution map was constructed with 19 AFLP markers spanning 2.73 cM. However, a severely reduced recombination rate in the region failed to delimit the *MXC3* locus within a 300-kbp interval defined by the overlapping BAC clones (Stirling et al. 2001). In this case, the most straightforward solution would be to shotgun sequence and assemble the whole BAC contig, which should not be a problem with current costs and throughput of sequencing.

4.5.3 Future Perspective: Overcoming Challenges to Map-Based Cloning in *Eucalyptus*

Given the challenges faced in positionally cloning a gene in *Eucalyptus*, efforts should be spent to enrich a reference microsatellite genetic map with hundreds, or even thousands, of candidate genes. Genes mapped and colocalized with QTLs or simply inherited phenotypes would make appealing positional candidate genes. This is true especially for those genes whose deduced function suggests that it could be the source of genetic variation in the trait in question. With the current availability of EST collections and the forthcoming genomic shotgun of *E. camaldulensis*, large numbers of gene sequences will be available to allow the design of PCR-based strategies or oligoarray genotyping (Borevitz et al. 2003) to carry out a massive gene-mapping effort (M. Kirst personal communication).

Finally, once a BAC clone carrying the target gene or a positional candidate gene is identified, an efficient transformation system of *Eucalyptus* will be required to carry out formal complementation tests. Although some transformation protocols have recently been developed for *Eucalyptus* (Kawazu et al. 2003; Tournier et al. 2003), all of these are still genotype dependent. While this is a limitation when planning to transform a specific elite clone, it should not represent a problem in the final stage of a map-based cloning work if the phenotype altered by the cloned gene can be accurately measured in the easily transformable genotype.

In concluding this section, it is clear that there are still several genomic and genetic tools that need to be developed or improved before one can seriously consider map-based cloning in *Eucalyptus*. While the genomic tools should come online in the next few years, the biggest challenge that deserves thinking about and planning still lies in the clear definition of a target gene or QTL and the development and phenotyping of the appropriate populations. QTL mapping and validation together with association mapping experiments will, in effect, be the fundamental information resources for any future map-based cloning effort in *Eucalyptus* and thus should receive continued and improved attention by the eucalypt genome community. The challenge thereafter will be to use this information to develop molecular breeding tools that can be successfully integrated into current eucalypt breeding programs.

4.6 Conclusions: Opportunities for Genome Research and Molecular Breeding in *Eucalyptus*

Eucalyptus tree species and their hybrids form the basis of the largest hardwood plantation crop in the world. These plantations produce the raw materials for multibillion-dollar processing and manufacturing industries based on wood fiber and timber. In some of these industries, the emphasis is shifting from traditional products such as pulp and paper to novel, lignocellulose-based polymers for use in future processing technologies. With the recent increase in oil-based energy prices, there is also renewed interest in more efficient use of the renewable biomass (e.g., lignin and hemicellulose) traditionally viewed as low-value byproducts of pulping. The idea of using the tree as a biorefinery for a range of secondary products (e.g., ethanol) will be made feasible as much by the new energy economy as by our ability to breed or genetically engineer trees that are much more suitable for such processing. Fast-growing eucalypt hybrids are already able to capture very large amounts of biomass per unit land. Further genetic improvement and domestication of eucalypt trees will provide a truly renewable source of timber, fiber, energy, and bioproducts for future generations. As discussed throughout this chapter, the completion of the *Eucalyptus* genome sequence and development of associated genomic mapping resources will be important endeavors in the molecular domestication of this important fiber crop. To conclude this chapter, we provide a brief summary of opportunities that exist for genome research and molecular breeding in *Eucalyptus*.

4.6.1 A Community Linkage Map

Research communities of model and crop plant species have benefited much from the availability of shared mapping resources such as recombinant inbred line (RIL) populations (Lister and Dean 1993) or doubled-haploid populations (Lu et al. 1996). Such populations have provided a stable source of DNA so that many researchers could contribute to the mapping of markers and genes. The opportunity now exists to generate a shared mapping pedigree for *Eucalyptus*, preferably using the *E. camaldulensis* or *E. grandis* clones donated for genome sequencing in interspecific crosses. The F_1 progeny (or later F_2 backcross progeny) can be immortalized by maintaining the plants as pruned hedges for future leaf sampling and DNA isolation. Such a resource will allow *Eucalyptus* researchers worldwide to contribute to the mapping of large numbers of genes and markers onto a shared community linkage map, which will provide a useful link between genetic linkage data and the genome sequence (see below).

4.6.2 An Integrated Physical and Genetic Linkage Map

The *E. camaldulensis* genome sequencing effort in Japan is proceeding without a genetic or physical map of the particular genotype (T. Hibino personal communication). This means that the genome sequence will eventually be made available as a large number of unordered contigs. An opportunity exists to generate a high-density (and high-resolution) genetic linkage map of *E. camaldulensis* and to use the same markers to anchor sequence (BAC) contigs onto the genetic linkage map. High-throughput marker systems such as AFLP could be used to genotype mapping progeny and BAC pools in order to obtain an integrated genetic and physical mapping framework (Klein et al. 2000). However, microarray-based marker systems such as Diversity Array Technology (DArT, Jaccoud et al. 2001; Lezar et al. 2004) or oligoarray markers (e.g., Borevitz et al. 2003) now hold the promise to simultaneously map thousands of anonymous or gene-based markers onto genetic and physical maps (M. Kirst personal communication). Such a high-resolution, integrated physical and genetic linkage map will facilitate map-based cloning of genes underlying QTLs. The *E. grandis* BAC library and BAC end sequences constructed by the *Genolyptus* project in Brazil (see previous section) will provide further opportunities for cross-linking of genetic and physical (sequence) data.

4.6.3 Comparative Genome Mapping

A generic genotyping chip (e.g., DArT array) and a set of highly transferable SSR markers will be useful for comparative genome mapping in *Eucalyptus*. As discussed earlier in the chapter, the first priority would be to construct comparative genetic maps of species in the sections *Maidenaria*, *Exsertaria*, and *Latoangulatae* in the subgenus *Symphyomyrtus*. The availability of comparative maps of the commercially important eucalypts will allow the integration of QTL information and the dissection of QTL multiallelism. Moreover, high-density genotyping (e.g., using microarrays) will provide the opportunity to study detailed, genomewide patterns of genome evolution and reproductive isolation in *Eucalyptus* (Myburg et al. 2004), which will greatly assist hybrid breeding in this genus. A genotyping chip with thousands of markers will also facilitate genomewide selection in intra- and interspecific pedigrees and will be particularly useful for advanced-generation hybrid breeding. It may even make backcross introgression feasible if combined with accelerated breeding techniques.

4.6.4 Association Genetics

The successful application of molecular breeding technologies in *Eucalyptus* will depend heavily on our ability to first demonstrate the breeding value of molecular polymorphisms (such as SNPs or SNP haplotypes) in well-designed association genetic studies. The first generation of association genetic studies in forest trees was based on the analysis of allelic diversity in typically fewer than 20 selected candidate genes. These studies have revealed that, although some candidate gene associations can be detected (e.g., Thumma et al. 2005), it is very difficult to predict an appropriate set of candidate genes for any trait of interest. We simply do not understand the distribution of molecular genetic variation in tree genomes well enough to predict the location of trait-altering polymorphisms. Instead, a much larger (and unbiased) set of genes and regulatory sequences have to be interrogated for allelic diversity and trait association. The most valuable associations may be present in genes (or their promoters) that would not necessarily be predicted to be candidates for a trait of interest. It will, however, be far too expensive for average *Eucalyptus* research groups to survey such large numbers of genes using current SNP discovery protocols based on allele sequencing.

Several alternatives to sequencing have been proposed to discover and type allelic polymorphism in a large set of individuals. One very promising technique is Ecotilling (Comai et al. 2004), an adaptation of the mutation detection technology used in Targeting Induced Local Lesions in Genomes (TILLING, Colbert et al. 2001). SNPs, small insertions and deletions, and microsatellites can be efficiently detected and typed in gene regions of up to 800 bp using this technique. It is also a low-cost technology that may provide an efficient way to survey natural variation in *Eucalyptus*. However, array-based genotyping (e.g., Borevitz et al. 2003; Hazen et al. 2005) may be the only approach that will allow sufficient throughput for association genetic analysis of thousands of genes at a time. Ultradense arrays of short (25 nt) oligonucleotides tiling gene and promoter regions may be useful to simultaneously discover and type SNPs and other polymorphisms in tree populations (Kirst 2004). Such arrays may even be useful for molecular breeding once marker-trait associations have been firmly established.

4.6.5 Integrative Genomics

The ability to measure the transcript levels of thousands of genes in mapping pedigrees has created the opportunity to study the genetic regulation of global gene expression patterns in *Eucalyptus* (Kirst et al. 2005). As discussed earlier in this chapter, transcript abundance data can be integrated with genetic linkage data to map expression QTLs (eQTLs) underlying transcript-level variation of individual genes. An eQTL represents a polymorphism in or near a gene (i.e., cis-acting), or in a transcription factor of the gene (i.e., trans-acting), that affects the expression level of the gene. Trans-acting polymorphisms in key regulatory genes are visualized by the colocalization (clustering) of eQTLs identified for the individual target genes of the putative transcription factor. When performed in F_2 interspecific pedigrees, eQTL mapping becomes a powerful tool to study differentiation in the regulation of gene expression among eucalypt species. Similarly, eQTL studies in intraspecific crosses will expose polymorphisms that underlie gene expression variation within species. Such polymorphisms may explain much of the phenotypic differentiation (within and between species) that is of interest to eucalypt breeders. Furthermore, eQTL-trait associations can be confirmed by the colocation of eQTLs and trait QTLs (Kirst et al. 2004), much like QTL-candidate gene colocation is used to identify positional candidate genes. Markers flanking eQTLs may therefore be valuable tools for marker-assisted breeding, once the breeding value of the eQTLs has been demonstrated.

eQTL studies in *Eucalyptus* are currently constrained by the high cost of microarray analysis and the availability of a whole-transcriptome arrays. Large numbers of arrays have to be performed in order to achieve adequate statistical confidence and power for eQTL detection. At US $300 to US $1,000 per hybridization (cDNA vs. oligoarray technologies) and at least one hybridization per progeny member, the cost of an eQTL mapping experiment will typically be more than US $100,000. Furthermore, eQTL experiments only provide data on gene expression variation in a specific tissue at a specific developmental stage. Whole-transcriptome (>30,000 gene) oligonucleotide arrays will become available for eQTL mapping as the *E. camaldulensis* genome-sequencing effort progresses (T. Hibino personal communincation). In the meantime, several private consortia will be able to synthesize large (>20,000 gene) arrays based on large EST collections.

Other forms of integrative genomics may also provide opportunities for genome research in *Eucalyptus*. For example, the increasing array of technologies for large-scale analysis of protein abundance may allow the integration of proteome and genome mapping data to identify polymorphisms that underlie variation in protein abundance, i.e., protein expression QTLs or pQTLs (also called protein quantity loci or PQLs, Amiour et al. 2003). Similarly, large-scale analysis of metabolite levels (metabolomics) can be integrated with genome mapping data to detect loci (mQTLs) underlying variation in metabolite levels. Variation in the proteome may be especially informative since transcript abundance is not sufficient to predict the structure, function, amount, and activity of the proteins in the cell. The complexity of the proteome is also much higher than that of the transcriptome. For instance, the human genome, which consists of ca. 30,000 genes, is expected to encode between 200,000 and 2 million proteins, mostly due to alternative splicing and posttranslational modification. Very little is known about the situation in plants, but it is assumed that plant proteome complexity will be of a similar order of magnitude (Rose et al. 2004).

The dissection of the genetic basis of the variation in individual protein amounts may prove to be a very powerful approach to select "candidate proteins," as illustrated first by studies on the effect of water stress in maize (Riccardi et al. 2004; Vincent et al. 2005). More recently, the interest and predictive potential of the PQL/candidate protein approach in forest trees was demonstrated by Plomion et al. (2004). The authors demonstrated the colocalization of the gene encoding glutamine synthetase, the corresponding PQL and a QTL for early height growth providing a strong indication that the enzyme is involved in the control of juvenile growth variation in maritime pine. The integration of transcriptome, proteome, metabolome, and genome mapping data clearly holds great promise for the identification of trait-linked markers for molecular breeding in *Eucalyptus*.

4.6.6 Comparative Genomics

The completion of the genome sequences of *Arabidopsis thaliana* (The Arabidopsis Genome Initiative 2000) and *Oryza sativa* (Yu et al. 2001; Goff et al. 2002), the recent public release of the complete genome sequence of *Populus trichocarpa* (http://genome.jgi-psf.org/poptr1/poptr1.home.html), the ongoing *E. camaldulensis* genome sequencing effort, and a possible DOE-funded *E. grandis* genome-sequencing effort will provide many excellent opportunities for comparative genomics research. Fundamental questions can be addressed regarding the evolution of woody and herbaceous growth forms. Comparative analysis of the *Eucalyptus* and poplar genomes will allow researchers to determine whether these two woody genera have evolved different solutions for challenges such as being large, long-lived organisms facing a multitude of biotic and abiotic stresses throughout their lifetimes.

Poplar was chosen to be sequenced as a model tree genome because it fulfilled a number of criteria that were required for a model system: (1) a relatively small genome size (500 Mbp), only 4 times larger than the genome of *Arabidopsis*, but 50 times smaller than the genome of pine; (2) diploid inheritance ($n = 19$); (3) facile clonal propagation; (4) rapid growth; and (5) the availability of an efficient transformation procedure via *Agrobacterium*. Although most eucalypt species have very similar characteristics such as relatively small genome sizes (ranging from 370 to 700 Mbp), diploid inheritance ($n = 11$), facile clonal propagation, and fast growth, the lack of efficient genetic transformation methods has hampered *Eucalyptus* from becoming the preferred model tree. Moreover, the huge commercial potential of eucalypts has fostered a situation in which access to genomic resources is restricted to a small number of private research consortia. These limitations may be overcome with the eventual public release of the *Eucalyptus* genome sequence and the development of public resources for genome research by the *Eucalyptus* Genome Network (EUCAGEN). Nevertheless, *Eucalyptus* research will benefit tremendously from model studies in *Arabidopsis* and poplar, which will help to focus hypothesis-driven research in *Eucalyptus* and expedite the development of molecular breeding tools for this important fiber crop.

References

Adams WT, Joly RJ (1980) Genetics of allozyme variants in loblolly pine. J Hered 71:33–40

Ahuja MR (2001) Recent advances in molecular genetics of forest trees. Euphytica 121:173–195

Amiour N, Merlino M, Leroy P, Branlard G (2003) Chromosome mapping and identification of amphiphilic proteins of hexaploid wheat kernels. Theor Appl Genet 108:62–72

Apiolaza LA, Raymond CA, Yeo BJ (2005) Genetic variation of physical and chemical wood properties of *Eucalyptus globulus*. Silvae Genet 54:160–166

Ashton DH (2000) Ecology of eucalypt regneration. In: Brown BN (ed) Diseases and Pathogens of Eucalypts. CSIRO, Collingwood, Australia, pp 35-46

Astorga R, Soria F, Basurco F, Toval G (2004) Diversity analysis and genetic structure of *Eucalyptus globulus* Labill. In: Borralho NMG, Pereira JS, Marques C, Coutinho J, Madeira M, Tomé M (eds) '*Eucalyptus* in a Changing World'. RAIZ, Instituto Investigação de Floresta e Papel, Aveiro, Portugal, pp 351–358

Barbour R, Potts BM, Vaillancourt RE, Tibbits WN, Wiltshire RJE (2002) Gene flow between introduced and native *Eucalyptus* species. New For 23:177–191

Barbour RC, Potts BM, Vaillancourt RE (2003) Gene flow between introduced and native *Eucalyptus*: exotic hybrids are establishing in the wild. Aust J Bot 51:429–439

Barbour RC, Potts BM, Vaillancourt RE (2005) Pollen dispersal from exotic eucalypt plantations. Conserv Genet 6:253–257

Bermúdez Alvite JD, Touza Vázquezm MC, Infante FS (2002) Manual da Madeira de Eucalipto Comum. CIS-Madera, Xunta de Galicia

Binkley D, Stape JL (2004) Sustainable management of eucalypt plantations in a changing world. In: Tomé M (ed) IUFRO Conference: *Eucalyptus* in a Changing World, 11-15 Oct 2004, RAIZ, Instituto Investigação de Floresta e Papel, Aveiro, Portugal

Boland DJ, Brooker MIH, Chippendale GM, Hall N, Hyland BPM, Johnston RD, Kleinig DA, Turner JD (1985) Forest Trees of Australia. Australian Government Publishing Service, Melbourne, Australia

Boland DJ, Brophy JJ, House APN (1991) *Eucalyptus* leaf oils. Inkata, Melbourne/Sydney, Australia

Borevitz JO, Liang D, Plouffe D, Chang HS, Zhu T, Weigel D, Berry CC, Winzeler E, Chory J (2003) Large-scale identification of single-feature polymorphisms in complex genomes. Genome Res 13:513–523

Borralho NMG (2001) The purpose of breeding is breeding for a purpose. In: Barros S (ed) Developing the Eucalypt of the Future. IUFRO International Symposium, 10-15 Sept 2001, INFOR, Chile

Borralho NMG, Dutkowski GW (1998) Comparison of rolling front and discrete generation breeding strategies for trees. Can J For Res 28:987–993

Borralho NMG, Cotterill PP, Kanowski PJ (1993) Breeding objectives for pulp production of *Eucalyptus globulus* under different industrial cost structures. Can J For Res 23:648–656

Botstein D, White RL, Skolnick M, Davis RW (1980) Construction of a genetic linkage map in man using restriction fragment length polymorphisms. Am J Hum Genet 32:314–331

Bowman D (1998) Tansley Review No. 101 - The impact of Aboriginal landscape burning on the Australian biota. New Phytol 140:385–410

Bradshaw HD, Stettler RF (1995) Molecular genetics of growth and development in Populus. 4. Mapping QTLs with large effects on growth, form, and phenology traits in a forest tree. Genetics 139:963–973

Brem RB, Yvert G, Clinton R, Kruglyak L (2002) Genetic dissection of transcriptional regulation in budding yeast. Science 296:752–755

Brondani RPV, Brondani C, Tarchini R, Grattapaglia D (1998) Development, characterization and mapping of microsatellite markers in *Eucalyptus grandis* and *E. urophylla*. Theor Appl Genet 97:816–827

Brondani RPV, Brondani C, Grattapaglia D (2002) Towards a genus-wide reference linkage map for *Eucalyptus* based exclusively on highly informative microsatellite markers. Mol Genet Genom 267:338–347

Brooker MIH (2000) A new classification of the genus *Eucalyptus* L'Her. (Myrtaceae). Aust Syst Bot 13:79–148

Brune A, Zobel BJ (1981) Genetic base populations, gene pools and breeding populations for *Eucalyptus* in Brazil. Silvae Genet 30:146–149

Bundock PC, Hayden M, Vaillancourt RE (2000) Linkage maps of *Eucalyptus globulus* using RAPD and microsatellite markers. Silvae Genet 49:223–232

Burley J, Kanowski PJ (2005) Breeding strategies for temperate hardwoods. Forestry 78:199–208

Burrows GE (2002) Epicormic strand structure in *Angophora*, *Eucalyptus* and *Lophostemon* (Myrtaceae) - implications for fire resistance and recovery. New Phytol 153:111–131

Butcher PA, Otero A, McDonald MW, Moran GF (2002) Nuclear RFLP variation in *Eucalyptus camaldulensis* Dehnh. from northern Australia. Heredity 88:402–412

Butcher PA, Williams ER (2002) Variation in outcrossing rates and growth in *Eucalyptus camaldulensis* from the Petford region, Queensland; evidence of outbreeding depression. Silvae Genet 51:6–12

Byrne M (1999) High genetic identities between three oil mallee taxa, *Eucalyptus kochii* ssp. *kochii* ssp. *plenissima* and *E. horistes*, based on nuclear RFLP analysis. Heredity 82:205–211

Byrne M, Macdonald B (2000) Phylogeography and conservation of three oil mallee taxa, *Eucalyptus kochii* ssp *kochii*, ssp *plenissima* and *E. horistes*. Aust J Bot 48:305–312

Byrne M, Moran GF, Tibbits WN (1993) Restriction map and maternal inheritance of chloroplast DNA in *Eucalyptus nitens*. J Hered 84:218–220

Byrne M, Moran GF, Murrell JC, Tibbits WN (1994) Detection and inheritance of RFLPs in *Eucalyptus nitens*. Theor Appl Genet 89:397–402

Byrne M, Murrell JC, Allen B, Moran GF (1995) An integrated genetic linkage map for *Eucalyptus* using RFLP, RAPD and isozyme markers. Theor Appl Genet 91:869–875

Byrne M, Marquezgarcia MI, Uren T, Smith DS, Moran GF (1996) Conservation and genetic diversity of microsatellite loci in the genus *Eucalyptus*. Aust J Bot 44:331–341

Byrne M, Murrell JC, Owen JV, Kriedemann P, Williams ER, Moran GF (1997a) Identification and mode of action of quantitative trait loci affecting seedling height and leaf area in *Eucalyptus nitens*. Theor Appl Genet 94:674–681

Byrne M, Murrell JC, Owen JV, Williams ER, Moran GF (1997b) Mapping of quantitative trait loci influencing frost tolerance in *Eucalyptus nitens*. Theor Appl Genet 95:975–979

Byrne M, Parrish TL, Moran GF (1998) Nuclear RFLP diversity in *Eucalyptus nitens*. Heredity 81:225–233

Campinhos E, Ikemori YK (1977) Tree improvement program of *Eucalyptus* spp.: preliminary results. In: Third World Consultation on Forest Tree Breeding. CSIRO, Canberra, Australia, pp 717-738

Campinhos E, Ikemori YK (1989) Selection and management of the basic population *Eucalyptus grandis* and *E. urophylla* established at Aracruz for the long term breeding programme. In: Matheson AC (ed) IUFRO Conference, Breeding Tropical Trees: Population Structure and Genetic Improvement Strategies in Clonal and Seedling Forestry. Oxford Forestry Institute, Oxford, UK and Winrock International, Arlington, VA, Pattaya, Thailand, pp 169–175

Cañas I, Soria F, López G, Astorga R, Toval G (2004) Genetic parameters for rooting trait in *Eucalyptus globulus* (Labill.). In: Tomé M (ed) IUFRO Conference: *Eucalyptus* in a Changing World, 11-15 Oct 2004, RAIZ, Instituto Investigação de Floresta e Papel, Aveiro, Portugal, pp 159–160

Carnero Diaz M, Martin F, Tagu D (1996) Eucalypt alpha-tubulin: cDNA cloning and increased level of transcripts in ectomycorrhizal root system. Plant Mol Biol 31:905–910

Carocha VJ, Melo M, Araújo JA, Borralho N, Marques CMP (2004) Candidate gene mapping in *E. globulus*. In: Tomé M (ed) IUFRO Conference: *Eucalyptus* in a Changing World, 11-15 Oct 2004, RAIZ, Instituto Investigação de Floresta e Papel, Aveiro, Portugal, pp 359-372

Castelo AT, Martins W, Gao GR (2002) TROLL - tandem repeat occurrence locator. Bioinformatics 18:634–636

Causse M, Rocher JP, Pelleschi S, Barriere Y, De Vienne D, Prioul JL (1995) Sucrose phosphate synthase: an enzyme with heterotic activity correlated with maize growth. Crop Sci 35:995–1001

Cauvin B, Potts BM (1991) Selection for extreme frost resistance in *Eucalyptus*. In: Schönau APG (ed) Intensive Forestry: The Role of Eucalypts. Proceedings of the IUFRO Symposium, P202-01 Productivity of Eucalypts, 2-6 Sept 1991. Southern African Institute of Forestry, Durban, South Africa, vol 1, pp 209–220

Chambers PGS, Borralho NMG (1997) Importance of survival in short-rotation tree breeding programs. Can J For Res 27:911–917

Chambers PGS, Potts BM, Tilyard PG (1997) The genetic control of flowering precocity in *Eucalyptus globulus* ssp. *globulus*. Silvae Genet 46:207–214

Chang YL, Tao Q, Scheuring C, Ding K, Meksem K, Zhang HB (2001) An integrated map of *Arabidopsis thaliana* for functional analysis of its genome sequence. Genetics 159:1231–1242

Chen M, Presting G, Barbazuk WB, Goicoechea JL, Blackmon B, Fang G, Kim H, Frisch D, Yu Y, Sun S, Higingbottom S, Phimphilai J, Phimphilai D, Thurmond S, Gaudette B, Li P, Liu J, Hatfield J, Main D, Farrar K, Henderson C, Barnett L, Costa R, Williams B, Walser S, Atkins M, Hall C, Budiman MA, Tomkins JP, Luo M, Bancroft I, Salse J, Regad F, Mohapatra T, Singh NK, Tyagi AK, Soderlund C, Dean RA, Wing RA (2002) An integrated physical and genetic map of the rice genome. Plant Cell 14:537–545

Chen X, Sullivan PF (2003) Single nucleotide polymorphism genotyping: biochemistry, protocol, cost and throughput. Pharmacogenomics J 3:77–96

Claverie M, Dirlewanger E, Cosson P, Bosselut N, Lecouls AC, Voisin R, Kleinhentz M, Lafargue B, Caboche M, Chalhoub B, Esmenjaud D (2004) High-resolution mapping and chromosome landing at the root-know nematode resistance locus Ma from Myrobalan plum using a large-insert BAC DNA library. Theor Appl Genet 109:1318–1327

Coe E, Cone K, McMullen M, Chen SS, Davis G, Gardiner J, Liscum E, Polacco M, Paterson A, Sanchez-Villeda H, Soderlund C, Wing R (2002) Access to the maize genome: an integrated physical and genetic map. Plant Physiol 128:9–12

Colbert T, Till BJ, Tompa R, Reynolds S, Steine MN, Yeung AT, McCallum CM, Comai L, Henikoff S (2001) High-throughput screening for induced point mutations. Plant Physiol 126:480–484

Comai L, Young K, Till BJ, Reynolds SH, Greene EA, Codomo CA, Enns LC, Johnson JE, Burtner C, Odden AR, Henikoff S (2004) Efficient discovery of DNA polymorphisms in natural populations by Ecotilling. Plant J 37:778–786

Costa e Silva J, Borralho NMG, Potts BM (2004) Additive and non-additive genetic parameters from clonally replicated and seedling progenies of *Eucalyptus globulus*. Theor Appl Genet 108:1113–1119

Costa e Silva J, Dutkowski GW, Borralho NMG (2005) Across-site heterogeneity of genetic and environmental variances in the genetic evaluation of *Eucalyptus globulus* trials for height growth. Ann For Sci 62:183–191

Cotterill P, Macrae S, Brolin A (1999) Growing eucalypt for high-quality papermaking fibres. Appita J 52:79–83

Coutinho TA, Wingfield MJ, Alfenas AC, Crous PW (1998) *Eucalyptus* rust: a disease with the potential for serious international implications. Plant Dis 82:819–825

Crisp M, Cook L, Steane D (2004) Radiation of the Australian flora: what can comparisons of molecular phylogenies across multiple taxa tell us about evolution of diversity in present-day communities? Philos Trans R Soc Lond B Biol Sci 359:1551–1571

Dale GT (2002) Salt tolerant eucalypts for commercial forestry: Progress and promise. In: Private Forestry - Sustainable, Accountable and Profitable, Session Paper 351. Proc of the

Australian Forest Growers, 2002 National Conference, 13-16 Oct 2002, Albany, Western Australia

Dale GT, Aitken KS, Sasse JM (2000) Development of salt-tolerant *E. camaldulensis* × *E. grandis* hybrid clones using phenotypic selection and genetic mapping. In: Proc QFRI/CRC-SPF Symp - Hybrid breeding and genetics of forest trees, Noosa, Australia, pp 227–233

Damerval C, Maurice A, Josse JM, de Vienne D (1994) QTL underlying gene product variation: A novel perspective for analysing regulation of genome expression. Genetics 137:289–301

de Assis TF (2000) Production and use of *Eucalyptus* hybrids for industrial purposes. In: Nikles DG (ed) Hybrid Breeding and Genetics of Forest Trees. Proc QFRI/CRC-SPF Symp, 9-14 April, Noosa, Dept of Primary Industries, Brisbane, Australia, pp 63–74

de Assis TF (2001) The evolution of technology for cloning *Eucalyptus* in a large scale. In: IUFRO Conf: Developing the Eucalypt of the Future, 10-15 Sept 2001, INFOR, Valdivia, Chile

de Assis TFd, Warburton P, Harwood C (2005) Artificially induced protogyny: an advance in the controlled pollination of *Eucalyptus*. Aust For 68:27–33

de Resende MDV, Thompson R (2004) Factor analytic multiplicative mixed models in the analysis of multi-environment forest field trials. In: Tomé M (ed) IUFRO Conference: *Eucalyptus* in a Changing World, 11-15 Oct 2004, RAIZ, Instituto Investigação de Floresta e Papel, Aveiro, Portugal, pp 123–129

de Vienne D, Leonardi A, Damerval C, Zivy M (1999) Genetics of proteome variation for QTL characterization: application to drought-stress responses in maize. J Exp Bot 50:291–302

Deng Z, Huang S, Ling P, Yu C, Tao Q, Chen C, Wendell MK, Zhang HB, Gmitter FG Jr (2001) Fine genetic mapping and BAC contig development for the citrus tristeza virus resistance gene locus in *Poncirus trifoliata* (Raf.). Mol Genet Genom 265:739–747

Devey ME, Delfino-Mix A, Kinloch, BB Jr, Neale DB (1995) Random amplified polymorphic DNA markers tightly linked to a gene for resistance to white pine blister rust in sugar pine. Proc Natl Acad Sci USA 92:2066–2070

Diatchenko L, Lau YFC, Campbell AP, Chenchik A, Moqadam F, Huang B, Lukyanov S, Lukyanov K, Gurskaya N, Sverdlov ED, Siebert PD (1996) Suppression subtractive hybridisation: a method for generating differentially regulated or tissue-specific cDNA probes and libraries. Proc Natl Acad Sci USA 93:6025–6030

Dillner B, Ljunger Å, Herud OA, Thune-Larsen E (1971) The breeding of *Eucalyptus globulus* on the basis of wood density, chemical composition and growth rate. Timber Bull Eur 23:120–151

Ding Y, Johnson MD, Colayco R, Chen YJ, Melnyk J, Schmitt H, Shizuya H (1999) Contig assembly of bacterial artificial chromosome clones through multiplexed fluorescence-labeled fingerprinting. Genomics 56:237–246

Ding Y, Johnson MD, Chen WQ, Wong D, Chen YJ, Benson SC, Lam JY, Kim YM, Shizuya H (2001) Five-color-based high-information-content fingerprinting of bacterial artificial chromosome clones using type IIS restriction endonucleases. Genomics 74:142–154

Doran JC, Bell RE (1995) Influence of genetic factors on yield of monoterpenes in leaf oils of *Eucalyptus camaldulensis* and implications for tree breeding. New For 8:363–373

Doughty RW (2000) The *Eucalyptus*. A natural and commercial history of the gum tree. The Johns Hopkins Univ Press, Baltimore, MD and London, UK

Downes GM, Hudson IL, Raymond CA, Dean GH, Michell AJ, Schimleck LR, Evans R, Muneri A (1997) Sampling plantation eucalypts for wood and fibre properties. CSIRO, Melbourne, Australia

Doyle JJ, Doyle JL (1990) Isolation of plant DNA from fresh tissue. Focus 12:13–15

Dumas P, Sun Y, Corbeil G, Tremblay S, Pausova Z, Kren V, Krenova D, Pravenec M, Hamet P, Tremblay J (2000) Mapping of quantitative trait loci (QTL) of differential stress gene expression in rat recombinant inbred strains. J Hypertens 18:545–551

Dutkowski GW, Potts BM (1999) Geographical patterns of genetic variation in *Eucalyptus globulus* ssp. *globulus* and a revised racial classification. Aust J Bot 46:237–263

Dutkowski GW, Costa e Silva J, Gilmour AR, Lopez GA (2002) Spatial analysis methods for forest genetic trials. Can J For Res 32:2201–2214

Egertsdotter U, van Zyl LM, MacKay J, Peter G, Kirst M, Clark C, Whetten R, Sederoff R (2004) Gene expression during formation of earlywood and latewood in loblolly pine: expression profiles of 350 genes. Plant Biol 6:654–663

Eldridge K, Davidson J, Harwood C, van Wyk G (1993) Eucalypt domestication and breeding. Clarendon, Oxford, UK

Evans R, Ilic J (2001) Rapid prediction of wood stiffness from microfibril angle and density. For Prod J 51:53–57

Fernandez J, Toro MA (2001) Controlling genetic variability by mathematical programming in a selection scheme on an open-pollinated population in *Eucalyptus globulus*. Theor Appl Genet 102:1056–1064

Ferreira ME, Grattapaglia D (1995) Introdução ao Uso de Marcadores RAPD e RFLP em Análise Genética. EMBRAPA - Cenargen, Brazil

Feuillet C, Boudet AM, Grima-Pettenati J (1993) Nucleotide sequence of a cDNA encoding cinnamyl alcohol dehydrogenase from *Eucalyptus*. Plant Physiol 103(4):1447

Florence RG (1996) Ecology and Silviculture of Eucalypt Forests. CSIRO, Collingwood, Australia

Foucart C, Paux E, Ladouce N, San-Clemente H, Grima-Pettenati J, Sivadon P (2006) Transcript profiling of a xylem vs phloem cDNA subtractive library identifies new genes expressed during xylogenesis in *Eucalyptus*. New Phytol 170:739–752

Franklin EC (1986) Estimation of genetic parameters through four generations of selection in *Eucalyptus grandis*. In: IUFRO Joint Meeting of Working Parties on Breeding Theory, Progeny Testing and Seed Orchards, 12-17 Oct 1986, Williamsburg, VA

Freeman JS, Milgate AW, Vaillancourt RE, Mohammed CL, Potts BM (2003) Genetic control of resistance to *Mycosphaerella* leaf disease in *Eucalyptus globulus*. Abstr 19th Genet Congr, 6-11July 2003, Melbourne, Australia, p 155

Fukuoka S, Namai H, Okuno K (1998) RFLP mapping of the genes controlling hybrid breakdown in rice (*Oryza Sativa* L.). Theor Appl Genet 97:446–449

Fullard KJ, Moran G (2003) Identification of frost tolerance genes in a *Eucalyptus* hybrid cross. Tree Biotechnology 2003, Umeå, Sweden

Gan S, Shi J, Li M, Wu K, Wu J, Bai J (2003) Moderate-density molecular maps of *Eucalyptus urophylla* S.T. Blake and *E. tereticornis* Smith genomes based on RAPD markers. Genetica 118:59–67

Gion J-M (2001) Etude de l'architecture génétique des caractères complexes chez l'*Eucalyptus*: des marqueurs anonymes aux gènes candidats. PhD thesis, University of Rennes I, France

Gion J-M, Rech P, Grima-Pettenati J, Verhaegen D, Plomion C (2000) Mapping candidate genes in *Eucalyptus* with emphasis on lignification genes. Mol Breed 6:441–449

Gion J-M, Cabannes E, Poitel M, Vigneron Ph (2005) Conservation of xylem preferentially expressed sequence tags in several subgenera of *Eucalyptus* L'Hérit. In: Plant & Animal Genome XIII conf, San-Diego

Glaubitz JC, Emebiri LC, Moran GF (2001) Dinucleotide microsatellite from *Eucalyptus sieberi*: Inheritance, diversity, and improved scoring of single-base differences. Genome 44:1014–1045

Goff SA, Ricke D, Lan TH, Presting G, Wang RL, Dunn M, Glazebrook J, Sessions A, Oeller P, Varma H, Hadley D, Hutchinson D, Martin C, Katagiri F, Lange BM, Moughamer T, Xia Y, Budworth P, Zhong JP, Miguel T, Paszkowski U, Zhang SP, Colbert M, Sun WL, Chen LL, Cooper B, Park S, Wood TC, Mao L, Quail P, Wing R, Dean R, Yu YS, Zharkikh A, Shen R, Sahasrabudhe S, Thomas A, Cannings R, Gutin A, Pruss D, Reid J, Tavtigian S, Mitchell J, Eldredge G, Scholl T, Miller RM, Bhatnagar S, Adey N, Rubano T, Tusneem N, Robinson R, Feldhaus J, Macalma T, Oliphant A, Briggs S (2002) A draft sequence of the rice genome (*Oryza sativa* L. ssp *japonica*). Science 296:92–100

Goicoechea M, Lacombe E, Legay S, Milhaevic S, Rech P, Jauneau A, Lapierre C, Pollet B, Verhaegen D, Chaubet-Gigot N, Grima-Pettenati J (2005) *EgMYB2*, a new transcriptional activator from *Eucalyptus* xylem, regulates secondary cell wall formation and lignin biosynthesis. Plant J 43:553–567

Goldman IL, Rocheford TR, Dudley JW (1994) Molecular markers associated with maize kernel oil concentration in an Illinois high protein X Illinois low protein cross. Crop Sci 34:908–915

Grattapaglia D (2004) Integrating genomics into *Eucalyptus* breeding. Gen Mol Res 3:369–379

Grattapaglia D, Bradshaw HD (1994) Nuclear DNA content of commercially important *Eucalyptus* species and hybrids. Can J For Res 24:1074–1078

Grattapaglia D, Sederoff R (1994) Genetic linkage maps of *Eucalyptus grandis* and *Eucalyptus urophylla* using a pseudo-testcross: mapping strategy and RAPD markers. Genetics 137:1121–1137

Grattapaglia D, Bertolucci FLG, Sederoff R (1995) Genetic mapping of QTLs controlling vegetative propagation in *Eucalyptus grandis* and *E. urophylla* using a pseudo-testcross mapping strategy and RAPD markers.Theor Appl Genet 90:933–947

Grattapaglia D, Bertolucci FLG, Penchel R, Sederoff R (1996) Genetic mapping of quantitative trait loci controlling growth and wood quality traits in *Eucalyptus grandis* using a maternal half-sib family and RAPD markers. Genetics 144:1205–1214

Grattapaglia D, Lourenço RT, Santos SN, Povoa A, Carocha V, Pappas GJ Jr, Brommonschenkel SH (2004) Fishing for complete lignification genes in a BAC library of *Eucalyptus grandis*. Abstr of the 50th Brazilian Genet Congr, p 1126. ISBN 85-89109-04-6

Greaves BL, Borralho NMG (1996) The influence of basic density and pulp yield on the cost of eucalypt kraft pulping: a theoretical model of tree breeding. Appita J 49:90–95

Greaves BL, Borralho NMG, Raymond CA (1997) Breeding objectives for plantation eucalypts grown for production of kraft pulp. For Sci 43:465–472

Greaves B, Hamilton M, Pilbeam D, Dutkowski G (2004a) Genetic variation in commercial properties of six and fifteen year-old *Eucalyptus globulus*. In: Tomé M (ed) IUFRO Conf: *Eucalyptus* in a Changing World, 11-15 Oct 2004, RAIZ, Instituto Investigação de Floresta e Papel, Aveiro, Portugal, pp 103-104

Greaves BL, Dutkowski GW, McRae TA (2004b) Breeding objectives for *Eucalyptus globulus* for products other than kraft pulp. In: Tomé M (ed) IUFRO Conf: *Eucalyptus* in a Changing World, 11-15 Oct 2004, RAIZ, Instituto Investigação de Floresta e Papel, Aveiro, Portugal, pp 175–180

Gregory SG, Sekhon M, Schein J, Zhao S, Osoegawa K, Scott CE, Evans RS, Burridge PW, Cox TV, Fox CA, Hutton RD, Mullenger IR, Phillips KJ, Smith J, Stalker J, Threadgold GJ, Birney E, Wylie K, Chinwalla A, Wallis J, Hillier L, Carter J, Gaige T, Jaeger S, Kremitzki C, Layman D, Maas J, McGrane R, Mead K, Walker R, Jones S, Smith M, Asano J, Bosdet I, Chan S, Chittaranjan S, Chiu R, Fjell C, Fuhrmann D, Girn N, Gray C, Guin R, Hsiao L, Krzywinski M, Kutsche R, Lee SS, Mathewson C, McLeavy C, Messervier S, Ness S, Pandoh P, Prabhu AL, Saeedi P, Smailus D, Spence L, Stott J, Taylor S, Terpstra W, Tsai M, Vardy J, Wye N, Yang G, Shatsman S, Ayodeji B, Geer K, Tsegaye G, Shvartsbeyn A, Gebregeorgis E, Krol M, Russell D, Overton L, Malek JA, Holmes M, Heaney M, Shetty J, Feldblyum T, Nierman WC,

Catanese JJ, Hubbard T, Waterston RH, Rogers J, de Jong PJ, Fraser CM, Marra M, McPherson JD, Bentley DR (2002) A physical map of the mouse genome. Nature 418:743–750

Griffin AR (1996) Genetically modified trees - the plantations of the future or an expensive distraction? Commonw For Rev For Rev 75:169–175

Griffin AR (2001) Deployment decisions - capturing the benefits of tree improvement with clones and seedlings. In: IUFRO Conf: Developing the Eucalypt of the Future, 10-15 Sept 2001, INFOR, Valdivia, Chile

Griffin AR, Burgess IP, Wolf L (1988) Patterns of natural and manipulated hybridisation in the genus *Eucalyptus* L'Herit. - a review. Aust J Bot 36:41–66

Groover A, Devey M, Fiddler T, Lee J, Megraw R, Mitchel-Olds T, Sherman B, Vujcic S, Williams C, Neale D (1994) Identification of quantitative trait loci influencing wood specific gravity in an outbred pedigree of loblolly pine. Genetics 138:1293–1300

Gupta PK, Rustgi S (2004) Molecular markers from the transcribed/expressed region of the genome in higher plants.Funct Integr Genom 4:139–62

Haldane JBS (1919) The combination of linkage values and the calculation of distances between loci of linked factors. J Genet 8:299–309

Harbard JL, Griffin AR, Espejo J (1999) Mass controlled pollination of *Eucalyptus globulus*: a practical reality. Can J For Res 29:1457–1463

Hardner CM, Potts BM (1995) Inbreeding depression and changes in variation after selfing *Eucalyptus globulus* subsp. *globulus*. Silvae Genet 44:46–54

Hardner CM, Potts BM (1997) Post-dispersal selection under mixed-mating in *Eucalyptus regnans*. Evolution 51:103–111

Hardner CM, Vaillancourt RE, Potts BM (1996) Stand density influences outcrossing rate and growth of open-pollinated families of *Eucalyptus globulus*. Silvae Genet 45:226–228

Hardner CM, Potts BM, Gore PL (1998) The relationship between cross success and spatial proximity of *Eucalyptus globulus* ssp. *globulus* parents. Evolution 52:614–618

Harkins DM, Johnson GN, Skaggs PA, Mix AD, Dupper GE, Devey ME, Kinloch BB, Neale DB (1998) Saturation mapping of a major gene for resistance to white pine blister rust in sugar pine. Theor Appl Genet 97:1355–1360

Hazen SP, Harmon FG, Pruneda-Paz FG, Schulz TF, Ecker JR, Borevitz JO, Kay SA (2005) Rapid linkage mapping in *Arabidopsis* by microarray genotyping: In: Proc Plant & Animal Genomes XIII Conf San Diego, W218

Hertzberg M, Aspeborg H, Schrader J, Andersson A, Erlandsson R, Blomqvist K, Bhalerao R, Uhlen M, Teeri TT, Lundeberg J, Sundberg B, Nilsson P, Sandberg G (2001) A transcriptional roadmap to wood formation. Proc Natl Acad Sci USA 98:14732–14737

Hill RS (2004) Origins of the southeastern Australian vegetation. Philos Trans R Soc B 359:1537–1549

Hill KD, Johnson LAS (1995) Systematic studies in the eucalypts 7. A revision of the bloodwoods, genus *Corymbia* (Myrtaceae). Telopea 6:185–504

Hill R, Truswell E, McLoughlin S, Dettmann M (1999) Evolution of the Australian flora: fossil evidence. In: George AS (ed) Flora of Australia, vol 1. Co-published by CSIRO and Australian Biological Resources Study, pp 251–320

Hingston AB, Potts BM (2005) Pollinator activity can explain variation in outcrossing rates within individual trees. Aust Ecol 30:319–324

Hingston AB, Potts BM, McQuillan PB (2004) The swift parrot, *Lathamus discolor* (Psittacidae), social bees (Apidae) and native insects as pollinators of *Eucalyptus globulus* ssp. *globulus* (Myrtaceae). Aust J Bot 52:371–379

Hittalmani S, Huang N, Courtois B, Venuprasad R, Shashidhar HE, Zhuang JY, Zheng KL, Liu, GF, Wang GC, Sidhu JS, Srivantaneeyakul S, Singh VP, Bagali PG, Prasanna HC, McLaren G, Khush GS (2003) Identification of QTL for growth- and grain yield-related traits in rice across nine locations of Asia. Theor Appl Genet 107:679–690

Holman JE, Hughes JM, Fensham RJ (2003) A morphological cline in *Eucalyptus*: a genetic perspective. Mol Ecol 12:3013–3025

Hoskins RA, Nelson CR, Berman BP, Laverty TR, George RA, Ciesiolka L, Naeemuddin M, Arenson AD, Durbin J, David RG, Tabor PE, Bailey MR, DeShazo DR, Catanese J, Mammoser A, Osoegawa K, de Jong PJ, Celniker SE, Gibbs RA, Rubin GM, Scherer SE (2000) A BAC-based physical map of the major autosomes of *Drosophila melanogaster*. Science 287:2271–2274

House SM (1997) Reproductive biology of eucalypts. In: Woinarski J (ed) Eucalypt Ecology: Individuals to Ecosystems. Cambridge University Press, Cambridge, UK, pp 30–56

Israelsson M, Eriksson ME, Hertzberg M, Aspeborg H, Nilsson P, Moritz T (2003) Changes in gene expression in the wood-forming tissue of transgenic hybrid aspen with increased secondary growth. Plant Mol Biol 52:893–903

Jaccoud D, Peng K, Feinstein D, Kilian A (2001) Diversity arrays: a solid state technology for sequence information independent genotyping. Nucleic Acids Res 29:e25

Jacobs MR (1981) Eucalypts for Planting. FAO, Rome

Janki-Ammal EK, Khosla SN (1969) Breaking the barrier to polyploidy in the genus *Eucalyptus*. Proc Ind Acad Sci (Bangalore) 20:248–249

Jansen RC, Stam P (1994) High resolution of quantitative traits into multiple loci via interval mapping. Genetics 136:1447–1455

Jones CJ, Edwards KJ, Castaglione S, Winfield MO, Sala F, Vandewiel C, Bredemeijer G, Vosman B, Matthes M, Daly A, Brettschneider R, Bettini P, Buiatti M, Maestri E, Malcevschi A, Marmiroli N, Aert R, Volckaert G, Rueda J, Linacero R, Vazquez A, Karp A (1997) Reproducibility testing of RAPD, AFLP and SSR markers in plants by a network of European laboratories. Mol Breed 3:381–390

Jones RC, McKinnon GE, Potts BM, Vaillancourt RE (2005) Genetic diversity and mating system of an endangered tree, *Eucalyptus morrisbyi*. Aust J Bot 53:366–377

Jones RC, Steane DA, Potts BM, Vaillancourt RE (2002) Microsatellite and morphological analysis of *Eucalyptus globulus* populations. Can J For Res 32:59–66

Jordan GJ, Potts BM, Kirkpatrick JB, Gardiner C (1993) Variation in the *Eucalyptus globulus* complex revisited. Aust J Bot 41:763–785

Junghans DT, Alfenas AC, Brommonschenkel SH, Oda S, Mello EJ, Grattapaglia D (2003) Resistance to rust (*Puccinia psidii* Winter) in *Eucalyptus*: mode of inheritance and mapping of a major gene with RAPD markers. Theor Appl Genet 108:175–180

Kanowski PJ, Borralho NMG (2004) Economics of tree improvement. In: Youngquist JA (ed) Encyclopedia of Forest Science. Elsevier, Oxford, pp 1561–1568

Karp CL, Grupe A, Schadt E, Ewart SL, Keane-Moore M, Cuomo PJ, Kohl J, Wahl L, Kuperman D, Germer S, Aud D, Peltz G, Wills-Karp M (2000) Identification of complement factor 5 as a susceptibility locus for experimental allergic asthma. Nat Immunol 1:221–226

Kawazu T, Doi K, Kondo K (2003) Process for transformation of mature trees of *Eucalyptus* plants. US Patent 6,563,024

Kearsey MJ, Farquhar AGL (1998) QTL analysis in plants: where are we now? Heredity 80:137–142

Kellison RC (2001) Present and future uses of eucalypts wood in the world. In: Barros S (ed) Developing the Eucalypt of the Future. IUFRO Int Symp, 10-15 Sept 2001, INFOR, Chile

Kershaw AP, Martin HA, McEwen MJRC (1994) The Neogene: a period of transition. In: Hill RS (ed) History of the Australian Vegetation: Cretaceous to Recent. Cambridge University Press, Cambridge, UK, pp 299-327

Kirst M (2004) Microarrays beyond gene expression: discovery of candidate genes for quantitative traits to large scale SNP genotyping. In: Li B, McKeand M (eds) Forest Genetics and Tree Breeding in the Age of Genomics: Progress and Future. IUFRO Joint Conf Div 2, 1-5 Nov 2005, Double Tree Guest Suites, Charleston, SC, p 341

Kirst M, Myburg AA, De Leon JP, Kirst ME, Scott J, Sederoff R (2004) Coordinated genetic regulation of growth and lignin revealed by quantitative trait locus analysis of cDNA microarray data in an interspecific backcross of *Eucalyptus*. Plant Physiol 135:2368–2378

Kirst M, Basten CJ, Myburg A, Zeng Z-B, Sederoff R (2005) Genetic architecture of transcript level variation in differentiating xylem of *Eucalyptus* hybrids. Genetics 169:2295–2303

Klein PE, Klein RR, Cartinhour SW, Ulanch PE, Dong JM, Obert JA, Morishige DT, Schlueter SD, Childs KL, Ale M, Mullet JE (2000) A high-throughput AFLP-based method for constructing integrated genetic and physical maps: progress toward a sorghum genome map. Genome Res 10:789–807

Kong A, Gudbjartsson DF, Sainz J, Jonsdottir GM, Gudjonsson SA, Richardsson B, Sigurdardottir S, Barnard J, Hallbeck B, Masson G, Shlien A, Palsson ST, Frigge ML, Thorgeirsson TE, Gulcher JR, Stefansson K (2002) A high-resolution recombination map of the human genome. Nat Genet 31:241–247

Krzywinski M, Wallis J, Gosele C, Bosdet I, Chiu R, Graves T, Hummel O, Layman D, Mathewson C, Wye N, Zhu B, Albracht D, Asano J, Barber S, Brown-John M, Chan S, Chand S, Cloutier A, Davito J, Fjell C, Gaige T, Ganten D, Girn N, Guggenheimer K, Himmelbauer H, Kreitler T, Leach S, Lee D, Lehrach H, Mayo M, Mead K, Olson T, Pandoh P, Prabhu AL, Shin H, Tanzer S, Thompson J, Tsai M, Walker J, Yang G, Sekhon M, Hillier L, Zimdahl H, Marziali A, Osoegawa K, Zhao S, Siddiqui A, de Jong PJ, Warren W, Mardis E, McPherson JD, Wilson R, Hubner N, Jones S, Marra M, Schein J (2004) Integrated and sequence-ordered BAC- and YAC-based physical maps for the rat genome. Genome Res 14:766–779

Kube PD, Raymond CA (2005) Breeding to minimise the effects of collapse in *Eucalyptus nitens*. For Genet 12:23–34

Kubisiak TL, Hebard FV, Nelson CD, Zhang JS, Bernatzky R, Huang H, Anagnostakis SL, Doudrick RL (1997) Mapping resistance to blight in an interspecific cross in the genus *Castanea* using morphological, isozyme, RFLP, and RAPD markers. Phytopathology 87:751–759

Ladiges PY (1997) Phylogenetic history and classification of eucalypts. In: Woinarski JCZ (ed) Eucalypt Ecology: Individuals to Ecosystems. Cambridge University Press, Cambridge, UK, pp 16–29

Ladiges PY, Udovicic F (2000) Comment on a new classification of the eucalypts. Aust Syst Bot 13:149–152

Ladiges PY, Udovicic F (2005) Comment on the molecular dating of the age of eucalypts. Aust Syst Bot 18:291–293

Ladiges PY, Udovicic F, Nelson G (2003) Australian biogeographical connections and the phylogeny of large genera in the plant family Myrtaceae. J Biogeogr 30:989–998

Lander ES, Botstein D (1989) Mapping mendelian factors underlying quantitative traits using RFLP linkage maps. Genetics 121:185–199

Lander ES, Green J, Abrahamson J, Baarlow A, Daly MJ (1987) MAPMAKER: an interactive computer package for constructing primary genetic linkage maps of experimental and natural populations. Genomics 1:174–181

Leon AJ, Andrade FH, Lee M (2003) Genetic analysis of seed-oil concentration across generations and environments in sunflower. Crop Sci 43:135–140

Lezar S, Myburg AA, Berger DK, Wingfield MJ, Wingfield BD (2004) Development and assessment of microarray-based DNA fingerprinting in *Eucalyptus grandis*. Theor Appl Genet 109:1329–1336

Lister C, Dean C (1993) Recombinant inbred lines for mapping RFLP and phenotypic markers in *Arabidopsis thaliana*. Plant J 4:745–750

Liu BH, Knapp SJ (1990) GMENDEL: a programme for Mendelian segregation and linkage analysis of individual or multiple progeny using log-likelihood ratios. J Herd 81:407

Lorieux M, Goffinet B, Perrier X, González de León D, Lanaud C (1995) Maximum-likelihood models for mapping genetic markers showing segregation distortion. 1. Backcross populations. Theor Appl Genet 90:73–80

Lourenço RT (2004) Genomic structure of three mega basespairs of genomic shotgun DNA of *Eucalyptus grandis*: nucleotide content, repetitive sequences and genes. MSc Thesis, Univ of Campinas, Brazil

Lu C, Shen L, Tan Z, Xu Y, He P, Chen Y, Zhu L (1996) Comparative mapping of QTLs for agronomic traits of rice across environments using a doubled haploid population. Theor Appl Genet 93:1211–1217

Luo MC, Thomas C, You FM, Hsiao J, Ouyang S, Buell CR, Malandro M, McGuire PE, Anderson OD, Dvorak J (2003) High-throughput fingerprinting of bacterial artificial chromosomes using the snapshot labeling kit and sizing of restriction fragments by capillary electrophoresis. Genomics 82:378–389

MacDonald MC, Borralho NMG, Potts BM (1997) Genetic variation for growth and wood density in *Eucalyptus globulus* ssp. *globulus* in Tasmania (Australia). Silvae Genet 46:236–241

Mangin B, Goffinet B, Rebaï A (1994) Constructing confidence intervals for QTL location. Genetics 138:1301–1308

Marques CM, Araujo JA, Ferreira JG, Whetten R, O'Malley DM, Liu BH, Sederoff R (1998) AFLP genetic maps of *Eucalyptus globulus* and *E. tereticornis*. Theor Appl Genet 96:727–737

Marques CM, Vasquez-Kool J, Carocha VJ, Ferreira JG, O'Malley DM, Liu BH, Sederoff R (1999) Genetic dissection of vegetative propagation traits in *Eucalyptus tereticornis* and *E. globulus*. Theor Appl Genet 99:936–946

Marques CM, Brondani RPV, Grattapaglia D, Sederoff R (2002) Conservation and synteny of SSR loci and QTLs for vegetative propagation in four *Eucalyptus* species. Theor Appl Genet 105:474–478

Marra MA, Kucaba TA, Dietrich NL, Green ED, Brownstein B, Wilson RK, McDonald KM, Hillier LW, McPherson JD, Waterston RH (1997) High throughput fingerprint analysis of large-insert clones. Genome Res 7:1072–1084

Martin B, Quillet J (1974) The propagation by cuttings of forest trees in the Congo. Bois et Forets des Tropiques 155:15–33

Martins-Corder MP, Lopez CR (1997) Isozyme characterization of *Eucalyptus urophylla* (S.T. Blake) and *E. grandis* (Hill ex Maiden) populations in Brazil. Silvae Genet 46:192–197

McDonald MW, Rawlings M, Butcher PA, Bell JC (2003) Regional divergence and inbreeding in *Eucalyptus cladocalyx* (Myrtaceae). Aust J Bot 51:393–403

McGowen MH, Wiltshire RJE, Potts BM, Vaillancourt RE (2001) The origin of *Eucalyptus vernicosa*, a unique shrub eucalypt. Biol J Linn Soc 74:397–405

McGowen MH, Potts BM, Vaillancourt RE, Gore P, Williams DR, Pilbeam DJ (2004) The genetic control of sexual reproduction in *Eucalyptus globulus*. In: Tomé M (ed) IUFRO Conference: *Eucalyptus* in a Changing World, 11-15 Oct 2004, RAIZ, Instituto Investigação de Floresta e Papel, Aveiro, Portugal, pp 104-108

McKinnon GE, Vaillancourt RE, Jackson HD, Potts BM (2001a) Chloroplast sharing in the Tasmanian eucalypts. Evolution 55:703–711

McKinnon GE, Vaillancourt RE, Tilyard PA, Potts BM (2001b) Maternal inheritance of the chloroplast genome in *Eucalyptus globulus* and interspecific hybrids. Genome 44:831–835

McKinnon GE, Vaillancourt RE, Steane DA, Potts BM (2004) The rare silver gum, *Eucalyptus cordata*, is leaving its trace in the organellar gene pool of *Eucalyptus globulus*. Mol Ecol 13:3751–3762

McKinnon GE, Potts BM, Steane DA, Vaillancourt RE (2005) Population and phylogenetic analysis of the cinnamoyl CoA reductase gene in *Eucalyptus globulus* Labill. (Myrtaceae). Aust J Bot 53:827–838

McPherson JD, Marra M, Hillier L, Waterston RH, Chinwalla A, Wallis J, Sekhon M, Wylie K, Mardis ER, Wilson RK, Fulton R, Kucaba TA, Wagner-McPherson C, Barbazuk WB, Gregory SG, Humphray SJ, French L, Evans RS, Bethel G, Whittaker A, Holden JL, McCann OT, Dunham A, Soderlund C, Scott CE, Bentley DR, Schuler G, Chen HC, Jang W, Green ED, Idol JR, Maduro VV, Montgomery KT, Lee E, Miller A, Emerling S, Kucherlapati, Gibbs R, Scherer S, Gorrell JH, Sodergren E, Clerc-Blankenburg K, Tabor P, Naylor S, Garcia D, de Jong PJ, Catanese JJ, Nowak N, Osoegawa K, Qin S, Rowen L, Madan A, Dors M, Hood L, Trask B, Friedman C, Massa H, Cheung VG, Kirsch IR, Reid T, Yonescu R, Weissenbach J, Bruls T, Heilig R, Branscomb E, Olsen A, Doggett N, Cheng JF, Hawkins T, Myers RM, Shang J, Ramirez L, Schmutz J, Velasquez O, Dixon K, Stone NE, Cox DR, Haussler D, Kent WJ, Furey T, Rogic S, Kennedy S, Jones S, Rosenthal A, Wen G, Schilhabel M, Gloeckner G, Nyakatura G, Siebert R, Schlegelberger B, Korenberg J, Chen XN, Fujiyama A, Hattori M, Toyoda A, Yada T, Park HS, Sakaki Y, Shimizu N, Asakawa S, Kawasaki K, Sasaki T, Shintani A, Shimizu A, Shibuya K, Kudoh J, Minoshima S, Ramser J, Seranski P, Hoff C, Poustka A, Reinhardt R, Lehrach H; International Human Genome Mapping Consortium (2001) A physical map of the human genome. Nature 409:934–41

McRae TA, Dutkowski GW, Pilbeam DJ, Powell MB, Tier B (2004a) Genetic evaluation using the TREEPLAN® System. In: McKeand S (ed) Forest Genetics and Tree Breeding in the Age of Genomics: Progress and Future, IUFRO Joint Conf Div 2, 1-5 Nov 2004, North Carolina State University, Charleston, SC

McRae TA, Pilbeam DJ, Powell MD, Dutkowski GW, Joyce K, Tier B (2004b) Genetic evaluation in Eucalypt breeding programs. In: Tomé M (ed) IURO Conf: *Eucalyptus* in a Changing World, 11-15 Oct 2004, RAIZ, Instituto Investigação de Floresta e Papel, Aveiro, Portugal, pp 189-190

Meddings RLA, McComb JA, Bell DT (2001) The salt-waterlogging tolerance of *Eucalyptus camaldulensis* × *E. globulus* hybrids. Aust J ExpAgric 41:787–792

Meyers BC, Scalabrin S, Morgante M (2004) Mapping and sequencing complex genomes: let's get physical! Nat Rev Genet 5:578–588

Michelmore RW, Paran I, Kesseli RV (1991) Identification of markers linked to disease resistance genes by bulked segregant analysis: a rapid method to detect markers in specific genomic regions by using segregating populations. Proc Natl Acad Sci USA 88:9828–9832

Milgate AW, Potts BM, Joyce K, Mohammed CL, Vaillancourt RE (2005) Genetic variation in *Eucalyptus globulus* for susceptibility to *Mycosphaerella nubilosa* and its association with tree growth. Aust Plant Pathol 34:11–18

Missiaggia AA, Grattapaglia D (2005) Genetic mapping of *Eef1*, a major effect QTL for early flowering in *Eucalyptus grandis*. Tree Genet Genom 1:79–84

Moran GF (1992) Patterns of genetic diversity in Australian tree species. New For 6:49–66

Moran GF, Bell JC (1983) *Eucalyptus*. In: Tanksley SD, Orton TJ (eds) Isozymes in Plant Genetics and Breeding. Elsevier, Amesterdam, pp 423–441

Moran GF, Thamarus KA, Raymond CA, Qiu D, Uren T, Southerton SG (2002) Genomics of *Eucalyptus* wood traits. Ann For Sci 59:645–650

Morgan TH (1910) Chromosomes and heredity. Am Nat 44:449–496

Morgante M, Salamini F (2003) From plant genomics to breeding practice. Curr Opin Biotechnol 14:214–219

Mullis KB, Faloona FA (1987) Specific synthesis of DNA *in vitro* via a polymerase-catalyzed chain reaction. Meth Enzymol 155:335–350

Myburg AA (2001) Genetic architecture of hybrid fitness and wood quality traits in a wide interspecific cross of *Eucalyptus* tree species. PhD thesis, North Carolina State University, Raleigh, NC: http://www.lib.ncsu.edu/theses/available/etd-20010723-175234/

Myburg AA (2004) The International *Eucalyptus* Genome Consortium (IEuGC): Opportunities and Resources for Collaborative Genome Research in *Eucalyptus*. In: Li B, McKeand S (eds) Forest Genetics and Tree Breeding in the Age of Genomics: Progress and Future, IUFRO Joint Conf of Division 2, Conf Proc, 1-5 Nov 2004, Double Tree Guest Suites, Charleston, SC

Myburg AA, Remington DL, O'Malley DM, Sederoff RR, Whetten RW (2001) High-throughput AFLP analysis using infrared dye-labeled primers and an automated DNA sequencer. Bio/Techniques 30:348–357

Myburg AA, Griffin AR, Sederoff RR, Whetten RW (2003) Comparative genetic linkage maps of *Eucalyptus grandis*, *Eucalyptus globulus* and their F_1 hybrid based on a double pseudo-backcross mapping approach. Theor Appl Genet 107:1028–1042

Myburg AA, Vogl C, Griffin AR, Sederoff RR, Whetten RW (2004) Genetics of postzygotic isolation in *Eucalyptus*: whole-genome analysis of barriers to introgression in a wide interspecific cross of *Eucalyptus grandis* and *E. globulus*. Genetics 166:1405–1418

Nehls U, Beguiristain T, Ditengou F, Lapeyrie F, Martin F (1998) The expression of a symbiosis-regulated gene in eucalypt roots is regulated by auxins and hypaphorine, the tryptophan betaine of the ectomycorrhizal basidiomycete *Pisolithus tinctorius*. Planta 207:296–302

Newcombe G, Bradshaw HD Jr (1996) Quantitative trait loci conferring resistance in hybrid poplar to leaf spot caused by *Septoria populicola*. Can J For Res 26:1943–1950

Newcombe G, Bradshaw HD, Chastagner GA, Stettler RF (1996) A major gene for resistance to *Melampsora medusae* f.sp. *deltoidae* in a hybrid poplar pedigree. Phytopathology 86:87–94

Orita M, Iwahana H, Kanazawa H, Hayashi K, Sekiya T (1989) Detection of polymorphisms of human DNA by gel electrophoresis as single-strand conformation polymorphisms. Proc Natl Acad Sci USA 86:2766–2770

Paterson AH, Lander ES, Hewitt JD, Peterson S, Lincoln SE, Tanksley SD (1988) Resolution of quantitative traits into Mendelian factors by using a complete linkage map of restriction fragment length polymorphisms. Nature 335:721–726

Paterson AH, Damon S, Hewitt JD, Zamir D, Rabinowitch HD, Lincoln SE, Lander ES, Tanksley SD (1991) Mendelian factors underlying quantitative traits in tomato: comparison across species, generations, and environments. Genetics 127:181–197

Patterson B, Gore PL, Potts BM, Vaillancourt RE (2004a) Advances in pollination technique for large-scale production of *Eucalyptus globulus* seed. Aust J Bot 52:781–788

Patterson B, Vaillancourt RE, Pilbeam DJ, Potts BM (2004b) Factors affecting outcrossing rates in *Eucalyptus globulus*. Aust J Bot 52:773–780

Paux E, Tamasloukht M, Ladouce N, Sivadon P, Grima-Pettenati J (2004) Identification of genes preferentially expressed during wood formation in *Eucalyptus*. Plant Mol Biol 55:263–80

Paux E, Carocha V, Marques C, Mendes de Sousa A, Borralho N, Sivadon P, Grima-Pettenati J (2005) Transcript profiling of *Eucalyptus* xylem genes during tension wood formation. New Phytol 167:89–100

Pelleschi S, Guy S, Kim JY, Pointe C, Mahe A, Barthes L, Leonardi A, Prioul JL (1999) *Ivr2*, a candidate gene for a QTL of vacuolar invertase activity in maize leaves. Gene-specific expression under water stress. Plant Mol Biol 39:373–380

Pinto GL, Loureiro J, Lopes T, Santos C (2005) Analysis of the genetic stability of *Eucalyptus globulus* Labill. somatic embryos by flow cytometry. Theor Appl Genet 109:580–587

Plomion C, Bahrman N, Costa P, Dubos C, Frigerio J-M, Gerber S, Gion J-M, Lalanne C, Madur D, Pionneau C (2004) Proteomics for genetics and physiological studies in forest trees: application in maritime pine. In: Kumar S, Fladung M (eds) Molecular Genetics and Breeding of Forest Trees, Haworth, Binghamton, NY, pp 53–80

Poke FS, Vaillancourt RE, Elliott RC, Reid JB (2003) Sequence variation in two lignin biosynthesis genes, cinnamoyl CoA reductase (*CCR*) and cinnamyl alcohol dehydrogenase 2 (*CAD2*). Mol Breed 12:107–118

Poke FS, Wright JA, Raymond CA (2004) Predicting extractives and lignin contents in *Eucalyptus globulus* using Near Infrared Reflectance Analysis. J Wood Chem Technol 24:55–67

Poke FS, Vaillancourt RE, Potts BM, Reid JB (2005) Genomic research in *Eucalyptus*. Genetica 125(1):79-101

Pole MS, Hill RS, Green N, Macphail MK (1993) The Oligocene Berwick Quarry flora - rainforest in a drying environment. Aust Syst Bot 6:399–427

Potts BM (2004) Genetic improvement of eucalypts. In: Burley J, Evans J, Youngquist JA (ed) Encyclopedia of Forest Science. Elsevier, Oxford, pp 1480–1490

Potts BM, Dungey HS (2004) Hybridisation of *Eucalyptus*: key issues for breeders and geneticists. New For 27:115–138

Potts BM, Jordan GJ (1994) Genetic variation in the juvenile leaf morphology of *Eucalyptus globulus* ssp. *globulus*. For Genet 1:81–95

Potts BM, Pederick LA (2000) Morphology, phylogeny, origin, distribution and genetic diversity of the eucalypts. In: Brown BN (ed) Diseases and Pathogens of Eucalypts. CSIRO, Collingwood, Australia, pp 11-34

Potts BM, Wiltshire RJE (1997) Eucalypt genetics and genecology. In: Wiliams J, Woinarski J (eds) Eucalypt Ecology: Individuals to Ecosystems. Cambridge University Press, Cambridge, UK

Potts BM, Barbour RC, Hingston AB, Vaillancourt RE (2003) Turner Review No. 6: Genetic pollution of native eucalypt gene pools - identifying the risks. Aust J Bot 51:1–25

Potts BM, Vaillancourt RE, Jordan GJ, Dutkowski GW, Costa e Silva J, McKinnon GE, Steane DA, Volker PW, Lopez GA, Apiolaza LA, Li Y, Marques C, Borralho NMG (2004) Exploration of the *Eucalyptus globulus* gene pool. In: Tomé M (ed) IUFRO Conf: *Eucalyptus* in a Changing World, 11-15 Oct 2004. RAIZ, Instituto Investigação de Floresta e Papel, Aveiro, Portugal, pp 46–61

Pound LM, Wallwork MAB, Potts BM, Sedgley M (2002a) Early ovule development following self- and cross-pollinations in *Eucalyptus globulus* Labill. ssp *globulus*. Ann Bot (London) 89:613–620

Pound LM, Wallwork MAB, Potts BM, Sedgley M (2002b) Self-incompatibility in *Eucalyptus globulus* ssp. *globulus* (Myrtaceae). Aust J Bot 50:365–372

Pound LM, Wallworth MAB, Potts BM, Sedgley M (2003) Pollen tube growth and early ovule development following self- and cross-pollination of *Eucalyptus nitens*. Sexual Plant Reprod 16:59–69

Prioul JL, Quarrie S, Causse M, De Vienne D (1997) Dissecting complex physiological functions through the use of molecular quantitative genetics. J Exp Bot 48:1151–1163

Pryor LD (1976) The Biology of Eucalypts. Edward Arnold, London

Pryor LD, Johnson LAS (1971) A Classification of the Eucalypts. Australian National University Press, Canberra, Australia

Pryor LD, Johnson LAS (1981) *Eucalyptus*, the universal Australian. In: Keast A (ed) Ecological Biogeography of Australia. Junk, The Hague, pp 499–536

Raga FR (2001) Perspectiva para el eucalipto Chileno. In: IUFRO Conf: Developing the Eucalypt of the Future, 10-15 Sept 2001. INFOR, Valdivia, Chile, pp 1–13

Raymond CA (2002) Genetics of *Eucalyptus* wood properties. Ann For Sci 59:525–531

Raymond CR, Apiolaza LA (2004) Incorporating wood quality and deployment traits in *Eucalyptus globulus* and *Eucalyptus nitens*. In: Carson M (ed) Plantation Forest Biotechnology for the 21st Century. Forest Research New Zealand, Rotorua, New Zealand, pp 87–99

Reddy KV, Rockwood DL (1989) Breeding strategies for coppice production in a *Eucalyptus grandis* base population with four generations of selection. Silvae Genet 38:148–151

Riccardi F, Gazeau P, Jacquemot MP, Vincent D, Zivy M (2004) Deciphering genetic variations of proteome responses to water deficit in maize leaves. Plant Physiol Biochem 42:1003–1011

Riesner DG, Steger U, Wiese M, Wulfert M, Heibey M, Henco K (1992) Temperature gradient electrophoresis for the detection of polymorphic DNA and for quantitative polymerase chain reaction. Electrophoresis 13:632–636

Rose JKC, Bashir S, Giovannoni JJ, Jahn MM, Saravanan RS (2004) Tackling the plant proteome: practical approaches, hurdles and experimental tools. Plant J 39:715–733

Rossetto M, Jezierski G, Hopper SD, Dixon KW (1999) Conservation genetics and clonality in two critically endangered eucalypts from the highly endemic south-western Australian flora. Biol Conserv 88:321–331

Rozefelds A (1996) *Eucalyptus* phylogeny and history: a brief summary. Tasforests 8:15–26

Saghai-Maroof MA, Solimen KM, Jorgensen RA, Allard RW (1984) Ribosomal DNA spacer-length polymorphisms in Barley: Mendelian inheritance, chromosomal location, and population dynamics. Proc Natl Acad Sci USA 81:8014–8018

Sale MM, Potts BM, West AK, Reid JB (1996) Relationships within *Eucalyptus* (Myrtaceae) using PCR-amplification and southern hybridisation of chloroplast DNA. Aust Syst Bot 9:273–282

Sanhueza R, Griffin AR (2001) FAMASA Fiber Yield Improvement Programme (F.Y.I.P.): 10 years experience breeding *Eucalyptus globulus*. In: Barros S (ed) Developing the Eucalypt of the Future. IUFRO Int Symp, 10-15 Sept 2001, INFOR, Chile

Sato S, Yamada N, Nakamoto S, Hibino T (2005) Expression profiling of the *Eucalyptus* transcription factor in differentiating xylem tissues. In: Plant & Animal Genome XIII conf, San-Diego, CA

Sax K (1923) The association of size differences with seed-coat pattern and pigmentation in *Phaseolus vulgaris*. Genetics 8:552560

Schadt EE, Monks SA, Drake TA, Lusis AJ, Che N, Colinayo V, Ruff TG, Milligan SB, Lamb JR, Cavet G, Linsley PS, Mao M, Stoughton RB, Friend SH (2003) Genetics of gene expression surveyed in maize, mouse and man. Nature 422:297–302

Schein J, Kucaba TA, Sekhon M, Smailus D, Waterston RH, Marra MA (2004) High-throughput BAC fingerprinting. In Zhao S, Stodolsky M (eds) Methods in Molecular Biology. Vol 255: Bacterial Artificial Chromosomes: Library Construction, Physical Mapping and Sequencing. Vol 1. Humana, Totowa, NJ, pp 143–156

Schimleck LR, Michell AJ, Vinden P (1996) Eucalypt wood classification by NIR spectroscopy and principal components analysis. Appita J 49:319–324

Shepherd M, Chaparro J, Dale G, Jefferson L, Duong H, Vogel H, Walsh J, Gibbings M, Teasdale R (1995) Mapping insect resistance and essential oil traits in a tropical *Eucalyptus* hybrid. CRC-THF IUFRO Conf - Eucalypt Plantations: Improving Fibre, Yield and Quality, 19-24 Feb 1995, Hobart, Tamania, pp 421–423

Shepherd M, Chaparro JX, Teasdale R (1999) Genetic mapping of monoterpene composition in an interspecific eucalypt hybrid. Theor Appl Genet 99:1207–1215

Shepherd M, Jones M (2005) Molecular markers in tree improvement: characterisation and use in *Eucalyptus*. In: Lörz H, Wenzel G (eds) Biotechnology in Agriculture and Forestry, Vol 55: Molecular Marker Systems in Plant Breeding and Crop Improvement. Springer, Berlin Heidelberg New York, pp 399–409

Shield E (2004) Silviculture for sawlogs - a review of the key elements, with special reference to *Eucalyptus grandis*. In: Tomé M (ed) IUFRO Conf: *Eucalyptus* in a Changing World, 11-15 Oct 2004. RAIZ, Instituto Investigação de Floresta e Papel, Aveiro, Portugal, pp 62–67

Shizuya H, Birren B, Kim UJ, MancinoV, SlepakT, TachiiriY, Simon M (1992) Cloning and stable maintenance of 300-kilobase-pair fragments of human DNA in *Escherichia coli* using an F-factor-based vector. Proc Natl Acad Sci USA 89:8794–8797

Skabo S, Vaillancourt RE, Potts BM (1998) Fine-scale genetic structure of *Eucalyptus globulus* ssp. *globulus* forest revealed by RAPDs. Aust J Bot 46:583–594

Smith S, Hughes J, Wardell-Johnson G (2003) High population differentiation and extensive clonality in a rare mallee eucalypt: *Eucalyptus curtisii* - conservation genetics of a rare mallee eucalypt. Conserv Genet 4:289–300

Snedden CL, Verryn SD (2004) A comparative study of predicted gains for selection from a cloned breeding population and the implications for deployment. In: Tomé M (ed) IUFRO Conf: *Eucalyptus* in a Changing World, 11-15 Oct, RAIZ, Instituto Investigação de Floresta e Papel, Aveiro, Portugal, pp 137–144

Soderlund C, Humphray S, Dunham A, French L (2000) Contigs built with fingerprints, markers, and FPC V4.7. Genome Res 10:1772–1787

Soria F, Borralho NMG (1998) The genetics of resistance to *Phoracantha semipunctata* attack in *Eucalyptus globulus* in Spain. Silvae Genet 46:365–369

Soria F, Basurco F, Toval G, Silio L, Rodriguez MC, Toro M (1998) An application of Bayesian techniques to the genetic evaluation of growth traits in *Eucalyptus globulus*. Can J For Res 28:1286–1294

Southerton S, Birt P, Porter G, Ford H (2004) Review of gene movement by bats and birds and its potential significance for eucalypt plantation forestry. Aust For 67:44–53

Specht RL (1996) The influence of soils on the evolution of the eucalypts. In: Adams MA (ed) Nutrition of Eucalypts. CSIRO, Melbourne, Australia

Stam P (1993) Construction of integrated genetic linkage maps by means of a new computer package: JoinMap. Plant J 3:739–744

Steane DA (2005) Complete nucleotide sequence of the chloroplast genome from the Tasmanian blue gum, *Eucalyptus globulus*. DNA Res 12(3):215-220

Steane DA, West AK, Potts BM, Overden JR, Reid JB (1991) Restriction fragment length polymorphisms in chloroplast DNA from six species of *Eucalyptus*. Aust J Bot 39:399–414

Steane DA, Vaillancourt RE, Russell J, Powell W, Marshall D, Potts BM (2001) Development and characterization microsatellite loci in *Eucalyptus globulus* (Myrtacea). Silvae Genet 50(2):89

Steane DA, Nicolle D, Vaillancourt RE, Potts BM (2002) Higher-level relationships among the eucalypts are resolved by ITS-sequence data. Aust Syst Bot 15:49–62

Steane DA, Jones RC, Vaillancourt RE (2005) A set of chloroplast microsatellite primers for *Eucalyptus* (Myrtaceae). Mol Ecol Notes 5:538–541

Stirling B, Newcombe G, Vrebalov J, Bosdet I, Bradshaw HD Jr (2001) Suppressed recombination around the MXC3 locus, a major gene for resistance to poplar leaf rust. Theor Appl Genet 103:1129–1137

Stokoe R, Shepherd M, Lee D, Nikles D, Henry R (2001) Natural inter-subgeneric hybridisation between *Eucalyptus acmenoides* Schauer and *Eucalyptus cloeziana* F. Muell (Myrtaceae) in south-eastern Queensland. Ann Bot (London) 84:563–570

Strauss SH, Lande R, Namkoong G (1992) Obstacles to molecular-marker-aided selection in forest trees. Can J For Res 22:1050–1061

Stuber CW, Lincoln SE, Wolff DW, Helentjaris T, Lander ES (1992) Identification of genetic factors contributing to heterosis in a hybrid from two elite maize inbred lines using molecular markers. Genetics 132:823–839

Sulston J, Mallett F, Staden R, Durbin R, Horsnell T, Coulson A (1988) Software for genome mapping by fingerprinting techniques. Comp Appl Biosci 4:125–132

Tanksley SD, Ganal MW, Martin GB (1995) Chromosome landing: a paradigm for map-based gene cloning in plants with large genomes. Trends Genet 11:63–68

Tao Q, Chang YL, Wang J, Chen H, Islam-Faridi MN, Scheuring C, Wang B, Stelly DM, Zhang HB (2001) Bacterial artificial chromosome-based physical map of the rice genome constructed by restriction fingerprint analysis. Genetics 158:1711–1724

Teulat B, Merah O, Souyris I, This D (2001) QTLs for agronomic traits from a Mediterranean barley progeny grown in several environments. Theor Appl Genet 103:774–787

Thamarus K, Groom K, Murrell J, Byrne M, Moran G (2002) A genetic linkage map for *Eucalyptus globulus* with candidate loci for wood, fibre and floral traits. Theor Appl Genet 104:379–387

Thamarus KA, Groom K, Bradley A, Raymond CA, Schimleck LR, Williams ER, Moran GF (2004) Identification of quantitative trait loci for wood and fibre properties in two full-sib pedigrees of *Eucalyptus globulus*. Theor Appl Genet 109:856–864

The Arabidopsis Genome Initiative (2000) Analysis of the genome sequence of the flowering plant *Arabidopsis thaliana*. Nature 408:796–815

Thumma R, F. NM, Evans R, Moran GF (2005) Polymorphisms in Cinnamoyl CoA Reductase (CCR) are associated with variation in microfibril angle in *Eucalyptus* spp. Genetics 171:1257–1265

Tibbits WN, Potts BM, Savva MH (1991) Inheritance of freezing resistance in interspecific F_1 hybrids of *Eucalyptus*. Theor Appl Genet 83:126–135

Toro MA, Silió L, Rodriguez MC, Soria F, Toval G (1998) Genetic analysis of survival to drought in *Eucalyptus globulus* in Spain. In: 6th World Congress on Genetics Applied to Livestock Production, Armidale, Australia, 27:499–502

Tournier V, Grat S, Marque C, El Kayal W, Penchel R, de Andrade G, Boudet AM, Teulieres C (2003) An efficient procedure to stably introduce genes into an economically important pulp tree (*Eucalyptus grandis* × *Eucalyptus urophylla*). Transgenic Res 12:403–411

Tragoonrung S, Kanizin V, Hayes PM, Blake TK (1992) Sequence tagged-site-facilitated PCR for barley genome mapping. Theor Appl Genet 84:1002–1008

Turnbull JW (1999) Eucalypt plantations. New For 17:37–52

Tuskan G, West D, Bradshaw HD, Neale D, Sewell M, Wheeler N, Megraw B, Jech K, Wiselogel A, Evans R, Elam C, Davis M, Dinus R (1999) Two high-throughput techniques for determining wood properties as part of a molecular genetics analysis of loblolly pine and hybrid poplar. Appl Biochem Biotechnol 77-9:55–65

Udovicic F, Ladiges PY (2000) Informativeness of nuclear and chloroplast DNA regions and the phylogeny of the eucalypts and related genera (Myrtaceae). Kew Bull 55:633–645

Vaillancourt RE, Jackson HD (2000) A chloroplast DNA hypervariable region in eucalypts. Theor Appl Genet 101:473–477

Vaillancourt RE, Potts BM, Manson A, Reid JB (1994) Detection of QTLs in a *Eucalyptus gunni* X *E. globulus* F_2 using a RAPD linkage map. In: Proc Int Wood Biotechnol Symp, 31 Aug-1 Sept, Hokutopia (Convention Hall) Tokyo, pp 63–70

Vaillancourt RE, Petty A, McKinnon GE (2004) Maternal inheritance of mitochondria in *Eucalyptus globulus*. J Hered 95:353–355

Varghese M, Nagarajan B, Nicodemus A, Bennet SSR, Subramanian K (2000) Hybrid breakdown in Mysore gum and need for genetic improvement of *Eucalyptus camaldulensis* and *E. tereticornis* in India. In: Nikles DG (ed) Hybrid Breeding and Genetics of Forest Trees. Proc QFRI/CRC-SPF Symp, 9-14 April 2000, Noosa, Queensland, Australia. Dept of Primary Industries, Brisbane, Australia, pp 519–525

Verhaegen D, Plomion C (1996) Genetic mapping in *Eucalyptus urophylla* and *E. grandis* using RAPD markers. Genome 39:1051–1061

Verhaegen D, Plomion C, Gion JM, Poitel M, Costa P, Kremer A (1997) Quantitative trait dissection analysis in *Eucalyptus* using RAPD markers. 1. Detection of QTL in interspecific hybrid progeny, stability of QTL expression across different ages. Theor Appl Genet 95:597–608

Vigneron P, Bouvet J (2000) Eucalypt hybrid breeding in Congo. In: Nikles DG (ed) Hybrid Breeding and Genetics of Forest Trees. Proc QFRI/CRC-SPF Symp, 9-14 April 2000, Noosa, Queensland, Australia, Dept of Primary Industries, Brisbane, Australia, pp 14–26

Vincent D, Lapierre C, Pollet B, Cornic G, Negroni L, Zivy M (2005) Water deficits affect caffeate O-methyltransferase, lignification, and related enzymes in maize leaves. A proteomic investigation. Plant Physiol 137:949–960

Vos P, Hogers R, Bleeker M, Reijans M, van de Lee T, Hornes M, Fritjters A, Pot J, Peleman J, Kuiper M, Zabeau M (1995) AFLP: a new technique for DNA fingerprinting. Nucleic Acids Res 23:4407–4414

Wagner DB, Furnier GR, Saghai-Maroof MA, Williams SM, Dancik BP, Allard RW (1987) Chloroplast DNA polymorphisms in lodgepole and jack pines and their hybrids. Proc Natl Acad Sci USA 84:2097–2100

Wallis JW, Aerts J, Groenen MA, Crooijmans RP, Layman D, Graves TA, Scheer DE, Kremitzki C, Fedele MJ, Mudd NK, Cardenas M, Higginbotham J, Carter J, McGrane R, Gaige T, Mead K, Walker J, Albracht D, Davito J, Yang SP, Leong S, Chinwalla A, Sekhon M, Wylie K, Dodgson J, Romanov MN, Cheng H, de Jong PJ, Osoegawa K, Nefedov M, Zhang H, McPherson JD, Krzywinski M, Schein J, Hillier L, Mardis ER, Wilson RK, Warren WC (2004) A physical map of the chicken genome. Nature 432:679–680

Wardell-Johnson GW, Williams JE, Hill KD, Cumming R (1997) Evolutionary biogeography and contemporary distribution of eucalypts. In: Woinarski JCZ (ed) Eucalypt Ecology: Individuals to Ecosystems. Cambridge University Press, Cambridge, pp 92–128

Waugh G (2004) Growing *Eucalyptus globulus* for high-quality sawn products. In: Tomé M (ed) IUFRO Conf: *Eucalyptus* in a Changing World, 11-15 Oct 2004. RAIZ, Instituto Investigação de Floresta e Papel, Aveiro, Portugal, pp 79–84

Wayne ML, McIntyre LM (2002) Combining mapping and arraying: an approach to candidate gene identification. Proc Natl Acad Sci USA 99:14903–14906

Wei X, Borralho NMG (1999) Objectives and selection criteria for pulp production of *Eucalyptus urophylla* plantations in South East China. For Genet 6:181–190

Welsh J, McClelland M (1990) Fingerprinting genomes using PCR with arbitrary primers. Nucleic Acids Res 18:7213–7218

Weng C, Kubisiak TL, Nelson CD, Stine M (2002) Mapping quantitative trait loci controlling early growth in a (longleaf pine × slash pine) × slash pine BC1 family. Theor Appl Genet 104:852–859

Whittock S, Steane DA, Vaillancourt RE, Potts BM (2003) Is *Eucalyptus* subgenus *Minutifructa* over-ranked? Trans R Soc South Aust 127:27–32

Wicking C, Williamson B (1991) From linked marker to gene. Trends Genet 7:288–293

Wilcox PL, Amerson HV, Kuhlman EG, Liu BH, O'Malley DM, Sederoff RR (1996) Detection of a major gene for resistance to fusiform rust disease in loblolly pine by genomic mapping. Proc Natl Acad Sci USA 93:3859–3864

Williams JE, Woinarski JCZ (1997) Eucalypt ecology: individuals to ecosystems. Cambridge University Press, Cambridge, UK

Williams JG, Kubelik AR, Livak KJ, Rafalski LA, Tingey SV (1990) DNA polymorphism amplified by arbitrary primers are useful as genetic markers. Nucleic Acids Res 18:6531–6535

Williams DR, Potts BM, Black PG (1999) Testing single visit pollination procedures for *Eucalyptus globulus* and *E. nitens*. Aust For 62:346–352

Williams E, Matheson A, Harwood C (2002) Experimental design and analysis for tree improvement. CSIRO, Melbourne, Australia

Wiltshire RJE (2004) Eucalypts. In: Burley J, Evans J, Youngquist JA (ed) Encyclopedia of Forest Science. Elsevier, Oxford, pp 1687–1699

Wu C, Sun S, Nimmakayala P, Santos FA, Meksem K, Springman R, Ding K, Lightfoot DA, Zhang HB (2004) A BAC- and BIBAC-based physical map of the soybean genome. Genome Res 14:319–26

Xu Z, van den Berg MA, Scheuring C, Covaleda L, Lu H, Santos FA, Uhm T, Lee MK, Wu C, Liu S, Zhang HB (2005) Genome physical mapping from large-insert clones by fingerprint analysis with capillary electrophoresis: a robust physical map of *Penicillium chrysogenum*. Nucleic Acids Res 33:e50

Yang JM, Park S, Kamdem DP, Keathley DE, Retzel E, Paule C, Kapur V, Han KH (2003) Novel gene expression profiles define the metabolic and physiological processes characteristic of wood and its extractive formation in a hardwood tree species, *Robinia pseudoacacia*. Plant Mol Biol 52:935–956

Yang JM, Kamdem DP, Keathley DE, Han KH (2004a) Seasonal changes in gene expression at the sapwood-heartwood transition zone of black locust (*Robinia pseudoacacia*) revealed by cDNA microarray analysis. Tree Physiol 24:461–474

Yang SH, van Zyl L, No EG, Loopstra CA (2004b) Microarray analysis of genes preferentially expressed in differentiating xylem of loblolly pine (*Pinus taeda*). Plant Sci 166:1185–1195

Yu J, Hu SN, Wang J, Li SG, Wong KSG, Liu B, Deng Y, Dai L, Zhou Y, Zhang XQ, Cao ML, Liu J, Sun JD, Tang JB, Chen YJ, Huang XB, Lin W, Ye C, Tong W, Cong LJ, Geng JN, Han YJ, Li L, Li W, Hu GQ, Huang XG, Li WJ, Li J, Liu ZW, Li L, Liu JP, Qi QH, Liu JS, Li L, Wang XG, Lu H, Wu TT, Zhu M, Ni PX, Han H, Dong W, Ren XY, Feng XL, Cui P, Li XR, Wang H, Xu X, Zhai WX, Xu Z, Zhang JS, He SJ, Zhang JG, Xu JC, Zhang KL, Zheng XW, Dong JH, Zeng WY, Tao L, Chen XW, He J, Liu DF, Tian W, Tian CG, Xia HG, Li G, Gao H, Li P, Chen W, Wang XD, Zhang Y, Hu JF, Wang J, Liu S, Yang J, Zhang GY, Xiong YQ, Li ZJ, Mao L, Zhou CS, Zhu Z, Chen RS, Hao BL, Zheng WM, Chen SY, Guo W, Li GJ, Liu SQ, Huang GY, Tao M, Wang J, Zhu LH, Yuan LP, Yang HM (2001) A draft sequence of the rice (*Oryza sativa* ssp *indica*) genome. Chin Sci Bull 46:1937–1942

Yvert G, Brem RB, Whittle J, Akey JM, Foss E, Smith EN, Mackelprang R, Kruglyak L (2003) Trans-acting regulatory variation in *Saccharomyces cerevisiae* and the role of transcription factors. Nat Genet 35:57–64

Zacharin RF (1978) Emigrant eucalypts: gum trees as exotics. Melbourne University Press, Melbourne, Australia

Zeng ZB (1994) Precision mapping of quantitative trait loci. Genetics 136:1457–1468

Zobel BJ, Sprague JR (1998) Juvenile Wood in Forest Trees, Springer, Berlin Heidelberg New York, p 300

5 Fagaceae Trees

Antoine Kremer[1], Manuela Casasoli[2], Teresa Barreneche[3], Catherine Bodénès[1], Paul Sisco[4], Thomas Kubisiak[5], Marta Scalfi[6], Stefano Leonardi[6], Erica Bakker[7], Joukje Buiteveld[8], Jeanne Romero-Severson[9], Kathiravetpillai Arumuganathan[10], Jeremy Derory[1], Caroline Scotti-Saintagne[11], Guy Roussel[1], Maria Evangelista Bertocchi[1], Christian Lexer[12], Ilga Porth[13], Fred Hebard[14], Catherine Clark[15], John Carlson[16], Christophe Plomion[1], Hans-Peter Koelewijn[8], and Fiorella Villani[17]

[1] UMR Biodiversité Gènes & Communautés, INRA, 69 Route d'Arcachon, 33612 Cestas, France, *e-mail*: antoine.kremer@pierroton.inra.fr
[2] Dipartimento di Biologia Vegetale, Università "La Sapienza", Piazza A. Moro 5, 00185 Rome, Italy
[3] Unité de Recherche sur les Espèces Fruitières et la Vigne, INRA, 71 Avenue Edouard Bourlaux, 33883 Villenave d'Ornon, France
[4] The American Chestnut Foundation, One Oak Plaza, Suite 308 Asheville, NC 28801, USA
[5] Southern Institute of Forest Genetics, USDA-Forest Service, 23332 Highway 67, Saucier, MS 39574-9344, USA
[6] Dipartimento di Scienze Ambientali, Università di Parma, Parco Area delle Scienze 11/A, 43100 Parma, Italy
[7] Department of Ecology and Evolution, University of Chicago, 5801 South Ellis Avenue, Chicago, IL 60637, USA
[8] Alterra Wageningen UR, Centre for Ecosystem Studies, P.O. Box 47, 6700 AA Wageningen, The Netherlands
[9] Department of Biological Sciences, University of Notre Dame, Notre Dame, IN 46556, USA
[10] Flow Cytometry and Imaging Core Laboratory, Benaroya Research Institute at Virginia Mason, 1201 Ninth Avenue, Seattle, WA 98101, USA
[11] UMR Ecologie des Forêts de Guyane, INRA, Campus agronomique BP 709, Avenue de France, 97387 Kourou, French Guyana
[12] Jodrell Laboratory, Royal Botanic Gardens, Kew, Richmond, Surrey TW9 3DS, UK
[13] Austrian Research Centre, 2444 Seibersdorf, Austria
[14] The American Chestnut Foundation Research Farms, 14005 Glenbrook Avenue, Meadowview, VA 24361, USA
[15] Department of Forestry, North Carolina State University, Box 8008, Raleigh, NC 27695-8008, USA
[16] The School of Forest Resources and Huck Institutes for Life Sciences, Pennsylvania State University, 323 Forest Resources Building, University Park, PA 16802, USA
[17] Istituto per l'Agroselvicoltura, CNR, V.le G. Marconi, 2 - 05010, Porano, Italy

5.1 Introduction

Worldwide, there are more than 1,000 species belonging to the Fagaceae. All Fagaceae species are woody plants and are spread throughout the northern hemisphere, from the tropical to the boreal regions. The family comprises seven genera (Govaerts and Frodin 1998), and the number of species is extremely variable among genera: *Castanea* (12), *Castanopsis* (100 to 200), *Chrysolepis* (2), *Fagus* (11), *Lithocarpus* (300), *Quercus* (450 to 600), *Trigonobalanus* (3). Oaks (*Quercus*), chestnuts (*Castanea*), and beeches (*Fagus*) are widely used in forestry for wood products over the three continents (Asia, Europe, and America) and are important economic species. Consequently, they have received more attention in forest genetic research than other genera. In addition to their cultivation in forestry, chestnuts are also used for their fruit production and have been partially domesticated for that purpose. *Castanopsis* and *Lithocarpus* are important ecological components of the Asian flora and have recently been investigated for their biological diversity (Cannon and Manos 2003). The remaining genera comprise only a very few species and for the time being have been studied mainly in botany and taxonomy.

Genetic research in Fagaceae has been restricted to the three genera of economic importance (*Castanea, Fagus*, and *Quercus*), although activities in phylogeny and evolutionary genetics have recently encompassed the whole family (Manos and Stanford 2001; Manos et al. 2001). Because of their long rotation times, breeding activities in the three main genera have been limited (Kremer et al. 2004). However, population differentiation has been investigated in a very large number of species, with the main aim of identifying geographic patterns or historical foot-

Genome Mapping and Molecular Breeding in Plants, Volume 7
Forest Trees
C. Kole (Ed.)

prints for molecular markers and phenotypic traits of forestry relevance. Population genetics has driven most of the research activities in molecular genetics and also genetic mapping, in contrast to other forest tree species where tree improvement has been the main goal. Genetic maps have been constructed in at least one species of *Quercus*, *Castanea*, and *Fagus*. Because of their low genetic divergence, it quickly became obvious that molecular markers could be easily transferred from *Quercus* to *Castanea* (and vice versa) but less easily to *Fagus*. These earlier findings led to further activities on comparative mapping across genera, especially between *Quercus* and *Castanea*.

5.1.1 Evolutionary Biology and Phylogeny of the Fagaceae

Fossil remains indicate that the Fagaceae appeared at the transition between the secondary and tertiary era. Remains of *Dryophyllum*, which is a fossil genus belonging to the Fagaceae, were discovered in layers belonging to the early Cretaceous (Jones 1986). Fossil remains that were unequivocally assigned to Fagaceae and dated to the Upper Eocene and Late Oligocene were found in North America (Herendeen et al. 1995) and Europe (Kvacek and Walther 1989). Differentiation of the various genera occurred during the mid Tertiary, and reported species of Fagaceae at the late Tertiary resemble already extant species. The oldest reported genera belonging to the Fagaceae occurred in Southeast Asia, where the extant species diversity is also the highest. The family originated from Southeast Asia and radiated toward Europe and America (Wen 1999; Xiang et al. 2000). Migration and major continental rearrangements contributed to disjunction and vicariance within the family, especially within *Quercus* (Manos and Stanford 2001). It is generally accepted that most major oak groups essentially evolved in the areas where they reside today (Axelrod 1983).

Phylogenetic investigations based on chloroplast or nuclear DNA data are poorly resolutive, suggesting that the differentiation into different genera was extremely rapid during the mid Tertiary (Manos and Steele 1997; Manos et al. 2001). All genera are usually clustered into a "starlike" dendrogram (polytomy), except *Fagus*, which diverged earlier from the common ancestor. However, there is a strong genomic similarity between *Quercus*, *Castanea*, *Lithocarpus*, and *Castanopsis*. Paleontological records suggest that *Quercus* and *Castanea* separated 60 million years ago (Crepet 1989). Interspecific separation within the genera *Quercus*, *Fagus*, and *Castanea* occurred between 22 and 3 million years ago as inferred from a molecular clock based on cpDNA divergence (Manos and Stanford 2001). The reduced genetic divergence among the different genera was recently confirmed by the results obtained in transferring molecular tools and markers among genera, as it is much more difficult to transfer microsatellite markers from *Quercus* to *Fagus* than it is from *Quercus* to *Castanea* (Steinkellner et al. 1997; Barreneche et al. 2004).

5.1.2 Ploidy, Karyotype, and Genome Size in Fagaceae

Reported karyotype studies in *Quercus*, *Lithocarpus*, *Castanopsis*, and *Castanea* (Mehra et al. 1972), in *Quercus* (D'Emerico et al. 1995), and *Fagus* (Ohri and Ahuja 1991) indicate that the number of chromosomes within the family is remarkably stable ($2n = 24$). Naturally occurring triploids have been mentioned occasionally in oaks (Butorina 1993; Naujoks et al. 1995). Extra chromosomes ($2n = 24+1$, 2 or 3) have been reported as consequences of irregular segregation in mitoses (Zoldos et al. 1998). Otherwise, C-banding comparisons have shown that the morphology of the chromosomes of *Fagus* (Ohri and Ahuja 1991) and *Quercus* (Ohri and Ahuja 1990) are quite similar.

The DNA content is variable across genera in the Fagaceae: the 2C DNA values varying from a low of 1.11 pg in *Fagus* to a high of 2.0 pg in *Quercus* species (Table 1). GC content on the other hand appears constant across genera (40%) and is similar to most higher plants (Table 1). All values reported in Table 1 were obtained by flow cytometric analysis of interphasic nuclei and are slightly higher than earlier assessments made with the Feulgen microdensitometry method (Ohri and Ahuja 1990). The 31 species in Table 1 represent a cross-section of the Fagaceae across the northern hemisphere. The two *Fagus* species, *Fagus grandifolia* and *F. sylvatica*, were quite similar in genome size (1.27 and 1.11 pg per 2C, respectively) and are at the lower range of genome sizes among the Fagaceae, suggesting that *Fagus* has either the most rudimentary genome or the most greatly reduced genome

Table 1. DNA content in 31 Fagaceae species determined by flow cytometric analysis

Species	2C nuclear DNA pg (mean value)	1C nuclear DNA Mbp	GC content (%)	Reference
Genus Castanea				
C. seguinii	1.57	755	–	Arumuganathan et al.*
C. sativa (1)	1.61	777	–	Arumuganathan et al.
C. sativa (2)	1.62	–	–	Brown and Siljak-Yakovlev (pers comm)
C. crenata	1.65	794	–	Arumuganathan et al.
C. mollissima	1.65	794	–	Arumuganathan et al.
C. dentata	1.67	803	–	Arumuganathan et al.
Genus Fagus				
F. grandifolia	1.27	610	–	Arumuganathan et al.
F. sylvatica	1.11	535	40	Gallois et al. 1999
Genus Quercus Subgen Erythrobalanus**				
Q. velutina	1.17	565	–	Arumuganathan et al.
Q. nuttallii	1.39	672	–	Arumuganathan et al.
Q. shumardii	1.47	709	–	Arumuganathan et al.
Q. nigra	1.52	735	–	Arumuganathan et al.
Q. rubra	1.58	762	–	Arumuganathan et al.
Q. palustris	1.60	774	–	Arumuganathan et al.
Q. coccinea	1.64	791	–	Arumuganathan et al.
Q. phellos	1.66	799	–	Arumuganathan et al.
Q. falcata	1.72	832	–	Arumuganathan et al.
Q. pagoda	1.75	843	–	Arumuganathan et al.
Q. imbricaria	1.81	871	–	Arumuganathan et al.
Genus Quercus Subgen Lepidobalanus**				
Q. bicolor	1.35	651	–	Arumuganathan et al.
Q. montana	1.49	719	–	Arumuganathan et al.
Q. robur	1.53	740	–	Arumuganathan et al.
Q. stellata	1.55	745	–	Arumuganathan et al.
Q. alba	1.59	766	–	Arumuganathan et al.
Q. macrocarpa	1.62	780	–	Arumuganathan et al.
Q. robur	1.84	885	42	Favre and Brown 1996
Q. pubescens	1.86	882	42.1	Favre and Brown 1996
Q. petraea	1.87	901	41.7	Favre and Brown 1996
Q. robur	1.88	–	39.4	Zoldos et al. 1998
Q. petraea	1.90	–	39.8	Zoldos et al. 1998
Q. pubescens	1.91	–	39.7	Zoldos et al. 1998
Genus Quercus Subgen Cerris**				
Q. acutissima	1.42	684	–	Arumuganathan et al.
Q. cerris	1.91	–	40.2	Zoldos et al. 1998
Q. suber	1.91	–	39.7	Zoldos et al. 1998
Genus Quercus Subgen Sclerophyllodrys				
Q. coccifera	2.00	–	40.4	Zoldos et al. 1998
Q. ilex	2.00	–	39.8	Zoldos et al. 1998

*Arumuganathan K, Schlarbaum SE, Carlson JE previously unpublished data (genome sizes are an average of three determinations of 2 to 4 individuals per species)

** According to Flora Europea (http://rbg-web2.rbge.org.uk/FE/fe.html)

among the Fagaceae. In addition, the small genome of *Quercus velutina* at 1.17 pg per 2C is essentially the same as that of the *Fagus* species, again suggesting a basal genome size of about 1.2 pg per 2C for the Fagaceae. Among the 24 *Quercus* species presented, the range of genome sizes is essentially continuous up to a maximum of 2.0 pg per 2C in *Q. coccifera* and *Q. ilex*. We looked for any indication that the interspecific variation in the observed genome sizes followed the botanical subdivisions within *Quercus*. We used here the classification into four distinct botanical subgenera from Flora Europaea (http://rbg-web2.rbge.org.uk/FE/fe.html). This classification corresponds to earlier botanical descriptions of Schwarz (1964) and Camus (1936-1954) and recent molecular analyses (Manos et al. 1999; Xu et al. 2005). The species that were investigated include representatives of all four subgenera: 12 species in *Erythrobalanus* (red oaks), seven species in *Lepidobalanus* (white oaks), three species in *Cerris*, and two in *Sclerophyllodrys*. The average 2C DNA contents were 2.0 pg for subgenus *Sclerophyllodris*, 1.75 pg for subgenus *Cerris*, 1.73 pg for subgenus *Lepidobalanus*, and 1.56 pg for subgenus *Erythrobalanus*. The two oak species with the largest genomes, *Q. coccifera* and *Q. ilex* (2.0 pg), are both evergreen species and are part of a disputed botanical subgenus (named *Sclerophyllodrys*, according to Schwarz 1964). This is intriguing, given that molecular phylogenetic analysis separates the evergreen species from the two sections of deciduous oaks (Manos and Steele 1997 and Xu et al. 2005), confirming their earlier subdivision in *Sclerophyllodrys* by Schwarz (1964).

Among the five *Castanea* species studied, genome sizes varied much less than among oaks, ranging only from 1.57 pg in *C. seguinii* to 1.67 pg per 2C in *C. dentata*. In fact DNA content varied as much within the chestnut species as between. For example, unrelated *C. seguinii* trees varied from 1.5 to 1.63 pg per 2C, while *C. sativa* varied from 1.57 to 1.65 pg per 2C (Arumuganathan et al. this study). Thus there may not be significant differences in average DNA content between *Castanea* species, and the range of average DNA content reported among species in Table 1 may just represent the natural variation in DNA content among *Castanea* individuals. The intraspecific variation in DNA content in *Quercus* was also as extensive as the amount of variation among the species. For example, the 2C DNA content varied between 1.88 pg and 2.0 pg among *Q. petraea* trees of the same populations (Zoldos et al. 1998), between 1.45 and 1.96 in *Q. pagoda*, and between 1.34 and 1.78 in *Q. macrocarpa* (Arumuganathan et al. this study). The intraspecific variation may be due in part to the occurrence of extra B chromosomes (Ohri and Ahuja 1990; Zoldos et al. 1998). While the range of DNA content among oak species appears to be greater than among chestnut and beech species, the magnitude of the differences among oak species may be related to experimental issues as well as biological ones. For example, the size estimates by Arumuganathan et al. (this study) were consistently smaller than those by Favre and Brown (1996) and Zoldos et al. (1998). One could speculate that the differences relate to the fact that Arumuganathan et al. (this study) studied primarily New World species, while the other two studies dealt exclusively with Old World species. However, the three groups report different genome sizes for *Q. robur* (1.53, 1.84, and 1.88 pg per 2C, respectively). Whether this discrepancy has a biological basis (the Arumuganathan et al. study sampled three trees of *Q. robur* "fastigiata," the "upright" horticultural variety) or resulted from experimental differences in sampling, internal size standards, and other methodologies is not clear.

In general, the genome sizes in the Fagaceae are only 3.5 to 6 times larger than the genome of *Arabidopsis* (0.32 pg; Bennett et al. 2003) and are within the size range of the sequenced rice and poplar genomes (both 1.0 pg; Brunner et al. 2004). Comparative genomics should thus be relatively efficient within the Fagaceae. Comparative genomics will lead to a better understanding of the extent to which the continuous range of DNA content is related to adaptive radiation of the species during evolution or is the result of overlapping ranges and interspecies hybridizations. Knowledge of the genome sizes reveals that genome-level comparisons between *Fagus sylvatica*, *Q. velutina*, *Q. coccifera*, and *Q. ilex* would be particularly informative and could illuminate the role of genome duplication in the evolution of the Fagaceae. When comparative genomics studies are extended to more species within the Fagaceae, it will be interesting to determine whether or not the broader range of genome sizes observed in *Quercus* relates to more extensive adaptations and specializations than exist among *Fagus* and *Castanea* species. Given the extensive natural populations of Fagaceae species that still exist across the northern hemisphere, such information will certainly provide insights into the ecology of temperate forest ecosystems.

5.2 Construction of Genetic Linkage Maps

5.2.1 Genetic Mapping in Forest Trees

PCR-based molecular markers and the two-way pseudotestcross strategy are useful tools for constructing genetic maps in forest trees (Grattapaglia and Sederoff 1994). These outbred species are characterized by long generation times, long life spans, and a high genetic load that often leads to significant inbreeding depression. Although all these elements hinder the development of the type of mapping populations normally used for genetic linkage mapping (for instance inbred lines and backcrosses), the high level of heterozygosity in forest species made two-generation full-sib pedigrees suitable populations for marker segregation analysis. Full-sib and half-sib crosses can, therefore, be used to construct single-tree genetic linkage maps thanks to dominant PCR-based molecular markers. Following this approach, called the two-way pseudotestcross strategy (Grattapaglia and Sederoff 1994), three types of segregation configurations can be obtained for dominant molecular markers in the mapping population: (1) male testcross markers, segregating in a 1:1 ratio and inherited from the male parent; (2) female testcross markers, segregating in a 1:1 ratio and inherited from the female parent; and (3) intercross markers, segregating in a 3:1 ratio and inherited from both parental trees. Male and female testcross markers are used to construct two independent single-tree genetic maps that are then aligned thanks to the intercross markers. RAPD (Williams et al. 1990) and AFLP (Vos et al. 1995) dominant molecular markers have been used most commonly to construct genetic linkage maps in forest tree species (Verhaegen and Plomion 1996; Marques et al. 1998; Arcade et al. 2000; Costa et al. 2000; Cervera et al. 2001), as their random distribution in the genome allows all chromosomes to be covered most efficiently.

The two-way pseudotestcross strategy was first applied in forest trees by Grattapaglia et al. (1995) to identify loci controlling quantitative trait loci (QTLs). In forest genetics, QTL analysis has been one of the most important applications of linkage mapping, and several studies reported successful QTL detections (Sewell and Neale 2000; van Buijtenen 2001).

5.2.2 Genetic Mapping Initiatives in Fagaceae

Oak Mapping Initiatives

European white oaks Starting in 1995, activities in genetic mapping were implemented in European white oaks at the INRA Research Centre in Bordeaux-Cestas (France). Motivations for genetic mapping in oaks were threefold: (1) the detection of genomic regions involved in species differentiation, (2) the detection of QTLs controlling traits of adaptive significance, and (3) the comparative analysis of genomic evolution in the Fagaceae. The whole mapping project is based on three pedigrees: one full-sib family of *Quercus robur* (3P × A4), one full-sib family of *Q. petraea* (QS28 × QS21), and one interspecific F_1 full-sib family *Q. robur* × *Q. petraea* (11P × QS29). An interspecific F_2 cross is planned as well. Given the objectives of the mapping experiments, the parents of the pedigrees were not selected for any particular criteria. The *Q. robur* parent trees originated from the southwest of France (INRA research station of Bordeaux-Cestas, and Arcachon) and the *Q. petraea* parents were from the central part of France (INRA research station of Orléans-Ardon). The controlled crosses were repeatedly done over successive years until 2004. From 200 to 1,000 seeds were obtained for each cross. The young seedlings were installed in a seedbed in a nursery, where they are raised as stool beds. Starting at age 5, the full-sibs were hedged every year at the ground level at the end of winter. Following the hedging, stump sprouts developing in spring were harvested and cut in 15- to 20-cm-long cuttings. These cuttings were then transplanted in field tests for phenotypic observations and further QTL detection. For the time being, only the *Q. robur* intraspecific cross has been fully exploited for genetic mapping and QTL detection. The clonal test of the full-sibs has now been planted in three different sites (two near Bordeaux, southwest France, and one near Nancy, northeastern France). The genetic mapping of *Q. robur* mapping was done on a sample of 94 offspring (pedigree 3P × A4), and the QTL detection on a sample of 278 offspring (replicated on average in five vegetative propagules).

Another mapping initiative for *Q. robur* was implemented in the Netherlands (ALTERRA, Wageningen). This mapping pedigree consists of 101 full-sibs (Bakker 2001). The sibs were screened by paternity analysis within an open-pollinated progeny set of 397 sibs collected on a single tree located in an urban area (Amsterdam). This tree was surrounded by three other oak trees within a radius of 10 m. One of these oak trees was selected to be the paternal tree. Paternity analysis revealed that 26% of the collected seeds were sired by this male parent. The selected seeds were germinated, grown individually in pots in a nursery (ALTERRA research station, Wageningen, the Netherlands), and measured for several morphological and physiological traits during the next 2 years (1999, 2000). The objective of this work was quite similar to the French initiative: detection of QTL controlling morphological and adaptive traits involved in species differentiation.

American red oaks Genetic mapping in northern red oak (*Quercus rubra* L.) was initiated at Purdue University (http://www.genomics.purdue.edu/forestry/; Romero-Severson 2003) and has continued at the University of Notre Dame. Using exclusion methods based on microsatellite polymorphisms (Aldrich et al. 2002, 2003a), a preliminary mapping population of 97 full-sibs was identified from the open-pollinated progeny of a single tree. The most likely male parent male was the closest conspecific. Recombination patterns revealed (Romero-Severson et al. 2003) six linkage groups (LGs) of three or more markers. A second acorn harvest from the same female parent yielded 462 full-sibs. The genetic map under construction now includes 15 microsatellites, 66 AFLP markers from the first set of progeny, and several hundred new AFLP markers from the second set of progeny. All of the potential pollen parents within 200 m of the female parent are being genotyped with all 15 microsatellite markers to eliminate any doubt over the full-sib status of the mapping population. The microsatellite markers used for genetic mapping are the same as those used for studies on interspecific gene flow (Aldrich et al. 2003b) and in northern red oak genetic diversity studies. No map has yet been published for *Q. rubra*. The long-term goal of the red oak mapping project is the detection of QTLs and genes controlling heartwood color and resistance to specific pests, specifically *Phytophthora ramorum*, the agent of sudden oak death.

Chestnut Mapping Initiatives

European chestnut Starting in 1998, a genetic mapping project for European chestnut (*Castanea sativa* Mill.) was implemented using a full-sib family obtained from a controlled cross performed between two highly differentiated trees originating from Turkey. Anatolia Peninsula was shown to be an important region for chestnut genetic diversity (Villani et al. 1991, 1992). As illustrated by these studies, a remarkably high level of genetic, morphological, and physiological differentiation was observed between two groups of chestnut populations coming from two phytogeographic regions, characterized by striking climatic differences: the Eurosiberian part of the peninsula in northeastern Anatolia (humid) and the Mediterranean region in western Anatolia (xeric). Common field experiments carried out at the experimental field site of Istituto di Biologia Agroambientale e Forestale, CNR (Porano, Italy), showed significant differences between these populations in growth rate, bud flush, and physiological parameters, related to the water use efficiency, allowing "drought-adapted" and "wet-adapted" ecotypes to be identified (Lauteri et al. 1997, 1999). Differences observed in the ecophysiological behavior suggested that Turkish chestnut populations are genetically adapted to contrasting environments, making them a suitable material to study the adaptive potential of this species.

The controlled cross was performed in 1998 between a female parent (Bursa) belonging to the "drought-adapted" type from western Turkey and a male parent (Hopa) belonging to the "wet-adapted" type from eastern Turkey. The parental individuals were 9 years old and were chosen according to their heterozygosity level at isozymes and high degree of variation in physiological traits. An F_1 full-sib family of 186 offsprings was obtained, and 96 F_1 individuals were used to construct two separate genetic linkage maps: a female or Bursa map and a male or Hopa map. The main objective of the project was to exploit the peculiar genetic and adaptive variation observed in these populations in order to identify the genomic regions affecting carbon isotope discrimination (related to the water use efficiency), bud phenology, and growth by means of QTL analysis.

American and Chinese chestnuts During the last century, American Chestnut, *Castanea dentata* (Marsch) Borkh, one of the most important timber and nut-producing tree species in eastern North America, was dramatically affected by a canker disease (chestnut blight) caused by *Cryphonectria parasitica*. American chestnut showed low levels of resistance to blight, whereas Asian chestnut species (*Castanea crenata* (Japanese chestnut) and *C. mollissima* (Chinese chestnut) exhibited higher levels of resistance to the disease. During the 1980s an important backcross breeding program was undertaken in the USA in order to obtain selected material combining the blight resistance of Asian chestnut and good timber qualities of American chestnut (Burnham et al. 1986).

In this context, a genetic map for chestnut was constructed. The main objective of this mapping project was to identify genomic regions involved in blight resistance. In addition, the map was also used to locate loci controlling morphological traits that differentiated both species. The mapping population was F_2 progeny derived from a cross between two *C. mollissima* × *C. dentata* F_1 hybrids. The female parent was the *C. mollissima* cultivar "Mahogany" and two different American chestnut trees from Roxbury, CT were used as male to create the F_1 hybrids. One hundred and two F_2 individuals were used for the map construction, and 185 individuals were assessed for resistance to *Cryphonectria parasitica*.

Beech Mapping Initiative

A genetic mapping project for European beech (*Fagus sylvatica*) has been implemented at the University of Parma (Italy) during the last 10 years (Scalfi et al. 2004). The objective was to dissect important adaptive traits and to identify their underlying QTLs to detect genomic regions involved in important quantitative traits such as growth, phenology, and water-use efficiency. The mapping pedigree consisted of a full-sib family comprising 143 offsprings. The family was the largest in a 4 × 4 diallel controlled cross performed in 1995 (Ceroni et al. 1997). The parents originated from a natural population located at high altitude in northern Italy (1,650 m altitude, just below the tree line).

5.2.3 Genetic Linkage Maps for *Quercus*, *Castanea*, and *Fagus*

Genetic Map of *Q. robur*

The first *Quercus* map was published in 1998 on *Q. robur* (Barreneche et al. 1998) (pedigree 3P × A4). Using the pseudotestcross mapping strategy, two maps were constructed comprising 307 markers (271 RAPD, 10 SCARs, 18 SSRs, 1 minisatellite, 6 isozymes, and 1 ribosomal DNA marker). Both maps provided 85 to 90% coverage of the *Q. robur* genome. Segregating markers could be aligned in 12 LGs, and the map size amounted to 893.2 cM for the paternal and 921.7 cM for the female map. These maps were further upgraded by the inclusion of new SSRs (Barreneche et al. 2004) and additional AFLP and STS. The upgrading is still ongoing and to date 854 markers (271 RAPD, 457 AFLP, 10 SCAR, 59 SSR, 49 EST, 1 minisatellite, 6 isozymes, and 1 ribosomal DNA marker) have been located (Table 2).

The Dutch *Q. robur* map (pedigree A1 × A2) was also constructed using the two-way pseudotestcross strategy (Bakker 2001). Two parental maps were first established comprising 18 SSR and 343 AFLP markers. The total lengths of the maternal and paternal maps were respectively 496 and 566 cM. Thirteen LGs were obtained (for 12 chromosomes) and the two maps could be partially merged using 58 "bridge" markers (2 LGs could not be aligned). One of the paternal LGs (LG 13, 27 cM) was highly dissimilar to the other LGs in terms of marker density. This LG contained almost half (48%) of all paternal markers and 22% of the segregating (heterozygote) markers. This marker-dense LG was homologous to one of the maternal LGs that remarkably was composed exclusively of 13 segregating markers. Congruence of LGs with the French map was based on the location of SSR markers (Sect. 5.3.1). The total map length of the integrated map was 659 cM, map density being one marker per 2.4 cM for the map without taking the exceptionally dense LG 13 into account.

Genetic Map of *Castanea sativa*

A first framework of the chestnut genetic linkage map was obtained with RAPD and ISSR markers (Casasoli et al. 2001). Few isozyme loci were integrated in this first version of the map. A total of 381 molecular markers segregated in the chestnut full-sib family covering a good portion of the chestnut genome (more than 70%). Intercross segregating markers allowed 11 of the 12 LGs identified to be aligned between the female and male maps. This original framework was then used to map AFLP markers and codominant locus-specific markers such as SSR- and EST-derived markers. Table 2 shows the number and type of molecular markers contained in the chestnut genetic linkage map. At present, 517 molecular markers have been mapped in chestnut covering 80% of its genome. The 12 LGs were aligned to obtain 12 consensus female and male LGs (chestnut linkage consensus groups are available at the Web site www.pierroton.inra.fr).

Table 2. Summary of genetic linkage maps of Quercus robur and Castanea sativa

Species (pedigree)	Number of LGs	Total number of marker loci	RAPD, ISSR, AFLP	SSR	STS	Isozymes	% of distorted markers	Total genetic distance (cM)	Genome saturation (%)	Total size (cM)	Ref.
Q. robur (3P × A4)	12	854	728	59	61	6	18	950	80	1200	Barreneche et al. 2004; and this study
Q. robur (A1 × A2)	13	361	343	18	–	–	17.5	659	64	1035	Bakker (2001); and this study
C. sativa (Bursa × Hopa)	12	517	427	39	46	5	10	865	82	1050	Casasoli et al. 2001; and this study
C. mollissima × *C. dentata* (Mahogany)	12	559	521	29	1	8	25	524	–	–	Kubisiak et al 1997; Sisco et al. 2005
F. sylvatica[a] (44 × 45)	12/11	138/124	128/113	10/6	9	–	23	844/971	78/82	1081/1185	Scalfi et al. 2004; Scalfi 2005

[a] The two numbers indicated correspond to the map of the female and male parent

Genetic Map of *Castanea mollissima*/*Castanea dentata*

The *C. mollissima*/*C. dentata* map was the first to be published in the *Fagaceae* (Kubisiak et al. 1997). At first a total of 241 markers, including 8 isozymes, 17 RFLPs, 216 RAPDs, were mapped in the F_2 family. Twelve LGs were identified, covering a genetic distance of 530.1 cM (corresponding to 75% of the chestnut genome). To saturate the map, additional markers were recently added to the initial map: 275 AFLP (Clark et al. 2001) and 30 STS (29 SSR and the 5SrDNA locus) (Sisco et al. 2005). To date, a total of 559 markers have been located. Relatively high levels of segregation distortion (more than 25%) have been reported in the *C. mollissima*/*C. dentata* family. Skewed segregation is a common feature in progenies resulting from interspecific crosses.

Genetic Map of *Fagus sylvatica*

The *Fagus* genetic linkage map was based on a total of 312 markers: 28 RAPDs, 274 AFLPs, and 10 SSRs. Two maps were constructed using the "double testcross" strategy. In the female map 132 markers were distributed in 12 LGs covering 844 cM. In the male parent only 11 LGs were identified, resulting in linkage relationships between 119 markers spanning over 971 cM (Table 2). The two maps cover about 78% and 82% of the *Fagus* genome. Using intercross markers (15 AFLP and 2 SSR) seven homologous LGs could be identified (Scalfi et al. 2004). Ten additional EST markers were then added to the map since its publication (Scalfi 2005).

5.3 Comparative Mapping between *Quercus*, *Castanea*, and *Fagus*

5.3.1 Mapping of Microsatellites in *Quercus robur*, *Castanea sativa*, *C. mollissima*, and *C. dentata*

Microsatellite markers, which are tandemly repeated units of 2 to 6 nucleotides evenly dispersed throughout plant genomes, have been sometimes used for comparative mapping studies (Marques et al. 2002). Amplification of orthologous SSR markers across phylogenetically related species depends largely on evolutionary distance and genome complexity of compared species (Powell et al. 1996). Usually, SSR cross-amplification is more efficient between closely related species with a low proportion of highly repeated sequences in their genome. Steinkellner et al. (1997) showed that microsatellite markers specifically developed in *Quercus* species were cross-amplified in chestnut. For these reasons microsatellites were supposed to be useful molecular markers for comparing *Q. robur* and *C. sativa* genetic linkage maps. To obtain orthologous markers for comparative mapping, SSR markers developed both in *Quercus* species and in *C. sativa* were therefore tested for cross-amplification and transferability between these two genera (Barreneche et al. 2004 and references therein) in a reciprocal way. We tested a total of 83 primer pairs: 53 developed in *Quercus* species and 30 in *C. sativa*. Primer pairs giving a strong amplification product were selected for mapping. Nineteen loci, 15 from oak and 4 from chestnut, were integrated into the two previously established genetic maps, allowing the first comparative mapping between LGs of the two species (Barreneche et al. 2004). Figure 1 shows the seven homeologous LGs identified by orthologous SSR and all microsatellite loci mapped in *Q. robur* and *C. sativa* genetic maps. These same SSR loci were used to align the European chestnut genetic linkage map with the *C. mollissima* × *C. dentata* interspecific map. Eleven of the 12 LGs of the two maps could be associated, nine LGs were aligned on the basis of pairs, triplets, or quadruplets of common markers, while three additional groups were matched using a single SSR marker (Sisco et al. 2005).

Overall, these findings showed that microsatellite markers could be cross-transferred between *Quercus* and *Castanea* genera and be used to recover orthologous markers for comparative mapping. Nevertheless, cross-transferability efficiency was low and the number of cross-transferred loci was not sufficient to link the 12 LGs of the two species. As expected, SSR loci were extremely useful for comparative mapping within the same genus (*Castanea*), but their cross-transferability efficiency decreased between different genera. SSR loci mapped both in *Q. robur* and *C. sativa* were sequenced in order to definitely demonstrate their orthology. Sequencing results clearly showed that both orthologous and paralogous loci could be recovered among the SSR cross-transferred between the two genera. Moreover, indels were sometimes observed within the flanking regions of the repeated motif. Therefore, although SSR loci can be cross-transferred between *Quercus* and *Cas-*

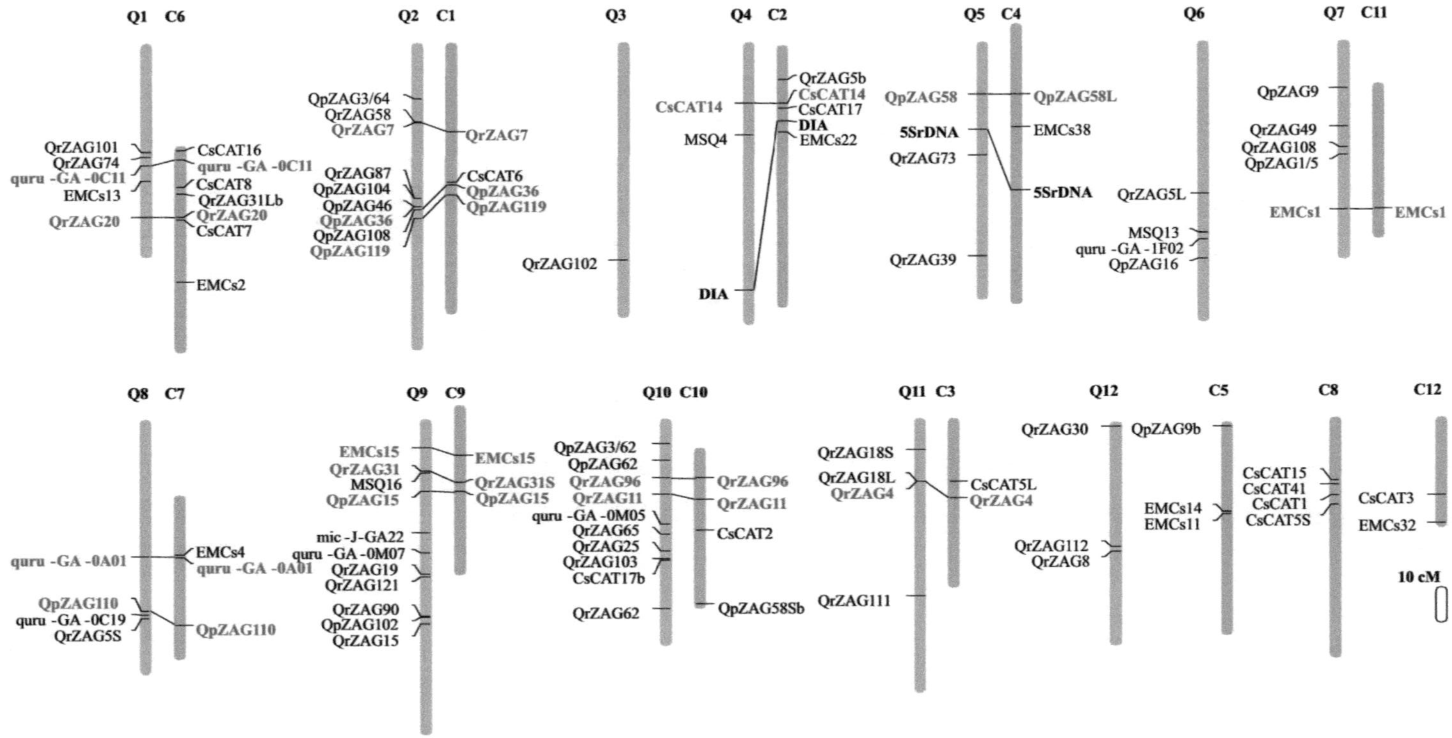

Fig. 1. Assignment between *Q. robur* and *C. sativa* based on orthologous microsatellites. Oak (Q, *green* on *left*, pedigree 3P × A4) and chestnut (C, *light blue* on *right*, pedigree Bursa × Hopa) LGs aligned using microsatellite markers. LGs are named as in Barreneche et al. (1998) and in Casasoli et al. (2001). Oak LGs are taken as reference and arranged in sequence from Q1 to Q12. Nine chestnut LGs, aligned with the corresponding oak LGs, are given on the *right*. The three remaining chestnut LGs are reported according to the oak LGs. Common orthologous SSR markers are shown in *red*. The EMCs1 marker was later shown to be a paralogous locus, and LG Q7 was not homeologous to LG C11 (Fig. 2). The figure, modified from Barreneche et al. (2004), was drawn using MapChart software (Voorrips 2002)

tanea genera, a sequence analysis is needed to demonstrate orthology and to avoid the risk of paralogy.

5.3.2
Mapping of EST-Derived Markers in *Q. robur* and *C. sativa*: Alignment of the 12 Linkage Groups between the Two Species

Several factors make EST (expressed sequence tag)-derived markers very useful for comparative mapping studies (Brown et al. 2001). First, ESTs are sequence fragments of coding regions; therefore sequence conservation among species is expected to be higher than that observed, for instance, in SSR loci. Second, ESTs correspond very often to genes of known function. This is of great interest because some ESTs colocalized with QTLs in a genetic linkage map may be putative positional candidate genes for a given trait. Finally, transcriptome analyses give rise to a high number of EST sequences that are the source of numerous EST-derived markers distributed throughout plant genomes. In oak, ESTs were developed by Derory et al. (2006) and Porth et al. (2005a). This gave the opportunity to exploit EST sequence information for marker design in order to complete the comparative mapping between *Q. robur* and *C. sativa* (Casasoli et al. 2006). About 100 EST sequences were selected from oak databases. Oak sequences were aligned with homologous sequences obtained from GenBank in order to design primer pairs for amplification in the most conserved regions of the sequence and assure a good cross-amplification efficiency in chestnut. A total of 82 primer pairs were designed. A proportion of about 70% produced by PCR a single and strong band both in oak and chestnut and 51 and 45 ESTs were mapped in oak and chestnut, respectively, using single strand conformation polymorphism (SSCP) and denaturing gradient gel electrophoresis (DGGE) approaches (Casasoli et al. 2006). These EST-derived markers, together with SSR markers previously mapped, provided 55 orthologous molecular markers that allowed the 12 LGs of *Q. robur* and *C. sativa* to be aligned. As shown in Fig. 2, from 2 to 7 common orthologous markers were mapped in the 12 homeologous pairs of LGs. Macrosynteny and macrocollinearity were well conserved between the two species. Few inversions, probably due to mapping errors, were observed. Although these data are still preliminary given the low number of common molecular markers mapped in the two species, no major chromosomal rearrangements have been identified, suggesting that oak and chestnut genomes are quite stable. Thus it appears likely that the "single genetic system" model of the grass genomes (Gale and Devos 1998) can also be applied to *Q. robur* and *C. sativa*. EST-derived markers were very easily transferred from oak to chestnut. About 50% of them contained intron-derived sequences. This increased the probability of detecting segregating polymorphisms useful for mapping in both oak and chestnut full-sib families. These markers proved to be ideal markers for comparative mapping within the Fagaceae family.

5.3.3
Mapping of Microsatellites and EST-Derived Markers in *Fagus sylvatica*, *Quercus robur*, and *Castanea sativa*

Success of transferability between *Fagus sylvatica*, *Quercus robur*, and *Castanea sativa* was lower. Although 86 SSR markers originally developed in other *Fagaceae* species were tested in *Fagus* (66 from *Quercus*, 20 from *Castanea*), only seven produced an interpretable banding pattern and only one marker from *Q. rubra* and one from *C. sativa* could be placed on the beech map (Scalfi 2005). One marker originally developed in *Fagus* gave good amplification also in *Quercus* and *Castanea* but was monomorphic in the crosses used for these species.

Similarly, 86 EST markers originally developed in *Quercus* were tested in beech, 46 coming from a budburst c-DNA library (Derory et al. 2006), 22 from osmotic stress response (Porth et al. 2005a), and 17 from hypoxia response cDNA-AFLP markers (C. Bodénès unpublished results). The success rate was higher than for microsatellites. In total 16 were polymorphic using various techniques (SSCP, DGGE, sequencing, CAPS, dCAPS), and 10 were finally mapped onto the beech map (Scalfi 2005).

Two markers (1T11 and 1T62) that mapped on *Quercus* and *Castanea* on LG 10 (Table 4) were mapped also on group 4 in *Fagus* with the help of a "bridge" marker (1T41): this can be considered as evidence of synteny between LG10 of *Q. robur* (3P*A4) and *C. sativa* with LG 4 of *Fagus*. For the two markers the sequence homology of *Fagus* with *Quercus* was 82% and 43%, respectively; the lower value is due to a large insertion in the beech sequence that was not present in the cDNA of *Quercus*. Eliminating the gap,

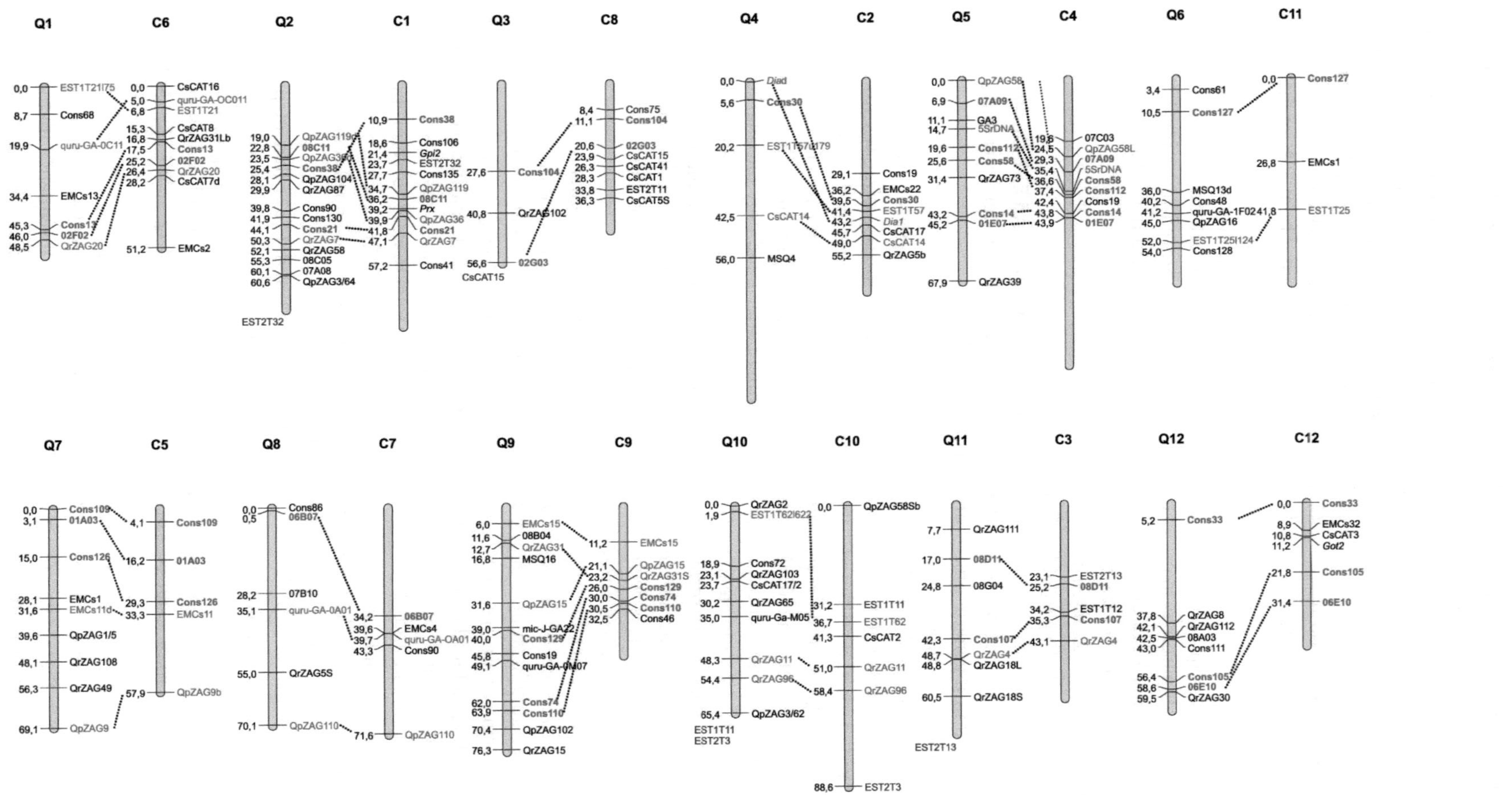

Fig. 2. Comparative mapping between *Q. robur* and *C. sativa.* The 12 homeologous LGs between *Q. robur* (Q, *green*, pedigree 3P × A4) and *C. sativa* (C, *light blue*, pedigree Bursa × Hopa). The orthologous molecular markers mapped in both species are shown in *red* (SSRs and EST-derived markers). A subsample of molecular markers of the oak and chestnut consensus genetic linkage maps (available at www.pierroton.inra.fr) is shown in this figure. Orthologous molecular markers mapped in a different oak cross (or showing a low mapping statistical support, Cons 75 in the oak LG Q3) are marked in *blue* below the LGs

the homology increased to 92%. Synteny could not be assessed for any other group since none had more than one marker mapped on it. For example, marker 2T32 mapped on LG 2 in *Quercus* was found linked to markers on LG 7-F in beech, but more than one comapping marker is needed to establish synteny.

5.3.4 Assignment of Linkage Groups Between *Quercus* and *Castanea*

Most genetic maps constructed within Fagaceae species comprised 12 LGs (Table 2). Cross-transferable SSR- and EST-derived markers made it possible to assign LGs among the four species *Q. robur*, *C. sativa*, *C. mollissima*, and *C. dentata* (Table 4). However, the assignment is still based on a limited number of markers per LG. Assignment was done by pairwise comparisons:

- Between the two *Q. robur* maps (3P × A4 and A1 × A2): 11 out of 12 LGs had at least two orthologous SSRs in common; the remaining LG was assigned by default.
- Between *Q. robur* (3P × A4) and *C. sativa* (Bursa × Hopa): between two and seven pairs of common markers (either SSR, isozymes, or EST) allowed the LGs to be assigned.
- Between the two chestnut maps (Bursa × Hopa and Mahogany): the assignment is still incomplete as only nine LGs could be assigned so far by at least two pairs of SSRs.

The results obtained so far need to be confirmed by further mapping experiments, based mainly on EST markers. They are also encouraging as suggested by the conservation of the macrosynteny and macro-collinearity that have so far been observed between the two most intensively studied species: *Q. robur* and *C. sativa*.

5.4 Genes Mapped in Oaks and Chestnut

Transcriptomic investigations and differential gene expression studies were implemented recently with the main aim of identifying genes that are involved in the adaptation of oak or chestnut trees to their environment. Gene expression was monitored for different traits, or tissues:

- Bud burst in oaks
- Hypoxia in oaks
- Osmotic stress in oaks
- Juvenile and mature shoots in oaks
- Blight infection in chestnuts

Various techniques were implemented for constructing expression profiles: cDNA-AFLP, SSH, and Quantitative RT-PCR. We will briefly summarize the experiments conducted and the functions of genes that were identified. Table 3 provides a list of EST-derived markers mapped in *Q. robur* and *C. sativa*. For each EST the accession number, amplification, sequencing, and mapping results are reported.

5.4.1 Bud Burst

Candidate genes for bud burst were identified in *Q. petraea* using SSH libraries, macroarray experiments, and RT-PCR. Three subtracted libraries (SSH method) were constructed, generating 801 ESTs derived from six developmental stages of bud burst. Expression patterns of these transcripts were monitored in apical buds during bud flushing in order to identify genes differentially expressed between the quiescent and active stage of bud development. After bioinformatic processing of the ESTs, macroarray experiments revealed a total of 233 unique transcripts exhibiting differential expression during the process, and a putative function was assigned to 65% of them (Derory et al. 2006). Cell rescue/defense-, metabolism-, protein synthesis-, cell cycle-, and transcription-related transcripts were among the most regulated genes. Reverse northern and RT-PCR showed that several genes exhibited contrasting expression between quiescent and swelling buds. Among this set of 233 unique transcripts, ca. 100 were selected and tentatively amplified and mapped in oak and chestnut, as previously described. In oak and chestnut, 51 and 45 ESTs were successfully mapped, respectively, using SSCP and DGGE approaches (Casasoli et al. 2006).

5.4.2 Hypoxia

Q. robur and *Q. petraea* exhibit different responses to hypoxia, the first one being more tolerant to waterlogged conditions. Hypoxia-induced genes were identified from vegetative copies of the two species

Table 3. List of genes mapped in both Quercus robur and Castanea sativa

EST Name[a]	Accession number	Reference for primer sequences and PCR protocols	Expected size (bp)[b]	Observed size	Identity[c]	Linkage group Q – C [d]	Functional category
1T11	CF369263	Porth et al. 2005b	555	555	99	10 – 10	Unknown
1T12	CF369264	Porth et al. 2005b	522	522	93.5*	nm – 3	Metabolism
1T21	CF369266	Porth et al. 2005b	338	770	94	1 – 6	Protein synthesis
1T25	CF369268	Porth et al. 2005b	187	187	97.2	6 – 11	Unknown
1T57	CF369273	Porth et al. 2005b	282	282	93.5	4 – 2	Transcription
1T62	CF369274	Porth et al. 2005b	346	705Q 600C	91.9	10 – 10	Metabolism
2T11	CF369278	Porth et al. 2005b	397	397Q 500C	89*	nm – 8	Cell rescue, defense and virulence
2T3	CF369283	Porth et al. 2005b	284	630	94.3	10 – 10	Unclassified (plasma membrane related?)
2T13	CF369280	Porth et al. 2005b	334	334	94.5	11 – 3	Metabolism
2T32	CF369284	Porth et al. 2005b	386	608Q 500C	93	2 – 1	Protein synthesis
01A03	CR627501	Casasoli et al. 2006	382	500	94.3	7 – 5	Protein synthesis
1,00E+07	CR627526	Casasoli et al. 2006	145	400Q 145C	86	5 [d] – 4	Transcription
02F02	CR627566	Casasoli et al. 2006	164	164	86.3	1 – 6	Transcription
02G03	CR627575	Casasoli et al. 2006	206	700	90.3	3 – 8	Hypothetical protein
06B07	CR627724	Casasoli et al. 2006	307	307	93.3	8 – 7	Hypothetical protein
6,00E+10	CR627745	Casasoli et al. 2006	333	700	94.5	12 – 12	Protein with binding function or cofactor requirement
07°08	CR627771	Casasoli et al. 2006	341	700	86.1*	2 – ni	Hypothetical protein
07°09	CR926157	Casasoli et al. 2006	252	300	95.7	5 – 4	Cellular transport
07B10	CR627781	Derory et al. 2006	360	360		8 – na	Transcription
07C03	CR627785	Casasoli et al. 2006	285	285	96.6*	ni – 4	Hypothetical protein

[a] STSs 08A01, 07B10, 08B04, Cons 86, and 08G04 have not been sequenced.

[b] The expected sizes were based on the knowledge of the EST sequence and primer design. The observed sizes were approximate because based on an electrophoresis on agarose gel. The unmapped amplified ESTs were either noninformative or mapping methods (SSCP or DGGE) have not been successfully optimized.

[c] Except for 01E07 and Cons 129, all STS sequences matched the same gene in both species using a BLASTX procedure. If STS was mapped or sequence was available for only one species, alignment has been done with the original oak EST (*). In 5 cases, sequence reaction did not work (-).

[d] We used a LOD threshold ³6.0 to map STS, except for those marked with e, for which 4.0 < LOD score < 6.0. ni: noninformative; nm: nonmapped. Q-C: Quercus robur (3P × A4)-Castanea sativa (Bursa × Hopa)

Table 3. (continued)

EST Name[a]	Accession number	Reference for primer sequences and PCR protocols	Expected size (bp)[b]	Observed size	Identity[c]	Linkage group Q – C [d]	Functional category
08°01	CR627918	Derory et al. 2006	210	500		3 [d] – nm	Metabolism
08°03	CR627920	Casasoli et al. 2006	454	454	94.1*	12[d] – ni	Protein with binding function or cofactor requirement
08B04	CR627933	Derory et al. 2006	327	327		9 – nm	Metabolism
08C05	CR627943	Casasoli et al. 2006	213	213	95.5*	2 – ni	Hypothetical protein
08C11	CR627947	Derory et al. 2006	316	316	94.4	2 – 1	Hypothetical protein
08D11	CR627958	Casasoli et al. 2006	343	700	88.4*	11 – 3	Metabolism
08G04	CR627986	Derory et al. 2006	393	1000		11 – na	Hypothetical protein
Cons 13	CR627506	Casasoli et al. 2006	301	301	89.5–93.1	1 – 6	Transcription
Cons 14	CR627508	Casasoli et al. 2006	243	1200	–	5 – 4	Protein synthesis
Cons 19	CR627517	Casasoli et al. 2006	178	300	81.3	9 – 2/4	Protein synthesis
Cons 21	CR627523	Casasoli et al. 2006	333	550	89.3	2 – 1	Protein synthesis
Cons 30	CR627541	Casasoli et al. 2006	424	1400Q 1500C	93.1	4 – 2	Hypothetical protein
Cons 33	CR627568	Casasoli et al. 2006	153	200Q 250C	95.3	12 [d] – 12	Hypothetical protein
Cons 38	CR627606	Casasoli et al. 2006	123	123	91.7	2 – 1 [d]	Energy
Cons 41	CR627646	Casasoli et al. 2006	443	500	90.7*	ni – 1	Cell rescue, defense, and virulence
Cons 46	CR627952	Casasoli et al. 2006	215	800C	–	na – 9	Cell rescue, defense, and virulence
Cons 48	CR627721	Casasoli et al. 2006	191	191Q 400C	–	6 – ni	Unknown
Cons 58	CR627732	Casasoli et al. 2006	255	500	92.7	5 – 4	Hypothetical protein
Cons 61	CR627776	Casasoli et al. 2006	260	1600Q	95.7*	6 – na	Cell rescue, defense, and virulence
Cons 68	CR627777	Casasoli et al. 2006	244	500Q	92.9*	1 – na	Metabolism
Cons 72	CR627907		312	1000Q 800C	90.9*	10 – ni	Cell cycle and DNA processing
Cons 74	CR627801	Casasoli et al. 2006	137	137	86.7	9 – 9	Cell rescue, defense, and virulence
Cons 75	CR627924	Casasoli et al. 2006	257	600	88*	ni – 8	Metabolism
Cons 86	CR627976	Casasoli et al. 2006	270	600		8 – nm	Unknown
Cons 90	CR628018	Casasoli et al. 2006	188	300Q 1200C	–	2 – 7	Cell rescue, defense, and virulence
Cons 104	CR627823	Casasoli et al. 2006	250	250	95.4	3 – 8	Hypothetical protein

Table 3. (continued)

EST Name[a]	Accession number	Reference for primer sequences and PCR protocols	Expected size (bp)[b]	Observed size	Identity[c]	Linkage group Q – C[d]	Functional category
Cons 105	CR627826	Casasoli et al. 2006	185	600	95.4	12 – 12	Metabolism
Cons 106	CR627828	Casasoli et al. 2006	326	326	91.2*	ni – 1	Energy
Cons 107	CR627830	Casasoli et al. 2006	272	900	92.6	11 – 3	Cell-type differentiation
Cons 109	CR627834	Casasoli et al. 2006	194	1200	100*	7 – 5	Cell rescue, defense, and virulence
Cons 110	CR627835	Casasoli et al. 2006	219	219	92.6	9 – 9	Metabolism
Cons 111	CR627837	Casasoli et al. 2006	219	219Q 600C	89.5*	12 – ni	Hypothetical protein
Cons 112	CR627839	Casasoli et al. 2006	171	171	93.4*	5 – 4	Transcription
Cons 126	CR628009	Casasoli et al. 2006	238	400	94.7	7 – 5	Protein synthesis
Cons 127	CR628014	Casasoli et al. 2006	289	289	94.3	6 – 11	Protein with binding function or cofactor requirement
Cons 128	CR628019	Casasoli et al. 2006	120	120	–	6 – ni	Energy
Cons 129	CR628021	Casasoli et al. 2006	210	500	79.6	9[d] – 9	Cell rescue, defense, and virulence
Cons 130	CR628241	Casasoli et al. 2006	190	190	91.9*	2 – ni	Energy
Cons 135	CR628167	Casasoli et al. 2006	115	200	100*	ni – 1	Hypothetical protein

Table 4. Homologous linkage groups (LGs) in genetic maps of *Quercus robur*, *Castanea sativa*, and *C. mollissima*/*C. dentata*

LG in *Quercus robur* [a] Pedigree 3P × A4	LG in *Quercus robur* [b] Pedigree A1 × A2	LG in *Castanea sativa* [c] Pedigree Bursa × Hopa	LG in *C. mollissima*/*C. dentata* [d] Pedigree Mahogany
1	1	6	H
2	2	1	A
3*	11*	8	C
4	4	2	K
5	5	4	E
6	6	11**	B**
7	7	5	I
8	10	*7*	*F*
9	8	9	L
10	3	10	D
11	9	*3*	*G*
12	12	12	J

Assignment of linkage groups was made by comparison within the following pairs: 3P × A4 and A1 × A2, 3P × A4 and Bursa × Hopa, Bursa × Hopa and Mahogany.
*) LG 3 in (3P × A4) and 11 in (A1 × A2) assigned by "default" (all other 11 LGs being assigned by at least 2 markers present in each species)
**) LG 11 in (Bursa × Hopa) and B in (Mahogany) assigned by "default" (all other 11 linkage group being assigned by at least two markers present in each species, except for pairs 7-F and 3-G where only one marker was common).)
The numbers or letters of linkage groups (LG) correspond to the following publications:
[a] Barreneche et al. (1998); Barreneche et al. (2004)
[b] Bakker (2001) and this study
[c] Casasoli et al. (2001)
[d] Kubisiak et al. (1997); Sisco et al. (2005)

grown in hydroponic conditions. Gene expression was monitored in seedlings raised under reduced oxygen (3%) applied for 24 h. RNA was extracted from root tips before (0 h time stress) and after oxygen reduction, following the protocol of Chang et al. (1993). Stress induction was validated by measuring alcohol dehydrogenase activity. Differentially expressed fragments were obtained by cDNA-AFLP, and 170 were sequenced and compared to databanks (C. Bodénès unpublished results).

5.4.3 Osmotic Stress

Osmotic stress induced genes were identified in a *Q. petraea* cell line grown under moderate stress (Porth et al. 2005a). Two subtraction libraries (SSH method) were established from callus cell cultures exposed to hyperosmotic stress for 1 h (indicated as 1T) and 2 d (2T), respectively. The differentially expressed ESTs were classified according to their putative functions. At least five of these gene products were assumed to be targets for stress tolerance in oak, e.g., betaine aldehyde dehydrogenase, two trans-acting transcription factors (one ABA-responsive, the other ABA-independent), a glutathione-S-transferase, and a heat shock cognate protein.

Seven genes were selected, based on their putative functions, to monitor their expression in vivo. Leaf tissue from hyperosmotically grown *Q. petraea* and *Q. robur* plantlets was harvested and investigated by RT-PCR at time intervals of 1, 6, 24, and 72 h. Indications of stress adaptation were found in *Q. petraea* based on up-regulation of certain genes related to protective functions, whereas in *Q. robur* down-regulation of those genes was evident (Porth et al. 2005a).

Segregating osmo-regulated loci were mapped to ten different LGs of *Quercus* (Porth et al. 2005b). By using orthologous primers, ten of the loci, including the four putatively water-stress tolerance related genes (1T57, 1T62, 2T11, and 2T13), were successfully amplified in *C. sativa*. Sequence analysis showed an identity of at least 90% (Table 3) with *Quercus*.

5.4.4
Differential Expression in Juvenile and Mature Oak Shoots

A gene named QRCPE (*Quercus robur* crown preferentially expressed) that is differentially expressed between mature and juvenilelike shoots was recently discovered in oaks (Gil et al. 2003). QRPCE accumulates in ontogenetically older organs of oak trees, although it is present in zygotic and somatic embryos but absent in callus cells. The encoded protein is small, contains a predicted N-terminal hydrophobic signal peptide that targets the protein to the cell wall, and is rich in glycine and histidine residues. In *C. sativa*, the QRCPE homolog is also expressed at different levels between adult and juvenilelike tissues.

5.4.5
Blight Infection in Chestnut

A cDNA clone showed similarity to a gene previously identified as encoding a cystatin. A protein shown to have antifungal activity in *C. sativa* (Pernas et al. 1998, 1999) was isolated from a cDNA library from stem tissues of *C. dentata* (Connors et al. 2001). The expression of this gene was verified by RT-PCR in healthy and diseased tissues of American chestnut (Connors et al. 2002). Amplification of a fragment of the gene in American and Chinese chestnuts and comparison of the sequences of the cloned amplification products revealed differences within the intron (SNPs or deletion). These differences could be used to locate the cystatin gene on the map of *C. mollissima*/*C. dentata* and to verify its putative colocalization with QTLs involved in blight resistance (Connors et al. 2002). However, cystatin did not map to any region known to be involved in resistance to chestnut blight.

5.5
QTL Detection

5.5.1
Phenotypic Traits Investigated

A common objective in genetic mapping in oak, chestnut, and beech is the detection of QTLs for adaptive traits, e.g., phenotypic traits that respond strongly to natural selection, and particularly to abiotic or biotic stresses. The interest in these traits lies in the issues raised by global change and the capacity of trees to respond to these challenges (Parmesan and Yohe 2003). This capacity depends on the level of genetic diversity for these traits and their underlying genes in natural populations. Knowledge of the genetic architecture of these traits (number and distribution of QTLs) is therefore of primary importance and has motivated research in QTL in conifers as well (Sewell and Neale 2000; van Buijtenen 2001).

In European oak, chestnut, and beech, the genetic control of three different adaptive traits, bud phenology, growth, and carbon isotope discrimination, were studied using a QTL approach (Casasoli et al. 2004; Scalfi et al. 2004; Scotti-Saintagne et al. 2004; Brendel et al. 2007). Bud phenology, growth, and carbon isotope discrimination (delta or Δ, which provides an indirect measure of plant water-use efficiency) are adaptive traits that show great phenotypic variation in natural populations of forest trees (Zhang and Marshall 1995; Tognetti et al. 1997; Lauteri et al. 1999; Hurme et al. 2000; Jermstad et al. 2001). Initiation and cessation of the growing seasons, defined through bud flush and bud set timing, have profound implications for adaptation of perennial plants to cold winter temperatures. Early flushing genotypes might be susceptible to spring frost damage. Likewise, bud set timing is related to the fall cold acclimation (Howe et al. 2000). Growth traits, such as annual height and diameter increments, are important components of plant vigor and biomass production, and they are profoundly influenced by abiotic and biotic stress occurrences during the growing season. In addition, they are relevant characteristics from an economic point of view and are often evaluated in breeding programs (Bradshaw and Stettler 1995). Carbon isotope discrimination (Δ) is a parameter related to the isotopic fractionation of carbon stable isotopes during the photosynthetic process (for review see Farquhar et al. 1989; Brugnoli and Farquhar 2000). Plant material is always enriched in ^{13}C with respect to the isotopic composition ($\delta^{13}C$) of atmospheric CO_2. This is particularly evident in C_3 plants where the fractionation effect mostly occurs during CO_2 diffusion from outside the leaf to the carboxylation sites into the chloroplasts, and during the carboxylation by ribulose 1,5-bisphosphate (RuBP) carboxylase. Due to its relationships with the diffusional path of photosynthetic gas exchange (for both CO_2 and water vapor in reverse directions) and with the photosynthetic substrate demand (CO_2 fixation by RuBP carboxylation activity), Δ has been theoretically predicted and empirically demonstrated to be inversely related to plant water-use efficiency (roughly the ratio of carbon gain to water losses; for deeper insights see Farquhar et al. 1989; Brugnoli and Farquhar 2000). Despite the complexity of this trait, significant heritabilities and low genotype × environment interactions have been found for Δ in crop species (Hall et al. 1994) encouraging the use of this parameter for breeding purposes.

5.5.2 Strategies and Methods Used for QTL Detection

In forest trees, QTLs for several traits have been detected, clearly showing the usefulness of this approach to dissect genomic regions controlling complex traits (Sewell and Neale 2000). With few exceptions (Brown et al. 2003; Jermstad et al. 2003), the size of segregating populations used in these studies is often small (150 to 200 individuals). Among factors influencing QTL detection power, small sample sizes and low trait heritability were shown to cause an overestimation of QTL effects and the underestimation of QTL number and to hamper detection of QTLs with low effects (Beavis 1995). For these reasons, a single QTL detection experiment does not give an exhaustive idea of the genetic architecture of a quantitative trait. One possible strategy to overcome these difficulties is to detect QTLs several times across different environments and developmental stages. In this way, environmental and temporal stability of QTLs can be verified and a more complete picture of genetic architecture of the complex trait can be drawn. Moreover, comparative QTL mapping between phylogenetically related species offers an important tool to validate QTLs from the evolutionary point of view. In oak and chestnut, a QTL-detection strategy based on multiple experiments across different environments and years has been performed to give an idea, as much as possible, of the complete genetic architecture of adaptive traits in both species. Afterwards, comparative QTL mapping for the three adaptive traits studied was carried out between the two species in order to identify genomic regions conserved through evolution controlling these traits.

In oak, QTL detection was done in both the French (3P*A4) and Dutch (A1*A2) *Q. robur* mapping pedigrees. In the French studies, phenotypic assessments were done over successive years using a clonal test planted with the vegetative copies of the full-sibs belonging to the pedigree. The phenotypic data obtained so far all originate from the two plantations installed in the southwest of France. The assessments first addressed the same three major adaptive traits as for chestnut: phenology, growth, and carbon isotope discrimination (Scotti-Saintagne et al. 2004; Brendel et al. 2007). In addition, leaf morphology characters (Saintagne et al. 2004) and the ability for vegetative reproduction by cutting propagation (Scotti-Saintagne et al. 2005) were assessed. The focus on leaf morphological traits is related to their use in species discrimination as shown by previous analyses (Kremer et al. 2002). The Dutch study focused on QTL detection for morphological and growth characters in one specific full-sib cross that was grown for two successive years in a nursery (Bakker 2001).

In European chestnut (*C. sativa*), bud flush, growth, and carbon isotope discrimination measurements were performed for 3 years: 2000, 2001, and 2002, corresponding to the growing seasons 2, 3, and 4 since seed germination. Bud set timing was scored only in 2002. During the three years, plants were grown in central Italy (Istituto di Biologia Agroambientale e Forestale, CNR, Porano, central Italy, 42° 43′ latitude, 500 m elevation) as previously reported. Details about phenotypic measurements are reported in Casasoli et al. (2004) and in Table 5.

In American (*C. dentata*) and Asian (*C. mollissima*) chestnut, the blight resistance response of F_2 progeny was assessed by using the agar-disk corkborer method (Griffin et al. 1983). During the growing season, each F_2 individual was inoculated with two different strains of *Cryphonectria parasitica*. Canker evaluations were made over two successive months. The mean canker sizes in each month for each isolate were used as relative measures of resistance. The degree of association between marker loci and blight resistance trait was investigated using successively single-locus or nonsimultaneous analysis of variance (ANOVA) models and multiple marker or simultaneous analysis of variance (ANOVA) models (Kubisiak et al. 1997). In European beech (*F. sylvatica*), leaf traits (size and shape) were assessed over 2 years, whereas growth and carbon isotope discrimination measurements were done only one year.

The MultiQTL software (Britvin et al. 2001, http://esti.haifa.ac.il/~poptheor) was used for QTL detection both in oak and chestnut in the French

Table 5. QTL data in oak, chestnut, and beech

Trait	Species (reference)	Pedigree	Number of offsprings phenotyped	Number of field plantations where the trait was assessed	Growing season(s) when the trait was assessed	Heritability or repeatability	Number of QTLs detected [f]	Range of variation of PEV (minimum – maximum)
Height growth	*Quercus robur* [a]	3P × A4	207	1	4 [a]	0.14–0.23 [c]	5	9.5–18.7
	Quercus robur [b]	A1 × A2	101	1	2	–	1	31.2
	Castanea sativa [c]	Bursa × Hopa	135–153	1	3	–	6	7.0–17.0
	Fagus sylvatica [d]	44 × 45	118	1	1	–	0	–
Bud burst	*Quercus robur* [a]	3P × A4	174–278	3	8, 4, 5	0.15–0.52 [c]	12	3.1–10.7
	Quercus robur [b]	A1 × A2	101	1	1	–	2	Trait not normally distributed. Nominal scale of 1–5
	Castanea sativa [c]	Bursa × Hopa	150–174	1	3	–	9	6.3–12.2
	Fagus sylvatica [d]	44 × 45	124	1	1	–	1	27.3
Delta	*Quercus robur* [e]	3P × A4	121–207	2	4, 5, 5	0.32–0.80 [c]	5	4.4–34.4
	Castanea sativa [c]	Bursa × Hopa	152–155	1	3	–	7	5.7–13.2
	Fagus sylvatica [d]	44 × 45	102	1	1	–	0	–
Bud set	*Castanea sativa* [c]	Bursa × Hopa	151	1	1	–	3	8.9–17.1
Diameter growth	*Quercus robur* [b]	A1 × A2	101	1	2	–	–	No significant QTL detected
	Castanea sativa [c]	Bursa × Hopa	136–153	1	3	–	4	5.9–10.3

[a] Scotti-Saintagne et al. (2004)
[b] Bakker (2001)
[c] Casasoli et al. (2004)
[d] Scalfi et al. (2004)
[e] Brendel et al. (2007)
[f] QTL detected at $p < 0.05$ at the genome level

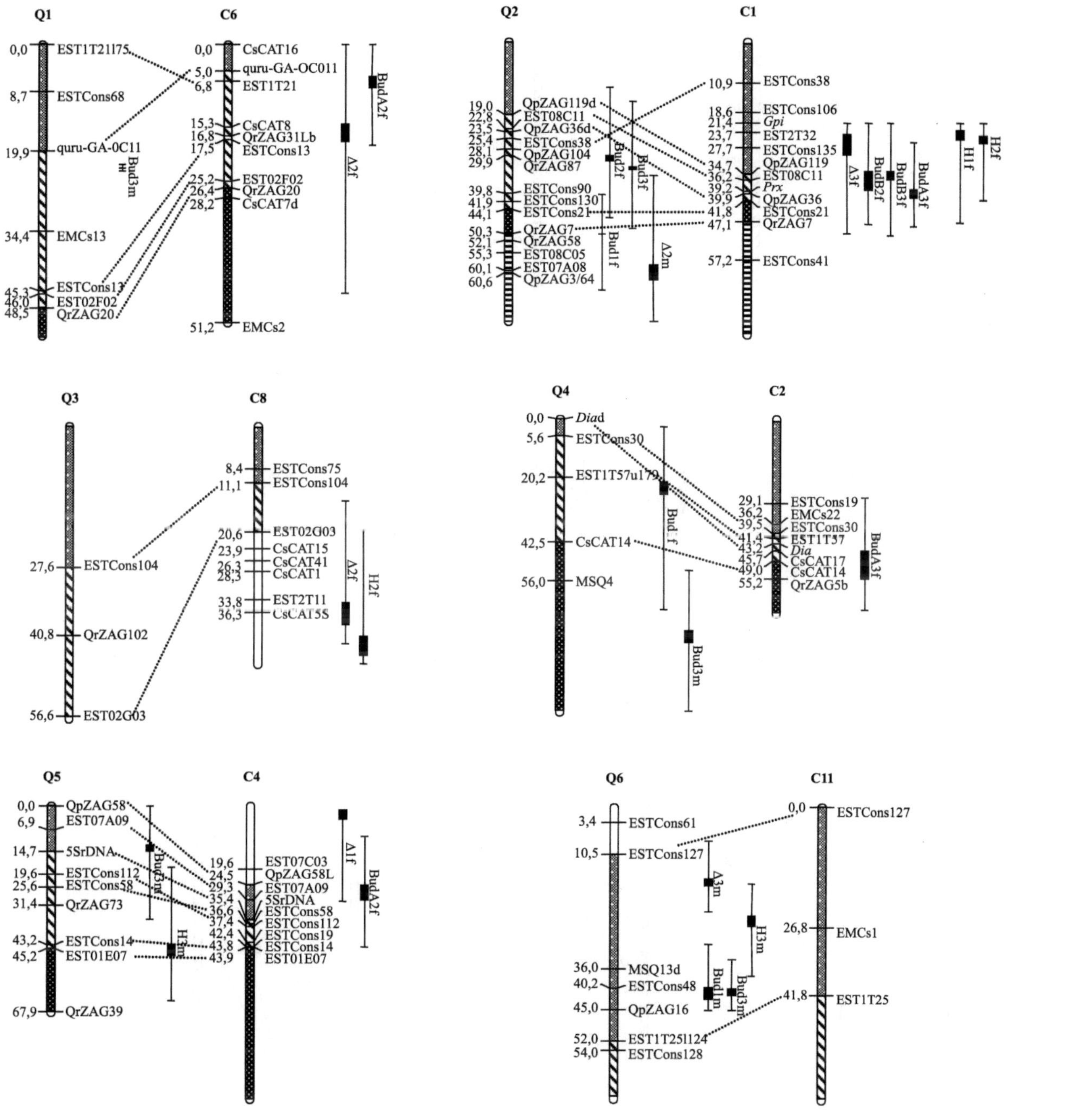

Fig. 3. Comparative QTL mapping between *Q. robur* and *C.sativa* (from Casasoli et al. 2006). Homeologous LGs between *Q. robur* (Q) and *C. sativa* (C) are named and ordered as in Figs. 1 and 2. Orthologous markers are linked by *dotted lines*. Common intervals between the two genomes identified by orthologous markers are filled with corresponding backgrounds in both oak and chestnut LGs. The figure was drawn using MapChart software (Voorrips 2002). Each QTL is represented on the right of the LG by its confidence interval (95% confidence intervals, *black line*) and the most probable position (Casasoli et al. 2006). QTLs were detected for three different phenotypic traits on the male (m) and female map (f): bud burst (Bud), total height (H), and carbon isotope discrimination (Δ). The phenotypic traits were observed over three seasons (indicated by subscripts 1 to 3). In oak, the date of bud burst was assessed as the date when the apical bud flushed. In chestnut, bud burst was assessed in two different ways: **A** = date of first observed unfolded leaf of a tree; **B** = date when 70% of buds showed an unfolded leaf BudA2f: QTL for bud burst assessed with method A during season 2 in female map. H2m: QTL for total height measured at season 2 and located on male map. Δ3f: QTL for carbon isotope discrimination assessed during season 3 and located on female map

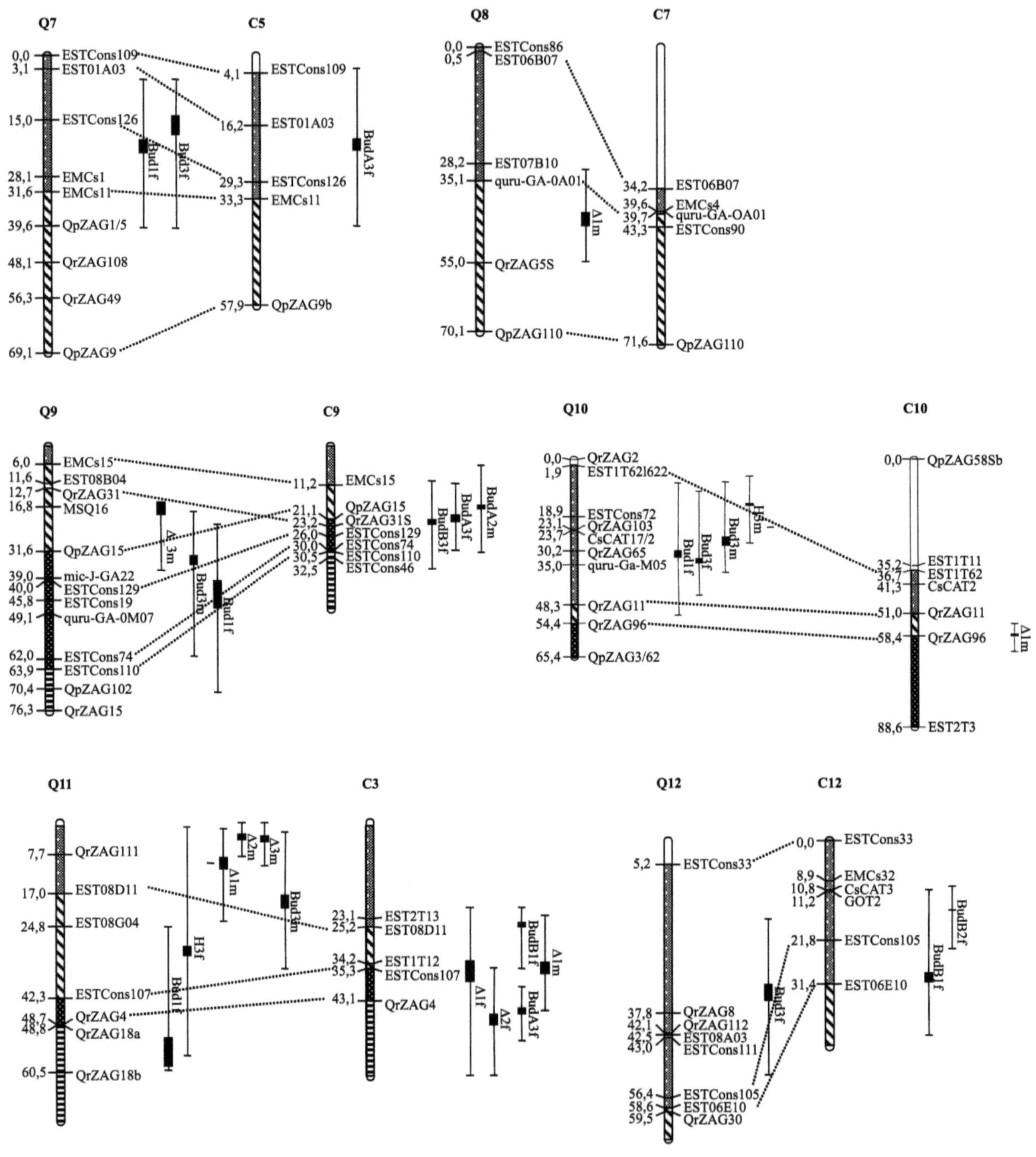

Fig. 3. (continued)

studies. This software was chosen for several reasons. First, the composite interval mapping was available (CIM, Jansen and Stam 1994; Zeng 1994); second, QTL significance thresholds could be computed by permutation (Churchill and Doerge 1994); and finally, confidence intervals for QTL position could be estimated by bootstrap (Visscher et al. 1996). The same statistical analysis was performed in oak and chestnut; the details are reported in Casasoli et al. (2004) and Scotti-Saintagne et al. (2004). The Dutch study used MapQTL 4.0 (http://www.kyazma.nl/index.php/mc.MapQTL) for QTL detection. For beech, QTL Cartographer 1.12 (Basten et al. 1994, 2002) and MultiQTL (Britvin et al. 2001) softwares were used to detect QTLs (Scalfi et al. 2004)

5.5.3 Number and Distribution of QTLs and Their Effects

The results of QTL detection are extremely heterogeneous across pedigrees and species. The survey of the results has been limited to traits that were assessed in at least two species (Table 5). Heterogeneity among the results is most likely related to the reduced size of the mapping pedigree. As already mentioned, with on average less than 200 sibs per pedigree, the contribution of a given QTL to the phenotypic variance of the trait is usually overestimated and exhibits a large sample variance (Beavis 1995). However, because the sampling efforts were similar and the phenotypic assessments were the same, a comparative analysis of QTL detection could be done between *Q. robur* and *C. sativa*.

The alignment of the 12 *Q. robur* and *C. sativa* LGs gives rise to a logical framework defined by common orthologous markers for comparing QTL location between the two species. Figure 3 shows the alignment of the 12 LGs, the common genomic regions identified by orthologous markers, and QTLs compared in oak and chestnut. Details about the definition of common genomic intervals and corresponding unique QTLs between the two species (i.e., more individual QTLs detected several times in the same genomic region in a single species) are reported in Casasoli et al. (2006). A total number of 34 common intervals were identified between the oak and chestnut genetic linkage maps thanks to the orthologous markers. Following the previously described criteria to declare unique QTLs, 13 and 10 unique QTLs were identified for timing of bud burst, 5 and 7 unique QTLs were identified for carbon isotope discrimination, and, finally, 5 and 6 unique QTLs for height growth were identified in oak and chestnut, respectively (Fig. 3). Among these unique QTLs, nine controlling timing of bud burst and two controlling height growth were colocated between the two species. No QTL involved in carbon isotope discrimination was colocated in the oak and chestnut map. Following Lin et al. (1995), the probability of obtaining these colocations by chance is $p = 0.0002$ in the case of timing of bud burst and $p = 0.20$ in the case of height growth. When QTL number and effects were compared for the three traits between the two species, a similar genetic architecture was observed for adaptive traits in oak and chestnut (Casasoli et al. 2006). From this simple comparison it was clear that adaptive traits are controlled by more loci of low and moderate than large effect in both species. Timing of bud flush was the trait showing the higher number of detected and stable QTLs. Despite this similar genetic architecture, most of the QTLs for bud flush were conserved, whereas only a few QTLs were conserved for height growth, and none for carbon isotope discrimination. The different conservation of QTLs may be explained taking into account differences for the three adaptive traits investigated in trait heritability values, QTL stability across experiments, and QTL-by-environment interactions. The striking conservation of QTLs for bud flush is very interesting from an evolutionary point of view. Although correspondence of QTLs does not imply correspondence of genes underlying the QTLs, as already reported in other species (Doust et al. 2004), these findings showed that loci controlling bud flush have remained highly polymorphic in both species. This high polymorphism of loci controlling bud flush, despite strong natural selection acting on this adaptive trait, may be explained with selection pressures able to maintain diversity over long evolutionary times (balancing, disruptive, or frequency-dependent selection) as discussed in Casasoli et al. (2006).

5.6 Conclusion

Mapping experiments in Fagaceae were hampered by various biological constraints that have limited research activities in this field. First, for most species, it was not possible to find adequate F_2 pedigrees that would allow us to screen the genome for QTLs of interest. This is somehow compensated by the high level of within-population diversity, which would allow segregation for QTLs of interest in F_1 pedigrees as well. Second, controlled crosses to obtain mapping F_1 pedigrees has been challenging in these species, and alternatives based on open-pollinated progeny screening using parentage analysis were implemented. Third, the development of mapping activities was restrained by the limited genomic resources available (genetic markers, ESTs) within this group of species. Despite these limitations, important progress has been made in the recent years as a result of international cooperation. Maps have been developed for each economically important genus (*Quercus, Castanea*, and *Fagus*), and the ongoing activities in comparative mapping suggest that there is a strong macrosynteny between phylogenetically close genera (*Quercus* and *Castanea*). For some traits, e.g., bud burst, there is even a strong conservation of the QTL position between the two genera. Extension of comparative mapping to *Fagus* might be more problematic as illustrated by difficulties described in this review. However, comparative mapping should be much easier with *Lithocarpus* and *Castanopsis*, as these genera are close to *Quercus* and *Castanea*. Furthermore, the genome of the Fagaceae is of small enough size (e.g., only 3.5 to 6 times larger than Arabidopsis) to make comparative genomics easily applicable to this family. These expectations should enhance research activities in genetics within a large group of ecologically and economically important species growing throughout the northern hemisphere.

Acknowledgement. The construction of the genetic linkage maps in European species *Quercus robur*, *Castanea sativa* and *Fagus sylvatica* was carried out with the financial support of the European Commission, DG Research (OAKFLOW, QLK5-2000-00960 for oaks; CASCADE, EVK2-1999-00065P for chestnut, and DYNABEECH, QLRT-1999-01210 for beech). The study on the Dutch *Q. robur* map was carried out with financial support from Programm 381 Functions of Nature, Forest and Landscape of the Dutch Ministry of Agriculture, Nature and Food quality. Jeremy Derory received a PhD grant from INRA to develop the EST used for the comparative mapping between *Quercus* and *Castanea*. The authors are grateful to Scott Schlarbaum (University of Tennessee-Knoxville) for providing the material for the genome size determination of different Fagaceae species, to Preston Aldrich, Kevin McAbee, David Chagne, Paolo Piovani, Weilin Sun, Michela Troggio, for their helpful contribution. Jeanne Romero thanks Antoine Kremer for suggesting to screen open-pollinated progenies by exclusion methods in order to identify full-sib progeny for the mapping.

References

Aldrich PR, Michler C, Sun W, Romero-Severson J (2002) Microsatellite markers for northern red oak (Fagaceae: *Quercus rubra*). Mol Ecol Notes 2:472–475

Aldrich PR, Jagtap M, Michler CH, Romero-Severson J (2003a) Amplification of north American red oak microsatellite markers in European white oaks and Chinese chestnut. Silvae Genet 52(3-4):176-179

Aldrich PR, Parker GR, Michler CH, Romero-Severson J (2003b) Whole-tree silvic identifications and the microsatellite genetic structure of a red oak species complex in an Indiana old-growth forest. Can J For Res 33:2228–2237

Arcade A, Anselin F, Faivre Rampant P, Lesage MC, Pâques LE, Prat D (2000) Application of AFLP, RAPD and ISSR markers to genetic mapping of European and Japanese larch. Theor Appl Genet 100:299–307

Axelrod DI (1983) Biogeography of oaks in the Arco-Tertiary province. Ann Mo Bot Gard 70:629–657

Bakker EG (2001) Towards molecular tools for management of oak forests, PhD Thesis. Alterra, Wageningen, The Netherlands

Barreneche T, Bodenes C, Lexer C, Trontin JF, Fluch S, Streiff R, Plomion C, Roussel G, Steinkellner H, Burg K, Favre JM, Glössl J, Kremer A (1998) A genetic linkage map of *Quercus robur* L. (pedunculate oak) based on RAPD, SCAR, microsatellite, minisatellite, isozyme and 5S rDNA markers. Theor Appl Genet 97:1090–1103

Barreneche T, Casasoli M, Russell K, Akkak A, Meddour H, Plomion C, Villani F, Kremer A (2004) Comparative mapping between *Quercus* and *Castanea* using simple-sequence repeats (SSRs). Theor Appl Genet 108:558–566

Basten CJ, Weir BS, Zeng ZB (1994) Zmap-a QTL cartographer. In: Smith C, Gavora JS, Benkel B, Chesnais J, Fairfull W, Gibson JP, Kennedy BW Burnside EB (eds) Proc 5th World Congress On Genetics Applied to Livestock Production: Computing Strategies and Software, Guelp, Ontario, Canada, 22:65–66

Basten CJ, Weir BS, Zeng ZB (2002) QTL cartographer, version 1.16. Department of Statistics, North Carolina State University, Raleigh, NC

Beavis WD (1995) The power and deceit of QTL experiments: lessons from comparative QTL studies. In: Proc 49th Annu Corn and Sorghum Industry Res Conf, ASTA, Washington, DC, pp 250–266

Bennett MD, Leitch IJ, Price HJ, Johnston JS (2003) Comparisons with *Caenorhabditis* (~100 Mb) and Drosophila (~175 Mb) using flow cytometry show genome size in Arabidopsis to be ~157 Mb and thus ~25% larger than the *Arabidopsis* Genome Initiative estimate of ~125 MB. Ann Bot 91:547–557

Bradshaw HD, Stettler RF (1995) Molecular genetics of growth and development in *Populus*. IV. Mapping QTLs with large effects on growth, form, and phenology traits in a forest tree. Genetics 139:963–973

Brendel O, Le Thiec D, Scotti-Saintagne C, Bodénès C, Kremer A, Jean-Marc Guehl JM, 2007 Quantitative trait loci controlling water use efficiency and related traits in *Quercus robur* L. Tree Genetics and Genomes (in press)

Britvin E, Minkov D, Glikson L, Ronin Y, Korol A (2001) MultiQTL, an interactive package for genetic mapping of correlated quantitative trait complexes in multiple environments, version 2.0 (Demo). In: Plant & Animal Genome IX Conf, San Diego

Brown GR, Kadel III EE, Bassoni DL, Kiehne KL, Temesgen B, van Buijtenen JP, Sewell MM, Marshall KA, Neale DB (2001) Anchored reference loci in loblolly pine (*Pinus taeda* L.) for integrating pine genomics. Genetics 159:799–809

Brown GR, Bassoni DL, Gill GP, Fontana JR, Wheeler NC, Megraw RA, Davis MF, Sewell MM, Tuskan GA, Neale DB

(2003) Identification of quantitative trait loci influencing wood property traits in loblolly pine (*Pinus taeda* L.). III. QTL Verification and candidate gene mapping. Genetics 164:1537–1546

Brugnoli E, Farquhar GD (2000) Photosynthetic fractionation of carbon isotopes. In: Leegood RC, Sharkey TD, von Caemmerer S (eds) Photosynthesis: Physiology and Metabolism. Advances in Photosynthesis, vol 9. Kluwer, Boston, pp 399–434

Brunner AM, Busov VB, Strauss SH (2004) Poplar genome sequence: functional genomics in an ecologically dominant plant species. Trends Plant Sci 9:49–56

Burnham CR, Rutter PA, French DW (1986) Breeding blight-resistant chestnuts. Plant Breed Rev 4:347–397

Butorina AK (1993) Cytogenetic study of diploid and spontaneous triploid oaks, *Quercus robur* L. Ann Sci For 50(Suppl 1):144s-150s

Camus A (1936-1954) Les chênes, Monographie du genre *Quercus* et Monographie du genre *Lithocarpus*. Encyclopédie Economique de Sylviculture. Vol. VI, VII, VIII. Académie des sciences, Paris, France

Cannon CH, Manos PS (2003) Phylogeography of the South Asian stone oaks (*Lithocarpus*). J Biogeogr 30:211–226

Casasoli M, Mattioni C, Cherubini M, Villani F (2001) A genetic linkage map of European chestnut (*Castanea sativa* Mill.) based on RAPD, ISSR and isozyme markers. Theor Appl Genet 102:1190–1199

Casasoli M, Pot D, Plomion C, Monteverdi MC, Barreneche T, Lauteri M, Villani F (2004) Identification of QTLs affecting adaptive traits in *Castanea sativa* Mill. Plant Cell Environ 27:1088–1101

Casasoli M, Derory J, Morera-Dutrey C, Brendel O, Porth I, Guehl J-M, Villani F, Kremer A (2006) Comparison of QTLs for adaptive traits between oak and chestnut based on an EST consensus map. Genetics 172:533- 546

Ceroni M, Leonardi S, Piovani P, Menozzi P (1997) Incrocio diallelico in faggio: obiettivi, metodologie, primi risultati. Monti e Boschi 48(1):46-51

Cervera MT, Storme V, Ivens B, Gusmao J, Liu BH, Hostyn V, Van Slycken J, Van Montagu M, Boerjan W (2001) Dense genetic linkage maps of three *Populus* species (*Populus deltoides*, *P. nigra* and *P. trichocarpa*) based on AFLP and microsatellite markers. Genetics 158:787–809

Chang S, Puryear J, Cairney J (1993) A simple and efficient method for isolating RNA from pine trees. Plant Mol Biol Rep 11(2):113-116

Churchill GA, Doerge RW (1994) Empirical threshold values for quantitative trait mapping. Genetics 138:963–971

Clark C, Kubisiak T, Lee B-C, O'Malley D, Sisco P (2001) AFLPs - towards a saturated genetic map for *Castanea*. In: Plant & Animal Genome IX Conf, San Diego

Connors BJ, Maynard CA, Powell WA (2001) Expressed sequence tags from stem tissue of the American chestnut, *Castanea dentata*. Biotechnol Lett 23 (17):1407-1411

Connors BJ, Laun NP, Maynard CA, Powell WA (2002) Molecular characterization of a gene encoding a cystatin expressed in the stems of American chestnut (*Castanea dentata*). Planta 215:510–513

Costa P, Pot D, Dubos C, Frigerio J-M, Pionneau C, Bodénès C, Bertocchi E, Cervera M-T, Remington D-L, Plomion C (2000) A genetic linkage map of Maritime pine based on AFLP, RAPD and protein markers. Theor Appl Genet 100:39–48

Crepet WL (1989) History and implications of the early North American fossil record of *Fagaceae*. In: Crane PR, Blackmore S (eds) Evolution, Systematics, and Fossil History of the *Hamamelidae*, vol 2: 'Higher' *Hamamelidae*'. Clarendon, Oxford, pp 45–66

D'Emerico S, Bianco P, Medagli P, Schirone B (1995) Karyotype analysis in *Quercus* ssp. (Fagaceae). Silvae Genet 44:66–70

Derory J, Léger P, Garcia V, Schaeffer J, Hauser M-T, Plomion C, Glössl J, Kremer A (2006) Transcriptome analysis of bud burst in sessile oak (*Quercus petraea* (Matt.) Liebl.). New Phytol 170:723–738

Doust AN, Devos KM, Gadberry MD, Gale MD, Kellogg EA (2004) Genetic control of branching in foxtail millet. Proc Natl Acad Sci USA 15:9045–9050

Farquhar GD, Ehleringer JR, Hubick KT (1989) Carbon isotope discrimination and photosynthesis. Annu Rev Plant Physiol Mol Biol 40:503–537

Favre JM, Brown S (1996) A flow cytometric evaluation of the nuclear content and GC percent in genomes of European oak species. Ann Sci For 53:915–917

Gale MD, Devos KM (1998) Comparative genetics in the grasses. Proc Natl Acad Sci USA 95:1971–1974

Gallois A, Burrus M, Brown S (1999) Evaluation of DNA content and GC percent in four varieties of *Fagus sylvatica* L. Ann For Sci 56:615–618

Gil B, Pastoriza E, Ballester A, Sanchez C (2003) Isolation and characterization of a cDNA from *Quercus robur* differentially expressed in juvenile-like and mature shoots. Tree Physiol 23:633–640

Govaerts R, Frodin DG (1998) World checklist and bibliography of *Fagales* (*Betulaceae, Fagaceae and Ticodendraceae*). Royal Botanical Garden, Kew, UK

Grattapaglia D, Sederoff R (1994) Genetic linkage maps of *Eucalyptus grandis* and *Eucalyptus urophylla* using a pseudo test-cross mapping strategy and RAPD markers. Genetics 137:1121–1137

Grattapaglia D, Bertolucci F, Sederoff R (1995) Genetic mapping of QTLs controlling vegetative propagation in *Eucalyptus grandis* and *E. urophylla* using a pseudo test-cross strategy and RAPD markers. Theor Appl Genet 90:933–947

Griffin GJ, Hebard FV, Wendt RW, Elkins JR (1983) Survival of American chestnut trees: Evaluation of blight resistance and virulence of *Endothia parasitica*. Phytopathology 73:1084–1092

Hall AE, Richards RA, Condon AG, Wright GC, Farquhar GD (1994) Carbon isotope discrimination and plant breeding. Plant Breed Rev 4:81–113

Herendeen PS, Crane PR, Drinman AN (1995) Fagaceous flowers, fruits and cupules from the Campanian (late Cretaceous) of Central Georgia, USA. Int J Plant Sci 156:93–116

Howe GT, Saruul P, Davis J, Chen THH (2000) Quantitative genetics of bud phenology, frost damage, and winter survival in an F_2 family of hybrid poplars. Theor Appl Genet 101:632–642

Hurme P, Sillanpää MJ, Arjas E, Repo T, Savolainen O (2000) Genetic basis of climatic adaptation in Scot pine by Bayesian quantitative trait locus analysis. Genetics 156:1309–1322

Jansen RC, Stam P (1994) High resolution of quantitative trait into multiple loci via interval mapping. Genetics 136:1447–1445

Jermstad KD, Bassoni DL, Jech KS, Wheeler NC, Neale DB (2001) Mapping of quantitative trait loci controlling adaptive traits in coastal Douglas-fir. I. Timing of vegetative bud flush. Theor Appl Genet 102:1142–1151

Jermstad KD, Bassoni DL, Jech KS, Ritchie GA, Wheeler NC, Neale DB (2003) Mapping of quantitative trait loci controlling adaptive traits in coastal Douglas fir. III. Quantitative trait loci-by-environment interactions. Genetics 165:1489–1506

Jones JH (1986) Evolution of the Fagaceaeae: the implication of foliar features. Ann Mo Bot Gard 73:213–229

Kremer A, Dupouey JL, Deans JD, Cottrell J, Csaikl U, Finkeldey R, Espinel S, Jensen J, Kleinschmit J, Van Dam B, Ducousso A, Forrest I, de Heredia UL, Lowe A, Tutkova M, Munro RC, Steinhoff S, Badeau V (2002) Leaf morphological differentiation between *Quercus robur* and *Quercus petraea* is stable across western European mixed oak stands. Ann For Sci 59:777–787

Kremer A, Xu LA, Ducousso A (2004) Genetics of oaks. In: Burley J, Evans J, Youngquist J (eds) Encyclopedia of Forest Sciences. Elsevier, Amsterdam, pp 1501–1507

Kubisiak TL, Hebard FV, Nelson CD, Zhang J, Bernatzky R, Huang H, Anagnostakis SL, Doudrick RL (1997) Molecular mapping of resistance to blight in an interspecific cross in the genus *Castanea*. Phytopathology 87:751–759

Kvaèek Z, Walther H (1989) Paleobotanical studies in *Fagaceae* of the European Tertiary. Plant Syst Evol 162:213–229

Lauteri M, Scartazza A, Guido MC, Brugnoli E (1997) Genetic variation in photo synthetic capacity, carbon isotope discrimination and mesophyll conductance in provenances of *Castanea sativa* adapted to different environments. Funct Ecol 11:675–683

Lauteri M, Monteverdi MC, Sansotta A, Cherubini M, Spaccino L, Villani F (1999) Adaptation to drought in European chestnut. Evidences from a hybrid zone and from controlled crosses between drought and wet adapted populations. Acta Hort 494:345–353

Lin YR, Schertz KF, Paterson AH (1995) Comparative analysis of QTLs affecting plant height and maturity across the Poaceae, in reference to an interspecific sorghum population. Genetics 141:391–411

Manos PS, Steele KP (1997) Phylogenetic analyses of "higher" *Hamamelidiae* based on plastid sequence data. Am J Bot 84:1407–1419

Manos PS, Doyle JJ, Nixon KC (1999) Phylogeny, biogeography, and processes of molecular differentiation in *Quercus* subgenus *Quercus* (*Fagaceae*). Mol Phylogenet Evol 12:333–349

Manos PS, Zhe-Kun Zhou, Cannon CH (2001) Systematics of Fagaceae: phylogenetic tests of reproductive trait evolution. Int J Plant Sci 162:1361–1379

Manos PS, Stanford AM (2001) The historical biogeography of *Fagaceae*: tracking the Tertiary history of temperate and subtropical forests of the northern hemisphere. Int J Plant Sci 162(6 Suppl):s77-s93

Marques CM, Araujo JA, Ferreira JG, Whetten R, O'Malley DM, Liu BH, Sederoff R (1998) AFLP genetic maps of *Eucalyptus globulus* and *E. tereticornis*. Theor Appl Genet 96:727–737

Marques CM, Brondani RPV, Grattapaglia D, Sederoff R (2002) Conservation and synteny of SSR loci and QTLs for vegetative propagation in four *Eucalyptus* species. Theor Appl Genet 105:474–478

Mehra PN, Hans AS, Sareen TS (1972) Cytomorphology of Himalayan Fagaceae. Silvae Genet 21:102–109

Naujoks G, Hertel H, Ewald D (1995) Characterisation and propagation of an adult triploid pedunculate oak (*Quercus robur*). Silvae Genet 44:282–286

Ohri D, Ahuja MR (1990) Giemsa C-banding in *Quercus* L. (oak). Silvae Genet 39:216–219

Ohri D, Ahuja MR (1991) Giemsa C-banding in *Fagus sylvatica* L., *Betula pendula* Roth and *Populus tremula* L. Silvae Genet 40:72–75

Parmesan C, Yohe G (2003) A globally coherent fingerprint of climate change impacts across natural systems. Nature 421:37–42

Pernas M, Sanchez-Monge R, Gomez L, Salcedo G (1998) A chestnut seed cystatin differentially effective against cysteine proteinase from closely related pests. Plant Mol Biol 38:1235–1242

Pernas M, Lopez-Solanilla E, Sanchez-Monge R, Salcedo G, Rodriguez-Palenzuela P (1999) Antifungal activity of a plant cystatin. Mol Plant-Micr Interact 12:624–627

Porth I, Koch M, Berenyi M, Burg A, Burg K (2005a) Identification of adaptation specific differences in the mRNA expression of sessile and pedunculate oak based on osmotic stress induced genes. Tree Physiol 25:1317–1329

Porth I, Scotti-Saintagne C, Barreneche T, Kremer A, Burg K (2005b) Linkage mapping of osmotic induced genes of oak. Tree Genet Genom 1:31–40

Powell W, Machray GC, Provan J (1996) Polymorphism revealed by simple sequence repeats. Trends Plant Sci 1:215–222

Romero-Severson J, Aldrich P, Mapes E, Sun W (2003) Genetic mapping in northern red oak (*Quercus rubra* L.). In: Plant & Animal XI Genome Conf, San Diego

Saintagne C, Bodénès C, Barreneche T, Pot D, Plomion C, Kremer A (2004) Distribution of genomic regions differentiating oak species assessed by QTL detection. Heredity 92:20–30

Scalfi M, Troggio M, Piovani P, Leonardi S, Magnaschi G, Vendramin GG, Menozzi P (2004) A RAPD, AFLP, and SSR linkage map, and QTL analysis in European beech (*Fagus sylvatica* L.) Theor Appl Genet 108:433–441

Scalfi M (2005) Studio della variabilita' genetica neutrale ed adattativa in *Fagus sylvatica* L. Tesi di Dottorato in Ecologia XVII ciclo. University of Parma, Italy

Schwarz O (1964) *Quercus*. In: Tutin TG, Heywood VH, Burges NA, Valentine DH, Walters SM, Webb DA (eds) Flora Europea. Vol 1: Lycopodiaceae to Platanaceae, Cambridge University Press, Cambridge, UK, pp 61–64

Scotti-Saintagne C, Bodénès C, Barreneche T, Bertocchi E, Plomion C, Kremer A (2004) Detection of quantitative trait loci controlling bud burst and height growth in *Quercus robur* L. Theor Appl Genet 109:1648–1659

Scotti-Saintagne C, Bertocchi E, Barreneche T, Kremer A, Plomion C (2005) Quantitative trait loci mapping for vegetative propagation in pedunculate oak. Ann For Sci 62:369–374

Sewell MM, Neale DB (2000) Mapping quantitative traits in forest trees. In: Jain SM, Minocha SC (eds) Molecular Biology of Woody Plants. Kluwer, Dordrecht, pp 407–423

Sisco PH, Kubisiak TL, Casasoli M, Barreneche T, Kremer A, Clark C, Sederoff RR, Hebard FV, Villani F (2005) An improved genetic map for *Castanea mollissima* / *Castanea dentata* and its relationship to the genetic map of *Castanea sativa*. Acta Hort 693:491–495

Steinkellner H, Lexer C, Turetschek E, Glössl J (1997) Conservation of (GA)n microsatellite loci between *Quercus* species. Mol Ecol 6:1189–1194

Tognetti R, Michelozzi M, Giovannelli A (1997) Geographical variation in water relations, hydraulic architecture and terpene composition of Aleppo pine seedlings from Italian provinces. Tree Physiol 17:241–250

Van Buijtenen H (2001) Genomics and quantitative genetics. Can J For Res 31:617–622

Verhaegen D, Plomion C (1996) Genetic mapping in *Eucalyptus urophylla* and *Eucalyptus grandis* using RAPD markers. Genome 39:1051–1061

Villani F, Pigliucci M, Benedettelli S, Cherubini M (1991) Genetic differentiation among Turkish chestnut (*Castanea sativa* Mill.) populations. Heredity 66:131–136

Villani F, Pigliucci M, Lauteri M, Cherubini M, Sun O (1992) Congruence between genetic, morphometric, and physiological data on differentiation of Turkish chestnut (*Castanea sativa*). Genome 35:251–256

Visscher PM, Thompson R, Haley CS (1996) Confidence intervals in QTL mapping by bootstrapping. Genetics 143:1013–1020

Voorrips RE (2002) MapChart: software for the graphical presentation of linkage maps and QTLs. J Hered 93:77–78

Vos P, Hogers R, Bleeker M, Reijans M, van de Lee T, Hornes M, Fritjers A, Pot J, Peleman J, Kuiper M, Zabeau M (1995) AFLP: a new concept for DNA fingerprinting. Nucleic Acids Res 23:4407–4414

Wen J (1999) Evolution of eastern Asian and eastern North American disjunct distributions in flowering plants. Annu Rev Ecol Syst 30:421–455

Williams JGK, Kubelik AR, Livak KJ, Rafalski JA, Tingey SV (1990) DNA polymorphisms amplified by arbitrary primers are useful as genetic markers. Nucleic Acids Res 18:6531–653

Xiang QY, Soltis DE, Soltis PS, Manchester SR, Crawford DJ (2000) Timing the Eastern Asia-Eastern North America floristic disjunction: molecular clock corroborates paleontological estimates. Mol Phylogenet Evol 15:462–472

Xu LA, Kremer A, Bodénès C, Huang MR (2005) Chloroplast DNA diversity and phylogenetic relationship of botanical sections of the genus *Quercus*. In: Proc Joint Meeting of IUFRO Working Groups on "Genetics of Quercus & Improvement and Silviculture of Oaks," 29 Sept-3 Oct 2003, Tsukuba, Japan

Zeng ZB (1994) Precision mapping of quantitative trait loci. Genetics 136:1457–1468

Zhang J, Marshall JD (1995) Variation in carbon isotope discrimination and photosynthetic gas exchange among populations of *Pseudotsuga menziesii* and *Pinus ponderosa* in different environments. Funct Ecol 9:402–412

Zoldos V, Papes D, Brown SC, Panaud O, Siljak-Yakovlev S (1998) Genome size and base composition of seven *Quercus* species: inter- and intra-population variation. Genome 4:162–168

6 Black Walnut

Charles H. Michler, Keith E. Woeste, and Paula M. Pijut

USDA Forest Service, Hardwood Tree Improvement and Regeneration Center at Purdue University, 715 West State Street, West Lafayette, IN 47907-2061, USA
e-mail: michler@purdue.edu

6.1 Introduction

6.1.1 Origin and History

Black walnut (*Juglans nigra* L.), also known as eastern black walnut or American walnut, is a fine hardwood species in the family Juglandaceae, section *Rhysocaryon* (Manning 1978). In general, *J. nigra* will not cross with species in the sections *Cardiocaryon* or *Trachycaryon*, but *J. nigra* will cross with *J. ailantifolia* (*Cardiocaryon*) (Williams 1990). *Juglans nigra* will also hybridize to some extent with other *Juglans* species (*Dioscaryon* and *Rhysocaryon*), and one hybrid is recognized: *J. nigra* × *J. regia* = *J.* x *intermedia* Carr. (USDA-NRCS 2004). Native to the deciduous forests of the eastern United States (USA), from Massachusetts to Florida and west to Minnesota and Texas, and occurring naturally in southern Ontario, Canada, black walnut is seldom found in pure stands, but rather in association with five mixed mesophytic forest cover types: sugar maple, yellow poplar, yellow poplar – white oak – northern red oak, beech – sugar maple, and silver maple – American elm (Williams 1990). Black walnut is a large tree and on good sites may attain a height of 30 to 38 m and diameter of 76 to 120 cm and can exceed 100 years of age (Williams 1990; Dirr 1998; USDA-NRCS 2004). Black walnut is shade intolerant, and control of competing vegetation is especially important in new plantations for the first 3 to 4 years. Black walnut grows best on moist, deep, fertile, well-drained, loamy soils, although it also grows quite well in silty clay loam soils or in good agricultural soils without a fragipan (Williams 1990; Cogliastro et al. 1997). These sites include coves, bottomlands, abandoned agricultural fields, and rich woodlands. Black walnut forms a deep taproot, wide-spreading lateral roots, and has been cultivated since 1686. A toxic chemical 'juglone' (5-hydroxy-1, 4-naphthoquinone), naturally occurring in the leaves, buds, bark, nut husks, and roots of black walnut, is a highly selective, cell-permeable, irreversible inhibitor of the parvulin family of peptidyl-prolyl *cis/trans* isomerases (PPIases) and functions by covalently modifying sullfhydryl groups in the target enzymes (Henning et al. 1998; Chao et al. 2001). Certain plants, especially tomato, apple, and several conifer species, are adversely affected (allelopathy; foliar yellowing, wilting, and even death) by being grown near the roots of black walnut trees (Goodell 1984; Dana and Lerner 1994). Horses can contract acute laminitis, an inflammation of the foot, when black walnut wood chips or sawdust is used for stall bedding or stables and paddocks are located too close to walnut trees (Galey et al. 1991). Historically, the bark of black walnut was used by several Native Americans, including the Cherokee, Delaware, Iroquois, and Meskwaki, in tea as a cathartic, emetic, or disease remedy agent, and chewed or applied for toothaches, snake bites, and headaches (Moerman 1998, 2003). *Caution: the bark should be used cautiously in medicine because it is poisonous.* The Cherokee, Chippewa, and Meskwaki also used the bark to make a dark brown or black dye (Moerman 1998, 2003). The Comanche pulverized the leaves of black walnut for treatment of ringworm, the Cherokee used leaves to make a green dye, and the Delaware used the leaves as an insecticide to dispel fleas (Moerman 1998, 2003). The nut meats were also a food source for Native Americans, and the nuts are still consumed today by people and are an important food source for wildlife.

6.1.2 Botany

Juglans nigra (section *Rhysocaryon*, Fig. 1) is the largest and the most valuable timber tree of the *Juglans*

Genome Mapping and Molecular Breeding in Plants, Volume 7
Forest Trees
C. Kole (Ed.)

Fig. 1. Form, leaf shape, bud characters, and flower and fruit morphology of *Juglans nigra* L.

species and is hardy to USDA hardiness zone range of four to nine (Dirr 1998). Black walnut is monoecious with male and female flowers maturing at different times (McDaniel 1956). Staminate catkins (5 to 10 cm) develop from axillary buds on the previous year's wood and appear as small, scaly, conelike buds, and the female flowers occur in two- to eight-flowered spikes borne on the current year's shoots (Brinkman 1974; Williams 1990; Flora of North America Editorial Committee 1993+; Dirr 1998). The female flowers more commonly appear first (protogyny) and flowering occurs with or shortly after the leaves. Because flowering is dichogamous, self-pollination is unlikely, which promotes outcrossing. The fruit is a drupe-like, furrowed nut enclosed in a thick, indehiscent yellowish-green husk that develops from a floral involucre (Brinkman 1974). Fruits are subglobose to globose, rarely ellipsoid, 3.5 to 8 cm, warty, with scales and capitate-glandular hairs (Flora of North America Editorial Committee 1993+). The fruit occur singly or in clusters of two to three and are edible, sweet, oily, and high in protein (Reid 1990). The nut is subglobose to globose, rarely ellipsoid, 3 to 4 cm, and very deeply longitudinally grooved, and the surface between the grooves is coarsely warty (Flora of North America Editorial Committee 1993+). Leaves are alternate, pinnately compound, 30 to 60 cm long, with 9 to 23 leaflets, nearly glabrous and somewhat lustrous dark green, pubescent and glandular beneath, with petioles 6.5 to 14 cm long covered with glandular hairs (Flora of North America Editorial Committee 1993+; Dirr 1998). Black walnut stems are stout, densely grey-downy, smooth, and reddish buff, and they have a chambered light brown pith (paler than that of butternut, *J. cinerea*) and a distinctly notched leaf scar. Terminal buds are ovoid or subglobose, 8 to 10 mm long, and weakly flattened (Flora of North America Editorial Committee 1993+). Lateral buds are smaller, often superposed, and greyish in color. *J. nigra* has a dark grey or brownish bark, deeply split into narrow furrows and thin ridges, and the ridges are chocolate in color when cut, forming a roughly diamond-shaped pattern. The sapwood of black walnut is nearly white, and the heartwood varies from light to dark brown. The wood is heavy, hard, strong, stiff, normally straight grained, and has good resistance to shock (Forests Products Laboratory 1999). The chromosome number of black walnut is $2n = 32$ (Woodworth 1930). Black walnut trees produce seed at about 12 years of age, with good seed crops occurring every 2 to 3 years (Brinkman 1974). Seeds of black walnut, like most *Juglans* spp., have a dormant embryo, but dormancy can be broken by fall sowing or by moist prechilling of seeds at 1 to 5 °C for 3 to 4 months (Brinkman 1974).

6.1.3 Economic Importance

Black walnut is one of the largest hardwood trees found in the USA and is valued economically and ecologically for its wood and edible nuts. Owners of quality black walnut wood receive high market prices, and the wood has many uses including furniture, veneer, cabinets, interior architectural woodwork, flooring, and gunstocks. Black walnut wood with figured grain gets even higher market prices. Curly and wavy figure can produce interesting characteristics in veneers, and these can arise from walnut butts, crotches, and burls. The nut is an important food source for wildlife and is also consumed by humans. The majority of black walnut trees occur in natural stands, with walnut plantations (ca. 13,800 ac) accounting for 1% of all the black walnut volume (ft^3) in the USA (Shifley 2004). There are 11 states that currently have the greatest volume of black walnut growing stock on timberland, and these include Missouri, Ohio, Iowa, Indiana, Illinois, Tennessee, West Virginia, Kansas, Pennsylvania, Virginia, and Michigan (Shifley 2004). Since the last (1997) comprehensive inventory and summary of the black walnut resource in the eastern USA, the number and volume of black walnut trees has increased, except in Michigan, Virginia, and Pennsylvania, where walnut volume is level or decreasing (Shifley 2004). In addition to the multimillion-dollar US market consumption of walnut wood, for the period 1999-2003, the USA exported walnut lumber to 67 countries (58,434.2 m^3; \$40,964,481; averages per year) and walnut logs to 49 countries (62,897 m^3; \$37,238,327; averages per year) (USDA-FAS 2004). Black walnut yields edible nuts that are used in baking (cookies, cakes, etc.) and ice cream products. The Hammons Products Company (Stockton, MO; http://www.black-walnuts.com) is the world's premier processor and supplier of American black walnuts for both food and industrial uses. Selection of black walnut trees for nut quality and production has developed slowly over the years, but over 700 cultivars have been named and the percent of edible kernel has improved to over 34% (Reid 1990; Reid et al. 2004). Black walnuts are low in saturated fats (3.4 g per 100 g edible nut), have zero cholesterol, and are high in polyunsaturated (35.1 g per 100 g edible nut) and monounsaturated fats (15 g per 100 g edible nut) (USDA-ARS 2004). Black walnuts are also a good source of protein (24.1 g per 100 g edible nut) and fiber (6.8 g per 100 g edible nut) containing low levels of sugar (1.1 g per 100 g edible nut) (USDA-ARS 2004). Ground black walnut shell (see Hammons Products Company, http://www.black-walnuts.com) is a hard, durable, nontoxic, biodegradable abrasive product used for blast cleaning and polishing. It is also used for industrial tumbling and deburring, as well as for uses in oil-well drilling, water filtration, and as explosive fillers.

6.2 Black Walnut Genetics

Black walnut genetic research has been focused on the practical improvement of the species for the production of timber or nuts. Although black walnut has a large native and commercial range, black walnut improvement has been largely a Midwestern preoccupation. The earliest recorded clonal selection of black walnut was the nut cultivar 'Thomas' (Corsa 1896, cited in Reid et al. 2004). New nut cultivars have been named, almost entirely by amateur breeders, throughout the 20th century. Now at least 700 such cultivars are recorded, although only a small number, perhaps 30 or 40, are suitable for inclusion in a contemporary nut-improvement program (Reid et al. 2004). The most important trait for selection for nut improvement is percent kernel. 'Thomas' averages 24% kernel, but the best of the newer selections routinely exhibit over 35% kernel. Other traits of interest include protandry, resistance to anthracnose (caused by *Gnomonia leptostyla* [Fr] Ces. & De Not.), high yield, low alternate bearing, and uniform fruit ripening. Protandry is important because black walnut, like all members of the genus *Juglans*, is dichogamous, and breeders would like to identify selections that can serve as pollenizers for the best nut-producing cultivars. There has been little published in refereed journals concerning the heritability or inheritance of important traits such as nut yield and percent kernel in black walnut, although these traits have been well studied in Persian walnut (*J. regia*) (Hansche et al. 1972). Observational studies related to almost every aspect of black walnut nut production have been published by scientists, hobbyists, and amateur breeders in the annual reports of the Northern Nut Growers, first published in 1910 and available through their Web site (http://www.nutgrowing.org).

Although single-species plantations of black walnut for timber production date from as long ago as the late 1800s, modern efforts at genetic research were initiated by Johnathan Wright and Leon Minckler in the early 1950s (Minckler 1952, 1953; Wright 1954). Subsequent genetic research can be loosely divided into studies of heritability and selection age, local and regional adaptation, methods in black walnut breeding, genetic variance as measured by neutral genetic markers and other markers, the use of hybrids to improve black walnut, and the development of general genetic and breeding resources. Each of these areas is briefly reviewed below.

6.2.1 Heritability of Important Traits and Selection Age

Height and diameter growth are the most studied traits in black walnut, perhaps because they are easily measured traits in young trees and because they give an overall impression of a tree's vigor. Over the long term, these traits also reflect a tree's adaptation to the site on which it is growing. As expected, heritability estimates for these basic growth traits are very high in the early generations of selection. All the studies cited below are based on progeny from open-pollinated trees growing either in forests or grafted into "clone banks." Rink and Clausen (1989) summarized the results of three progeny tests after 13 years, finding that the narrow sense heritability for height was about 0.41, but that there were significant family × site interactions. This heritability estimate was similar to or slightly lower than those summarized in other studies (Beineke 1989). Kung et al. (1974) found h^2 for height growth of about 0.4 based on a study of twinned seedlings, and Rink (1984), Rink and Kung (1995), and Beineke (1974) indicated a value slightly higher ($h^2 = 0.55$), but in any case there is sufficient evidence that breeders can expect to make rapid progress in this trait in the first few generations of selection.

The heritability of diameter growth is in the same range as height ($h^2 = 0.35$ to 0.65) based on studies of clones (Beineke and Stelzer 1991), twins (Kung et al. 1974), and open-pollinated families (Rink and Kung 1995). Woeste (2002) found a somewhat lower value ($h^2 = 0.28$) based on a 35-year-old progeny test. As was true for height, the heritability of this trait was near zero at outplanting, but increased over the first 10 years and did not appear to stabilize until about age 15 (Hammitt 1996; Rink 1997). Tree form is more difficult to measure than height or diameter; nevertheless, estimates of the heritability of this trait also range from 0.4 (Beineke 1989) to 0.5 (Beineke and Stelzer 1991). Other important traits are less well studied, but heritability estimates have been published for foliation date ($h^2 = 0.92$), defoliation date (0.73), sweep (0.32), number of crooks (0.24), branch angle (0.20), and branch number (0.41) (all reported in Beineke 1974). Black walnut geneticists have studied the heritability of phenological traits because of their role in the adaptation of seed sources to local environments (see below). Heritability estimates for multiple stems (0.18), leaf drop date (0.13; note the discrepancy between this value and that reported by Beineke), insect damage (0.27), and leaf angle (0.32) were reported by Bey (1970). Walnut anthracnose, described above, is the most important foliar disease of black walnut. Trees heavily infested with anthracnose appear to senesce and defoliate earlier than more resistant genotypes. Anthracnose resistance may be highly heritable (Funk et al 1981; Woeste and Beineke 2001), and this could be important in extending the growing season, and thus the growth, of black walnut, but attempts to associate anthracnose resistance with growth are not conclusive (Todhunter and Beineke 1984). The heritability of important wood-quality traits such as heartwood formation, heartwood color, and wound recovery is expensive to determine as these traits require destructive sampling of mature trees. Nelson (1976), Rink (1987a), and Woeste (2002) all report that heartwood area has a moderate to high heritability ($h^2 > 0.4$), perhaps because this trait is strongly associated with tree vigor (Woeste 2002). Rink (1987a) was unable to find any genetic component to heartwood color.

Because the rotation age for black walnut is more than 60 years, selection of juvenile trees for their anticipated rotation-age value is an essential part of any improvement program for the species. Juvenile-mature correlations and rank correlations for height and diameter growth based on progeny tests were reported by McKeand et al. (1979), Rink (1984), Beineke (1989), and Rink and Kung (1995). In general, these studies found family selection for the high heritability traits of height and diameter growth can begin by about age 8 years, but that within-family selection should be delayed until after age 12. The optimal age for selection depends on thinning schedules and site quality, as these factors influence selection intensity, intertree competition, and trait heritability (Kung 1973).

6.2.2 Local and Regional Adaptation

Even mature black walnut trees bear a relatively small number of large seeds, and trees often only bear in alternate years. Seed-germination rates fall to near zero if seeds are stored more than 2 years (Beineke 1989). Consequently, seed from improved sources is typically expensive and subject to local shortages in supply, and provenance trials to measure the effects of long-distance seed movement have been a fundamental part of black walnut improvement. These provenance trials were established to characterize regional genetic variability and to determine the relationship between the latitude of seed sources and tree growth. The underlying rationale was that trees from southern sources would leaf out earlier and lose their leaves later than trees from more northern latitudes. This, in general, was what Bey (1970) concluded. Longer growing seasons, in turn, would translate into faster growth. Frost injury, which can cause poor form, and dieback from winter injury were considered potential drawbacks to the use of southern seed sources. Early reports (Bey and Williams 1975) indicated that trees from provenances to the south of the planting site would perform well compared to local sources. Bey (1979, 1980) and Bresnan et al. (1992, 1994) refined this analysis. The results from a large number of provenance studies have been summarized with a general guideline often called the "200 mile rule," i.e., plant seeds from sources 200 miles (322 km) to the south for optimal growth. In fact, the recommendations found in the reports of Bey and others were far more nuanced, recognizing that planting site climate is not strictly dependent on latitude. The use of grafted trees to evaluate site effects and adaptability is only now under way (Woeste and McKenna 2004).

6.2.3 Methods in Black Walnut Breeding

The publication "Genetics of Black Walnut" (Funk 1970) reviewed much of what was known at the time on the subject, including sections on micropropagation, seed orchard design, and the use of hybrids, topics that continue to be the subject of research. At the time Funk wrote his synopsis, Beineke was developing his own ideas for clonal seed orchards of black walnut (Beineke and Lowe 1969) as a means to improve the availability of improved seed (Beineke 1982). Beineke was the first to document the effects of inbreeding on black walnut seedling growth (Beineke 1972). He believed that wild stands of black walnut were genetically depauperate and that the species was suffering from dysgenic selection. Beineke published a justification of his approach and his own synopsis of black walnut breeding 13 years after Funk (Beineke 1983). Beineke, a fierce proponent of clonal forestry, became the first person in the USA to patent a tree (Beineke 1980).

Black walnut is not an easy species to vegetatively propagate (Coggeshall and Beineke 1997a, b), and the expense inherent in clonal approaches to improvement made Beineke's proposed improvement methods (McKeand and Beineke 1980; Lowe and van Buijtenen 1986) theoretically attractive but difficult and expensive in practice. Rink (1987b) focused on the practical problems related to seed production, including fertilization and thinning of seed orchards. A comparison of improved seed from seed orchards with elite clonal and nursery-run stock published by Hammitt (1997) indicated that the (patented) clones and improved sources were superior to 1-0 common nursery stock, especially with respect to form.

6.2.4 Genetic Variance Measured by Neutral and Other Markers

By the late 1980s, it became possible to determine the mating system parameters for black walnut using allozyme systems (Rink et al. 1989). The goal of the research was to determine the level of inbreeding and outcrossing of black walnut in native stands and the allocation of genetic variance at marker loci among and within populations. Forest fragmentation, overharvesting, and dysgenic selection were thought by many to have led to inbreeding and low genetic variance (Beineke 1972), but Rink found that black walnut had a high outcrossing rate (about 90%) and was highly heterozygous based on eight loci (Rink et al. 1989). The 26 maternal trees in Rink et al.'s 1989 study were significantly more heterozygous than their progeny, and Rink suggested this might have been caused by selection against inbred progeny. These findings would be substantiated by later research (Rink et al. 1994; Busov et al. 2002). These later studies also showed that the mating parameters of black walnut seed orchards were similar to those found in wild stands, and that as much as 94% of the variance in the isozyme marker loci was distributed within populations, i.e., that very little differentiation among populations could be detected at the level of allozymes. Fjellstrom (1993) found high levels of heterozygosity in black walnut using RFLPs, and unpublished research by one of the authors (Woeste) indicates that black walnut is highly heterozygous at microsatellite loci (Woeste et al. 2002) as well.

6.2.5 Black Walnut Hybrids

Luther Burbank and others noticed the unusual vigor demonstrated by interspecies hybrids within the genus *Juglans*. Burbank's "Royal" hybrid (*J. hindsii* × *J. nigra*) has been cited by others as an exemplar of the phenomenon (Wright 1966). The "Royal" hybrid has been impractical in the Midwest because of poor cold tolerance, but in milder climates it is fruitful and appears to have excellent potential as a timber tree in California (Forde and McGranahan 1996) and the Pacific Northwest. Hybrids between the *J. nigra* × *J. regia* F_1 known as *J.* x *intermedia* Carr. have attracted the most interest in the United States (Wright 1966; Funk 1970), and Europe (Hussenforfer 1999; Germain 2004). Bey (1969) reported that researchers at the USDA Forest Service research unit at Carbondale, IL had begun to collect *Juglans* species with the intention of producing and testing hybrids. Funk (1970) summarized the hybridization research of McKay (1965) and others who found that, while *J.* x *intermedia* had variable vigor, it was always nearly sterile, and all attempts to restore fertility through the production of amphidiploids failed. The clonal deployment of *J.* x *intermedia* Carr. has not yet been pursued, although the possible uses of seedling hybrid rootstock for timber production in the Midwestern USA was discussed by Woeste and McKenna (2004). They cite the potential of *J. nigra* × *J. major* hybrids as a rootstock for drier sites. While seed availability previously limited research into the use of hybrids in *J. nigra* improvement (Bey 1969), *Juglans* hybrid seedlings can be produced on a commercial scale; seedling Paradox (*J. hindsii* × *J. regia*) rootstocks are routinely used in the Persian walnut industry in California (Forde and McGranahan 1996). The identity and parentage of *Juglans* hybrids can be verified using inter-simple sequence repeats (Potter et al. 2002).

6.2.6 General Genetic and Breeding Resources

There is no catalog of black walnut germplasm in the United States, but there are abundant resources. Germplasm collections or seed orchards of black walnut are maintained by at least one state agency (Vallonia Tree Nursery, http://www.in.gov/dnr/forestry/index.html?http://www.in.gov/dnr/forestry/nursery/&2), by universities (University of Missouri Horticultural and Agroforestry Research Center, New Franklin, MO; Purdue University Department of Forestry and Natural Resources, West Lafayette, IN; Southern Illinois University Department of Forestry, Carbondale, IL), private organizations such as the Arbor Day Foundation (Nebraska City, NE); state chapters of the Northern Nut Growers (e.g., Nebraska Nut Growers), and by at least two federal agencies in the United States (USDA Forest Service Hardwood Tree Improvement and Regeneration Center, West Lafayette, IN and the National Clonal Germplasm Repository, Davis, CA). A small number of *J. nigra* clones is also maintained by breeding programs in Europe (Germaine 2004).

There are few tools available for molecular genetic research in black walnut. There are no published, publicly available resources such as cDNA or genomic libraries; however, the National Center for Biotechnology Information (http://www.ncbi.nlm.nih.gov/) Web site has records for 88 nucleotide sequences including microsatellite loci, 19 proteins, and 22 population genetic data sets. This resource is likely to expand in the future.

6.3 Tissue Culture and Genetic Transformation

Important methods for capturing genetic gains from breeding programs include the use of vegetative propagation, both through rooted cuttings or grafting of elite clones and tissue culture. Beineke (1983) described optimum variables for grafting that included storage conditions for scion wood and rootstock condition. Although grafting success can be very high, this method is also extremely labor and resource intensive and not highly desirable for mass propagation. To date, success from rooted cutting propagation has been minimal, if not recalcitrant for most researchers (Coggeshall and Beineke 1997a, b). More promise has come from research on rooting of microshoots (Van Sambeek et al. 1990, 1997; Khan 1995; Long et al. 1995) as well as somatic embryogenesis (Neuman et al. 1993; Preece et al. 1995).

Genetic transformation can be used in lieu of conventional breeding when the desire is to either introduce a gene from another species or to circumvent long generation times as found in tree species such as black walnut. Only one report has been published to date on the successful genetic transformation of black walnut (Bosela et al. 2004). In this research, somatic embryo lines were established from cotyledons of immature zygotic embryos. From 16 embryo lines, transgenic callus was regenerated following infection with disarmed strains of *Agrobacterium tumefaciens* carrying transgenes for β-glucuronidase (GUS) and kanamycin resistance (uidA). GUS expression assays showed that the methods deployed resulted in high rates of gene transfer in all but two lines tested. When kanamycin was used for selection, the majority of the secondary embryos were transgenic. If these secondary embryos were kept on selection media, the chimeric embryos would regenerate fully transgenic embryos. These methods are now being used to generate transgenic lines with genes for herbicide resistance, insect resistance, and flowering control.

6.4 Future Scope of Works

To date, no genetic maps have been made for black walnut. Molecular markers previously developed have been used to begin to characterize population including estimation of genetic variation both in small stands and across large portions of the natural range. This type of genetic information was needed to aid in management of the genetics in fragmented stands and to guide movement of seed and seedlings in the Midwest USA. Once genetic maps have been developed, we will be able to target genes for cloning as well as to begin to identity quantitative trait loci, which could lead to marker-assisted breeding on an operational level.

References

Beineke WF (1972) Recent changes in the population structure of black walnut. In: Proc 8th Central States For Tree Improv Conf, Columbia, MO, pp 43–45

Beineke WF (1974) Inheritance of several traits in black walnut clones. Department of Forestry and Conservation, Ag Exp Stn, Purdue Univ, Stn Bull No 38, West Lafayette, IN

Beineke WF (1980) Plant patent pp 4542 issued June, 1980 Distinct variety of black walnut [*Juglans nigra*] tree [Excellent timber quality, fast growing early nut bearing] United States Patent. [Washington DC, The Office]: http://patft.uspto.gov/netacgi/nph-Parser?Sect1=PTO2&Sect2=HITOFF&p=1&u=%2Fnetahtml%2Fsearch-bool.html&r=0&f=S&l=50&TERM1=walnut&FIELD1=&co1=AND&TERM2=beineke&FIELD2=&d=ptxt). Accessed March 21, 2005

Beineke WF (1982) New directions in genetic improvement: grafted black walnut plantations. In: Black Walnut for the Future: Third Walnut Symp: Proc 12th Annu Mtg Walnut Council 1981; West Lafayette, IN. Gen Tech Rep NC74. St. Paul, MN: US Dept Agric, Forest Service, NC For Expt Stn, pp 64–68

Beineke WF (1983) Genetic improvement of black walnut for timber production. Plant Breed Rev 1:236–266

Beineke WF (1989) Twenty years of black walnut genetic improvement at Purdue University. North J Appl For 6:68–71

Beineke WF, Lowe WJ (1969) A selection system for superior black walnut trees and other hardwoods. Proc 10th South Tree Improv Conf, Houston, TX, 10:27–32

Beineke W, Stelzer HE (1985) Genetic improvement of black walnut: Is it working? Annu Rep North Nut Grow Assoc 76:26–31

Beineke WF, Stelzer HE (1991) Genetic variation and heritability estimates in black walnut clones at different ages. Indiana Acad Sci Proc 99:137–140

Bey C (1969) The Forest Service Black Walnut Genetics Project: a progress report. In: Schreiner EJ (ed) Proc 15th NE Forest Tree Improv Conf; Morgantown, WV, US Dept Agric For Serv North East For Expt Stn, Upper Darby, PA, pp 31–34

Bey CF (1970) Geographic variation for seed and seedling characters in black walnut. USDA For Serv Res Note NC101, p 4

Bey CF (1979) Geographic variation in *Juglans nigra* in the Midwestern United States. Silvae Genet 28:132–135

Bey CF (1980) Growth gains from moving black walnut provenances northward. J For Oct:640-645

Bey CF, Williams RD (1975) Black walnut trees of southern origin growing well in Indiana. Indiana Acad Sci Proc 84:122–127

Bosela MJ, Smagh GS, Michler CH (2004) Genetic transformation of black walnut (*Juglans nigra*). In: Michler CH, Pijut PM, Van Sambeek J, Coggeshall M, Seifert J, Woeste K, Overton R (eds) Black Walnut in a New Century, Proc 6th Walnut Council Res Symp, Gen Tech Rep NC-243, US Dept Agric, For Serv, North Central Res Stn, pp 45–58

Bresnan D, Geyer WA, Lynch KD, Rink G (1992) Black walnut provenance performance in Kansas. North J Appl For 9:41–43

Bresnan DF, Rink G, Diebel KE, Geyer WA (1994) Black walnut provenance performance in seven 22-year-old plantations. Silvae Genet 43:246–252

Brinkman KA (1974) *Juglans* L. Walnut. In: Schopmeyer CS (tech coord) Seeds of Woody Plants in the United States. USDA For Serv Agric Handbook 450, Washington, DC, pp 454–459

Busov VB, Rink G, Woeste K (2002) Allozyme variation and mating system of black walnut (*Juglans nigra* L.) across the central hardwood region of the United States. For Genet 9:319–327

Chao SH, Greenleaf AL, Price DH (2001) Juglone, an inhibitor of the peptidyl-prolyl isomerase Pin1, also directly blocks transcription. Nucleic Acids Res 29(3):767-773

Coggeshall MV, Beineke WF (1997a) Black walnut vegetative propagation: the challenge continues. Annu Rep North Nut Grow Assoc 88:83–92

Coggeshall MV, Beineke WF (1997b) Black walnut vegetative propagation: the challenge continues. In: Van Sambeek, JW (ed) Knowledge for the Future of Black Walnut: 5th Black Walnut Symp, Gen Tech Rep NC-191, US Dept Agric, For Serv, North Central Res Stn, St Paul, MN, pp 70–77

Cogliastro A, Gagnon D, Bouchard A (1997) Experimental determination of soil characteristics optimal for the growth of ten hardwoods planted on abandoned farmland. For Ecol Manage 96(1-2):49-63

Corsa WP (1896) Nut culture in the United States. Special report to the Secretary of Agriculture. US Dept Agric, Div of Pomology, Govt Printing Office, Washington, DC

Dana MN, Lerner BR (1994) Black walnut toxicity. HO-193, Purdue University Coop Extension Service, West Lafayette, IN, pp 1–2

Dirr MA (1998) Manual of woody landscape plants: their identification, ornamental characteristics, culture, propagation and uses, 5th edn. Stipes, Champaign, IL, pp 500–502

Fjellstrom RG (1993) Genetic diversity of walnut (*Juglans* L.) species determined by restriction fragment length polymorphisms, PhD dissertation, University of Califiornia-Davis

Flora of North America Editorial Committee (1993+) Flora of North America North of Mexico, 7+ vols. New York and Oxford, http://www.eFloras.org

Forde HI, McGranahan GH (1996) Walnuts. In: Janick J, Moore JN (eds) Fruit Breeding, vol III: Nuts. Wiley, New York, pp 241–273

Forest Products Laboratory (1999) Wood Handbook-Wood as an Engineering Material. Gen Tech Rep FPL-GTR-113. USDA Forest Service, Forest Products Lab, Madison, WI, p 9

Funk DT (1970) Genetics of black walnut. USDA Forest Service Res Paper WO10, p 13

Funk DT, Nealy D, Bey CF (1981) Genetic resistance to anthracnose of black walnut. Silvae Genet 30:115–117

Galey FD, Whiteley HE, Goetz TE, Kuenstler AR, Davis CA, Beasley VR (1991) Black walnut (*Juglans nigra*) toxicosis: a model for equine laminitis. J Comp Pathol 104:313–326

Germaine E (2004) Inventory of Walnut Research, Germplasm and References. Food and Agric Org United Nations, FAO Regional Office for Europe, REU Tech Series 66, Rome, p 264

Goodell E (1984) Walnuts for the northeast. Arnoldia 44(1):3-19

Hammitt WE (1996) Growth differences among patented walnut grafts and selected seedlings 12 years after establishment. In: Van Sambeek JW (ed) Knowledge for the Future of Black Walnut. Proc 5th Black Walnut Symp, Gen Tech Rep NC-191. US Dept Agric, For Serv, North Central For Expt Stn, St. Paul, MN, pp 63–68

Hansche PE, Beres V, Forde HI (1972) Estimates of quantitative genetic properties of walnut and their implications for cultivar improvement. J Am Soc Hort Sci 97:279–285

Henning L, Christner C, Kipping M, Schelbert B, Rucknagel KP, Grabley S, Kullertz G, Fischer G (1998) Selective inactivation of parvulin-like peptidyl-prolyl *cis/trans* isomerases by juglone. Biochem 37:5953–5960

Hussendorfer E (1999) Identification of natural hybrids *Juglans* × *intermedia* CARR- using isoenzyme gene markers. Silvae Genet 48:50–52

Khan SB (1995) Micropropagation of adult black walnut (*Juglans nigra* L.). MS thesis, South IL University, Carbondale IL

Kung FH (1973) Development and use of juvenile mature correlations in a black walnut tree improvement program. Southern Tree Improv Conf 12:243–249

Kung FH, Bey CF, Larson JL (1974) Nursery performance of black walnut twins. In: Garrett PW (ed) Proc 22nd Northeastern For Tree Improv Conf: US Dept Agric For Serv Northeastern For Expt Stn Upper Darby, PA, pp 184–190

Long LM, Preece, JE, Van Sambeek, JW (1995) Adventitious regeneration of *Juglans nigra* L (eastern black walnut). Plant Cell Rep 14:799–803

Lowe WJ, van Buijtenen JP (1986) The development of a sublining system in an operational tree improvement program.

In: Proc IUFRO Conf on Breeding Theory, Progeny Testing and Seed Orchards, Williamsburg, VA, pp 98–106

Manning WE (1978) The classification within the Juglandaceae. Annu Mo Bot Gard 65:1058–1087

McDaniel JC (1956) The pollination of Juglandaceae varieties - Illinois observations and review of earlier studies. Annu Rep North Nut Grow Assoc 47:118–132 (published 1957)

McKay JW (1965) Progress in black × Persian walnut breeding. Annu Rep North Nut Grow Assoc 56:76–80

McKeand E, Beineke WF (1980) Sublining for half-sib breeding populations of forest trees. Silvae Genet 29:14–17

McKeand SE, Beineke WF, Todhunter, MN (1979) Selection age for black walnut progeny tests. Proc North Central Tree Improv Conf, Madison, WI, 1:68–73

Minckler L (1952) What do foresters hope to accomplish in forest genetics. J For 50:871–872

Minckler LS (1953) Recent advances in the field of forest genetics. Illinois Acad Sci Trans 46:56–62

Moerman DE (1998) Native American ethnobotany. Timber, Portland, OR, 927 p

Moerman DE (2003) Native American ethnobotany: a database of foods, drugs, dyes and fibers of Native American peoples, derived from plants. University of Michigan-Dearborn, http://herb.umd.umich.edu

Nelson ND (1976) Gross influences on heartwood formation in black walnut and black cherry trees. US Dept Agric For Serv Res Pap FPL-268

Neuman MC, Preece JE, Van Sambeek JW, Gaffney GR (1993) Somatic embryogenesis and callus production from cotyledon explants of eastern black walnut. Plant Cell Tiss Org Cult 32:9–18

Potter D, Gao F, Baggett S, McKenna J, McGranahan G (2002) Defining the sources of paradox: DNA sequence markers for North American walnut (*Juglans* L.) species and hybrids. Sci Hort 94:157–170

Preece JE, McGranahan GH, Long LM, Leslie CA (1995) Somatic embryogenesis in walnut (*Juglans regia*). In: Jain SM, Gupta PK, Newton RJ (eds) Somatic embryogenesis in woody plants. Kluwer, Dordrecht, 2:99–116

Reid W (1990) Eastern black walnut: potential for commercial nut producing cultivars. In: Janick J, Simon JE (eds) Advances in New Crops. Timber, Portland, OR, pp 327–331

Reid W, Coggeshall MV, Hunt KL (2004) Cultivar evaluation and development for black walnut orchards. In: Michler CH, Pijut PM, Van Sambeek J, Coggeshall M, Seifert J, Woeste K, Overton R (eds) Black walnut in a new century. Proc 6th Walnut Council Res Symp, Gen tech Rep NC-243, US Dept Agric, For Serv, North Central Res Stn, St Paul, MN, pp 18–24

Rink G (1984) Trends in genetic control of juvenile black walnut height growth. For Sci 30:821–827

Rink G (1987a) Heartwood color and quantity variation in a young black walnut progeny test. Wood Fiber Sci 19:93–100

Rink G (1987b) Practical strategies of black walnut improvement. Walnut Council Bull 14:3–5

Rink G (1989) Recent advances in walnut tree improvement. In: Phelps JE (ed) The continuing quest for quality. Proc 4th Black Walnut Symp, Walnut Council, Indianapolis, IN, pp 140–144

Rink G (1997) Genetic variation and selection potential for black walnut timber and nut production. In: van Sambeek J (ed) Knowledge for the Future of Black Walnut. Proc 5th Black Walnut Symp, Gen Tech Rep NC-191US Dept Agric For Serv North Cent For Expt Stn, St Paul, MN, pp 58–62

Rink G, Clausen KE (1989) Site and age effects on genotypic control of juvenile *Juglans nigra* L. tree height. Silvae Genet 38:17–21

Rink G, Kung FH (1995) Age trends in genetic control of *Juglans nigra* L. height growth. In: Gottschalk KW, Fosbroke SL (eds) Proc 10th Cent Hardwood For Conf, Morgantown, WV, pp 247–255

Rink G, Carroll ER, Kung FH (1989) Estimation of *Juglans nigra* L. mating system parameters. For Sci 35:623 627

Rink G, Zhang G, Jinghua Z, Kung FH, Carroll ER (1994) Mating parameters in *Juglans nigra* L. seed orchard similar to natural population estimates. Silvae Genet 43:261–263

Shifley SR (2004) The black walnut resource in the United States. In: Michler CH, Pijut PM, Van Sambeek J, Coggeshall M, Seifert J, Woeste K, Overton R (eds) Black walnut in a new century. Proc 6th Walnut Council Res Symp, Gen Tech Rep NC-243, US Dept Agric, For Serv, North Central Res Stn, St Paul, MN, p 188

Todhunter MN, Beineke WF (1984) Effect of anthracnose on growth of grafted black walnut. Plant Dis 68:203–204

USDA-ARS (2004) US Department of Agriculture, Agricultural Research Service, USDA nutrient data lab, USDA national nutrient database for standard reference, release 17 (http://www.nal.usda.gov/fnic/foodcomp)

USDA-FAS (2005) Foreign Agricultural Service Export Commodity Aggregations, Dept of Commerce, US Census Bureau, Foreign Trade Statistics (http://www.fas.usda.gov/ustrade)

USDA-NRCS (2004) The PLANTS Database, v. 3.5 (http://plants.usda.gov). National Plant Data Center, Baton Rouge, USA

Van Sambeek JW, Preece JE, Lindsay TL, Gaffney GR (1990) In vitro studies on black walnut embryo dormancy. Annu Rep North Nut Grow Assoc 80:55–59

Van Sambeek JW, Lambus LJ, Khan SB, Preece JE (1997) In vitro establishment of tissues from adult black walnut. In: Van Sambeek, JW (ed) Knowledge for the future of black walnut: 5th Black Walnut Symp, Gen Tech Rep NC-191, US Dept Agric, For Serv, North Central Res Stn, pp 78–92

Williams RD (1990) *Juglans nigra* L., Black walnut. In: Burns RM, Honkala BH (tech coords) Silvics of North America, vol 2: Hardwoods. USDA For Serv Agric Handbook 654, Washington, DC, pp 386–390

Woeste KE (2002) Heartwood production in a 35-year-old black walnut progeny test. Can J For Res 32:177–181

Woeste KE, Beineke WF (2001) An efficient method for evaluating black walnut for resistance to walnut anthracnose in field plots and the identification of resistant genotypes. Plant Breed 120:454–456

Woeste KE, McKenna JR (2004) Walnut genetic improvement at the start of a new century. In: Michler CH, Pijut PM, Van Sambeek J, Coggeshall M, Seifert J, Woeste K, Overton R (eds) Black walnut in a newcentury. Proc 6th Walnut Council Res Symp, Gen Tech Rep NC-243, US Dept Agric, For Serv, North Central Res Stn, St Paul, MN, pp 9–17

Woeste K, Burns R, Rhodes O, Michler C (2002) Thirty polymorphic nuclear microsatellite loci from black walnut. J Hered 93:58–60

Woodworth RH (1930) Meiosis of microsporogenesis in the Juglandaceae. Am J Bot 17(9):863-869

Wright JW (1954) Preliminary report on a study of races in black walnut. J For 52:673–675

Wright J (1966) Breeding better timber varieties. In: Black Walnut Culture. US Dept Agric For Serv, North Central For Expt Stn, St Paul, MN, pp 53–57

7 Douglas-Fir

J. E. Carlson[1], A. Traore[2], H. A. Agrama[3], and K. V. Krutovsky[4]

[1] The School of Forest Resources and Huck Institutes for Life Sciences, Pennsylvania State University, 323 Forest Resources Building, University Park, PA 16802, USA
e-mail: jec16@psu.edu
[2] The Schatz Center for Tree Molecular Genetics, Pennsylvania State University, University Park, PA 16802, USA
[3] Rice Research Ext. Center, University of Arkansas & DB National Rice Research Center, USDA, 2890 Hwy 130 E., P.O. Box 1090, Stuttgart, AR 72160, USA
[4] Department of Forest Science, Texas A&M University, College Station, TX 77843-2135, USA

7.1 Introduction

Douglas-fir, *Pseudotsuga menziesii* var. *menziesii* (Mirb.) Franco, is a long-lived tree of the Pinaceae family native to North America. It has an extensive natural range stretching from central California north to Alaska for the coastal variety *menziesii* and in the interior Rockies from central Mexico north to British Columbia for the interior variety *glauca* (Alexander 1985, 1988). Douglas-fir is a very important tree species for the timber and landscape and Christmas tree industry in North America. Its adaptation to a wide variety of environments from extremely dry to moist sites and sea level to very high elevations (10,500 ft, 3,200 m) makes it ideal as a plantation species for both timber and Christmas tree production. Because of its economic importance, major investments have been made in Douglas-fir tree breeding and genetic studies (Johnson at el. 2000; Hamann et al. 2004; Howe 2006; Howe et al. 2006). Therefore, there is considerable interest in understanding the genetic structure of Douglas-fir at the genome level.

Like many conifer trees, Douglas-fir has a large and complex genome that contains a large proportion of repeated DNA sequences (Kriebel 1993). The genome size of Douglas-fir has been estimated to be 19 pg per haploid nucleus, which is about average for the Pinaceae (5.8 to 32.2 pg per haploid genome) (Murray 1998; Murray et al. 2004) although 100-fold larger than *Arabidopsis thaliana* (0.18 pg) (Bennett and Smith 1991). Douglas-fir is unique among the Pinaceae family in having a diploid chromosome number of $2n = 26$. All other members of the family have a chromosome number of $2n = 24$.

7.2 Genetic Marker Development for Douglas-Fir Genome Mapping

The history of genome mapping in Douglas-fir can be divided into three phases. In the first phase (before the mid-1970s) markers used in Douglas-fir genetic studies consisted mostly of morphological markers, such as yellow foliage, white cotyledon, nonwhite cotyledonous lethals, virescence, dwarfs, and mottled and curly needles (Piesch and Stettler 1971; Sorensen 1971, 1973). The small number of these markers, which were often controlled by single genes, made them inappropriate for the deeper genomic study that is necessary for a better understanding of a large and complex genome such as the Douglas-fir genome. From the mid-1970s to the end of the 1980s a second phase in Douglas-fir genetic-marker development and usage began with the development of monoterpene genetic markers and allozymes or isozymes markers. Unlike morphological markers that were limited in number, numerous monoterpene markers were developed and used to study diversity, pest resistance, genetic drift, and migration history in Douglas-fir (Zavarin and Snajberk 1973; von Rudloff 1975; Snajberk and Zavarin 1976; von Rudloff and Rehfeldt 1980; Critchfield 1984; Hanover 1992; Jermstad et al. 1994). Isozymes are simple multiallelic codominant markers that have been extensively used for various Douglas-fir genetic studies including diversity (Yeh and O'Malley 1980; Merkle and Adams 1987), mat-

Genome Mapping and Molecular Breeding in Plants, Volume 7
Forest Trees
C. Kole (Ed.)

ing system (Neale et al. 1984; Nakamura and Wheeler 1992), seed orchard efficiency (Adams 1983), and linkage studies (El Kassaby et al. 1982; Adams et al. 1990). However, isozymes could not deliver the large number of genetic markers needed to construct high-density genomic maps for Douglas-fir. This problem was solved by the development of DNA-based molecular markers such as restriction fragment length polymorphisms or RFLPs (Botstein et al. 1980), random amplified polymorphic DNA or RAPD (Williams et al. 1990), and simple sequence repeats or SSRs (Tautz 1989; Beckman and Soller 1990; Morgante and Olivieri 1993).

The application of RFLP, RAPD, and SSR markers to Douglas-fir genome mapping constitutes the third phase of its genetic-marker development (Jermstad et al.1994, 1998; Agrema and Carlson 1995; Krutovskii et al 1998; Amarasinghe and Carlson 2002; H. Agrama, J. Broome, J. Woods, J. E. Carlson unpublished; Krutovsky et al. 2004; Slavov et al. 2004). RFLP and RAPD markers were the first of the DNA molecular markers to be widely and successfully used in Douglas-fir genome mapping (Jermstad et al. 1994, 1998; Agrema and Carlson 1995; Krutovskii et al. 1998; Amarasinghe and Carlson 2002; Krutovsky et al. 2004). RFLPs can provide a large number of polymorphic markers for the development of genetic linkage maps. In tree species RFLPs were used to construct genetic linkage maps in *Pinus taeda* (Devey et al. 1994; Groover et al. 1994), *Populus tremuloides* (Liu and Furnier 1993), *Cryptomeria japonica* (Mukai et al. 1995), and *Pseudotsuga menziesii* (Jermstad et al. 1998, 2003. RFLPs provide orthologous markers for aligning the linkage information from each parental data set to produce a sex-averaged map (Devey et al. 1994). RFLP-based genetic maps in conifers proved to be difficult and time-consuming endeavors, however, as RFLP analysis relies on reproducibly detecting single-copy loci via Southern hybridization, and requires large amounts of highly pure DNA from species with large genomes, such as conifers.

The RAPD approach, on the other hand, provides an alternative for genetic analysis with conifers that circumvents factors that complicate a standard RFLP approach. RAPD markers are quick and efficient means of identifying DNA-sequence-based polymorphisms at a large number of loci using small amounts of DNA. In Tulsieram et al. (1992), the application of the RAPD technique of Williams et al. (1990) with conifers was first described. However, RAPD markers also have their limitations for genome mapping, as discussed below, including dominance, in which only one allele generates an amplification product while the remaining alleles are those that failed to initiate amplification.

Simple sequence repeats (SSRs), also known as microsatellites, are a class of molecular markers based on tandem repeats of short (2–6 bp) DNA sequences (Litt and Lutty 1989). The microsatellite motifs AG, AC, and ATG were found to be the most abundant in Douglas-fir and several other conifer tree species among di-, tri-, and tetranucleotide SSRs. PCR primers were designed from flanking sequences in 102 of the SSR clones, of which 50 primer pairs (for 10 AC-repeat and 40 AG-repeat microsatellites) produced robust amplification products (Amarasinghe and Carlson 2002). Variability at these SSR loci was confirmed with 24 unrelated Douglas-fir trees, and Mendelian segregation was demonstrated with progeny sets from full-sib populations. Allele frequencies for the 48 polymorphic loci varied from 0.017 to 0.906, with a mean allele frequency of 0.250. In a second study, another 22 highly variable SSR markers were developed in Douglas-fir (*P. menziesii*) from five SSR-enriched genomic libraries (Slavov et al. 2004). Of these, 15 PCR primer pairs amplified a single codominant locus, while seven primer pairs amplified two loci. The Mendelian inheritance of all 22 SSRs was confirmed via segregation analyses in several Douglas-fir families. SSR markers have great potential in Douglas-fir genome mapping but so far have only been used as a complement to other markers in established maps (Slavov et al. 2004) and in comparative genomics (Krutovsky et al. 2004).

7.3 Genetic Linkage Map Construction

7.3.1 Map Construction Using RFLP Markers

RFLP markers are simple Mendelian markers that are based on variation at the nucleotide-sequence level. They have the advantage and power of being codominant DNA markers and can be developed in relatively large numbers to cover the genome. Disadvantages of RFLP markers can include lengthy development time and the large amount of DNA required for the assay. In addition, the complex and large size of the Douglas-fir

Table 1. Summary of genetic linkage and QTL mapping studies in Douglas-fir

Marker #	Marker density, cM	Genetic distance, cM	Linkage groups	Mapping population	Software used	QTLs	Reference
RAPD (153) RAPD (132)	n/a	2335 2143	13 13	Hybrid cross	JOINMAP 3.0	Wood density, tree growth	H. Agrama, J. Broome, J. Woods, J. E. Carlson unpublished
RAPD (121) RAPD (140)	27.9 26.0	2487 2598	17 16	Intervarietal F_1 hybrids	MAPMAKER	Cold hardiness, growth rate, bud set, bud flush, biomass, height, branching	K. V. Krutovsky, S. S. Vollmer, J. Y. Wu, F. C. Sorensen, W. Th. Adams and S. H. Strauss unpublished
RFLP (126), RAPD (15), STS (1)	7.5	1062	17	3 generations, outbred	JOINMAP 2.0	Bud flush, cold hardiness, QTL × Env.	Jermstad et al. 1998, 2001, 2003
RAPD (201) RAPD (238)	10.1 12.9	2600 3000	16 18	intervarietal F_1 hybrids	GMENDEL	none	Krutovskii et al. 1998
RFLP (172), ESTP (101), RAPD (77), SSR (20), STS (4), Isozyme (2)	n/a	1859	22	3 generations, outbred	JOINMAP 2.0	n/a	Krutovsky et al. 2004

genome makes it difficult to detect single- or low-copy loci. Unlike the nuclear genome, organelle genomes are as compact in Douglas-fir as they are in other plant species, and RFLP markers have been successfully employed to study inheritance and variation in chloroplast (Neal et al. 1986; Tsai and Strauss 1989; Ali et al. 1991; Ponoy et al. 1994) and mitochondrial (Marshall and Neal 1992) DNA in Douglas-fir. However, despite the difficulties with nuclear RFLPs, they have also been successfully used to study ribosomal gene number variability (Strauss and Tsai 1988). Later RFLPs were used in the construction of the first relatively complete genetic map for Douglas-fir (Jermstad et al. 1994, 1998). In the construction of that map, Jermstad et al. (1994, 1998) used a mapping population made of a three-generation outbred pedigree (Table 1). The source of RFLP probes were from Douglas-fir cDNA and genomic probes and loblolly pine (*Pinus taeda* L.) cDNA probes. A total of 151 RFLP loci were identified in the first study by Jermstad (et al. 1994). In a follow-up study (Jermstad et al. 1998), they reported the construction of a sex-averaged genetic linkage map in coastal Douglas-fir that comprised a total of 141 markers organized into 17 linkage groups (LGs) that covered 1,062 centiMorgans (cM). The marker data were analyzed using the JOINMAP version 2.0 mapping software (Stam and Oojen 1995). In this case the map included 15 RAPD and 1 PCR-amplified phytochrome locus as well as 125 RFLP markers. Including the 15 RAPD markers along with the RFLP markers increased the density of the derived linkage map. Despite these successes, RFLP markers have had limited application in conifers due to the large number of multigene families that they reveal (Kinlaw and Neale 1997).

7.3.2 Map Construction Using RAPD Markers

RAPD genetic markers are simple to use and can be developed and used without substantial investment. In Douglas-fir, RAPDs have been used to identify heritable markers in F_1 families (Carlson et al. 1991), to select parents for mapping studies (Carlson et al. 1994), and to generate genetic linkage maps

(Jermstad et al. 1994, 1998; Agrema and Carlson 1995; Krutovskii et al. 1998). The advent of RAPD markers permitted linkage maps to be generated quickly by individual investigators, which was a significant advance in plant genomics at the time. Maps constructed entirely of RAPD markers were also reported for loblolly pine (Grattapaglia et al. 1991; Sewell et al. 1999), white spruce (Tulsieram et al. 1992), Arabidopsis (Reiter et al. 1992), peach (Chaparro et al. 1994), and numerous other species.

One of the limitations of RAPDs is that they are dominant markers, which reduces the number of genotypes that can be resolved. However, in Douglas-fir and other conifers, that limitation was overcome by the availability of haploid DNA from the megagametophyte tissue contained in the seed. The combination of readily available haploid tissue from the large numbers of seeds that individual trees produce, the convenience of using open-pollinated seeds vs. full-sib populations, and the potential for a large number of markers permitted the first high-density linkage maps in Douglas-fir to be constructed quickly and inexpensively using only RAPD markers (Agrema and Carlson 1995; Krutovskii et al. 1998; H. Agrama, J. Broome, J. Woods, and J. E. Carlson unpublished and Fig. 1). These maps were constructed using a pseudotestcross strategy (Grattapaglia and Sederoff 1994) in which a small subset of seed from one tree was first screened to identify loci segregating in a 1:1 Mendelian manner, and those preselected loci were then used to conduct full-scale segregation analysis for map construction. This pseudotestcross strategy with haploid megagametophyte DNA came to be known as single-tree linkage mapping (Carlson et al. 1991). Enough loci can be identified in testcross configuration to construct RAPD-marker-based linkage maps simply by following segregation among haploid megagametophyte DNAs without needing to conduct crosses first (Agrema and Carlson 1995; Krutovskii et al. 1998; H. Agrama, J. Broome, J. Woods, and J. E. Carlson unpublished and Fig. 1). This is possible because of the high level of heterozygosity and null alleles in Douglas-fir and because most trees in Douglas-fir breeding programs are no more than two generations removed from the wild. RAPDs are best suited to single-tree linkage mapping. Although each seed yields only a limited amount of DNA, RAPDs require less template DNA than do other marker systems.

Although there are now additional PCR-based marker systems available (below), RAPD markers still have a role in genome mapping in Douglas-fir for the following reasons: (1) The long generation time of trees precludes establishing new mapping pedigrees quickly, which means that single-tree maps constructed with RAPDs will continue to be useful. (2) Single-tree maps are consistent with most tree-improvement programs, which rely on open-pollinated progeny tests in which the genetic value of female parent trees are determined. (3) The low level of linkage disequilibrium in Douglas-fir (Krutovksy and Neale 2005) means that markers will often not be directly transferable from a quantitative trait locus (QTL) in one pedigree to other genotypes, regardless of the DNA marker system employed, and use of less expensive markers like RAPDs will allow associations to be found among more genotypes and marker-assisted selection (MAS) to be applied on a particular basis. (4) The virtually unlimited number of RAPD markers and the rapidity with which maps can be constructed with RAPDs should make it more feasible to compare the architecture of chromosomes and of complex genetic traits among genotypes within a species with RAPDs than with other marker systems.

An example of a Douglas-fir single-tree linkage map is shown in Fig. 1 (H. Agrama, J. Broome, J. Woods, and J. E. Carlson unpublished). This is one of two maps constructed in 2 months as an undergraduate student project for two parent trees from the British Columbia Tree Improvement Program entirely with RAPD markers and JOINMAP software (Stam and van Oojen 1995). The analysis yielded 13 LGs with a total of 153 and 132 mapped loci for 2,335 cM and 2,143 cM map distance for the trees DF69 and DF60, respectively (map shown for clone DF60). The markers on these maps were relatively well spaced and thus suitable for QTL analysis (see below). In another example, Krutovskii et al. (1998) produced single-tree RAPD linkage maps for two hybrids between the interior and coastal varieties, using haploid DNA from 80 seeds from each tree. They obtained 201 segregating loci combined into 16 LGs for one map and 238 segregating loci combined into 18 LGs for the other (Table 1). These maps covered 2,600 and 3,000 cM, respectively, and the two maps had enough in common among their LGs so that the total number of unique groups between the two hybrids was also close to the haploid chromosome number of 13.

Despite the conveniences of RAPD markers, the system also has its challenges and disadvantages. RAPD markers are notoriously difficult to reproduce

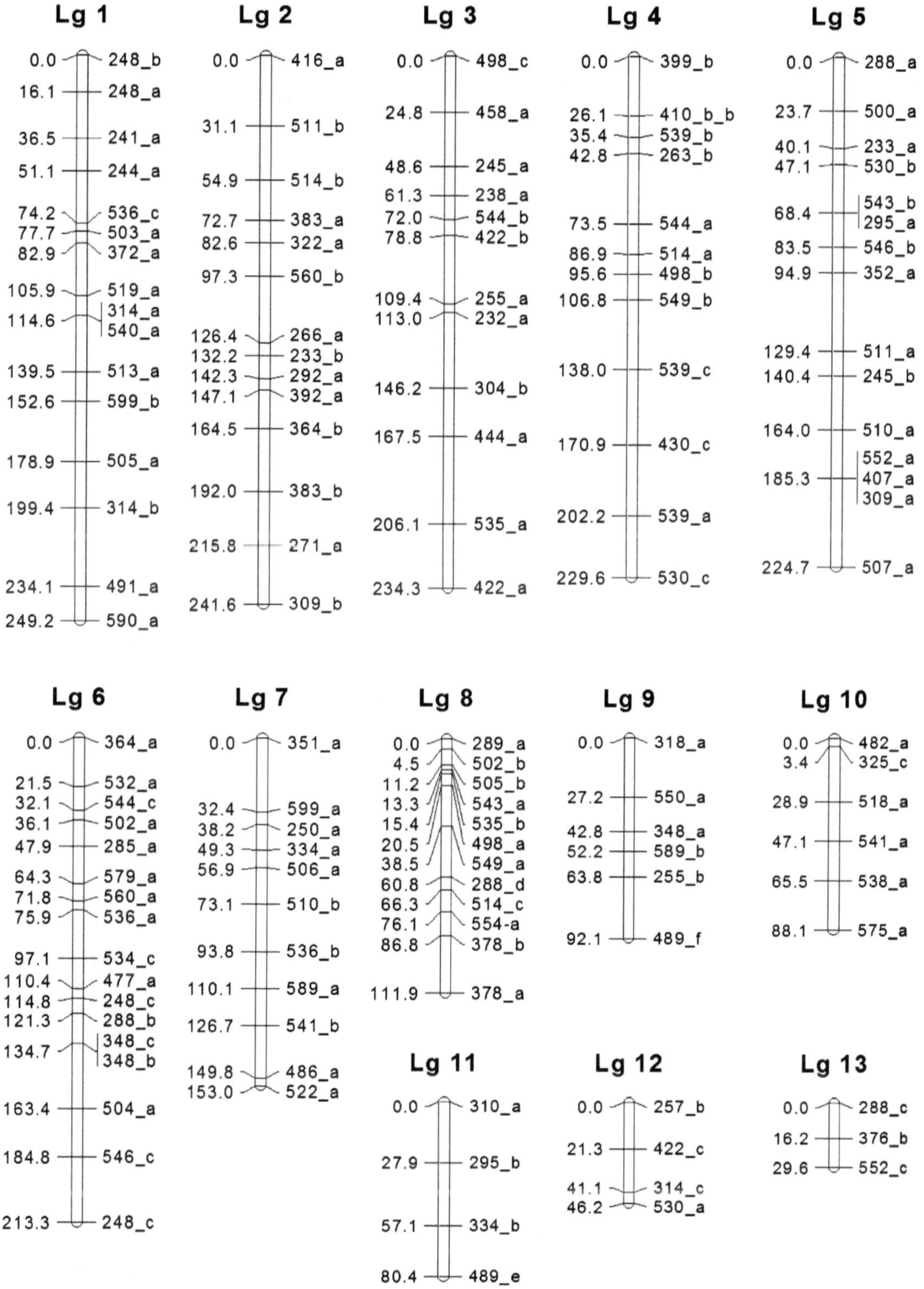

Fig. 1. Random amplified polymorphic DNA (RAPD) marker-based single-tree linkage map of Douglas-fir clone DF60 using haploid megagametophyte (seed) DNA. The map contains 132 markers, listed to the *right* of the map position, with map distances in centiMorgan listed on the *left*. The map encompasses 13 LGs, covering a total of 2,143 cM. The data were analyzed and the map was constructed using the software program JoinMap version 3.0 (Stam and van Oojen 1995). Two sets of data were used to assign loci to different groups of male and female parents using a LOD score threshold of 3.0

between research groups, making it difficult to independently confirm and compare results. Although a highly heterozygous species like Douglas-fir produces many RAPD bands with typical 1:1 Mendelian segregation, during linkage analysis a high percentage of the loci show segregation distortion (12% in H. Agrama, J. Broome, J. Woods, J. E. Carlson unpublished, and 10 to 18% in Krutovskii et al. 1998). For these reasons, RAPDs are not suitable for defining chromosomes, for developing a framework set of loci that are transferable among genotypes, or for comparative genomics among species. Thus better genetic marker systems continue to be sought.

7.3.3 SSRs and other new PCR-Based Markers for Douglas-Fir Genome Mapping

A great deal of effort has been put into the development of SSR markers for Douglas-fir (Amarasinghe and Carlson 2002; Slavov et al. 2004). Twenty SSR markers were incorporated in a preexisting Douglas-fir linkage map (Slavov et al. 2004), which then aided in comparisons of Douglas-fir and loblolly pine maps (Krutovsky et al. 2004). However, the 72 SSR markers reported to date for Douglas-fir are not sufficient in number for map construction alone and unfortunately often suffer from the same multilocus detection problems that many of the earlier RFLP markers did. Thus, more recently marker development has focused on expressed sequence tag polymorphism (ESTP) markers and single nucleotide polymorphisms (SNPs, below). ESTP markers have been developed for Douglas-fir by Krutovsky et al. (2004) and together with SSR markers have been added to the RFLP and RAPD markers in preexisting linkage maps (Jermstad et al. 1998). The most recently published genetic map of Douglas-fir consists of 376 markers, including 172 RFLP, 77 RAPD, 2 isozyme, 20 SSR, 4 sequence tagged site (STS), and 101 expressed sequence tag (EST) markers (Krutovsky et al 2004). This map is organized into 22 LGs that have three or more linked markers and spans 1,859 cM (Table 1). When enough markers are mapped, the number of LGs should coalesce into 13, corresponding to the 13 chromosome pairs in Douglas-fir. It would be valuable to map additional SSR and EST markers (i.e., create a high-density map) so that (1) all 13 LGs can be defined and (2) the locations of more genes can be determined.

7.4 QTL Mapping and Marker-Assisted Selection (MAS)

The variety of DNA markers available in Douglas-fir has led to maps of greater density, accuracy, and reliability that have been applied in the detection of QTL for adaptive and economically important traits of (Wheeler et al. 2005; Jermstad et al. 2001a, b, 2003; Agrema and Carlson 1995; H. Agrama, J. Broome, J. Woods, and J. E. Carlson unpublished; K. V. Krutovsky, S. S. Vollmer, J. Y. Wu, F. C. Sorensen, W. Th. Adams, and S. H. Strauss unpublished). Most of the traits of economic and ecological importance in forest tree species like Douglas-fir are governed by several genes, making them too complex to study by traditional Mendelian genetics. The construction of genetic linkage maps has made it possible to detect QTLs, to place their location on genetic linkage maps, and to track them in tree-improvement programs. In Douglas-fir, genetic linkage maps have been used to detect QTL for wood density and growth (Agrema and Carlson 1995; H. Agrama, J. Broome, J. Woods, and J. E. Carlson unpublished), cold hardiness (Wheeler et al. 2005), and spring bud flush (Jermstad et al. 2001a, b), and to study QTL-by-environment interactions (Jermstad et al. 2003). QTLs for several adaptive traits, including cold hardiness, growth rate and rhythm (bud set and bud flush), biomass, branching pattern, and shoot phenology, have also been mapped in two F_1 hybrids between the coastal and interior varieties or races of Douglas-fir (*Pseudotsuga menziesii* [Mirb.] Franco var. *menziesii* and var. *glauca*, respectively) using RAPD markers (K. V. Krutovsky, S. S. Vollmer, J. Y. Wu, F. C. Sorensen, W. Thomas Adams, and S. H. Strauss unpublished). More than 1,000 offspring have been genotyped at more than 130 RAPD loci, and phenotypes have been determined for 61 quantitative traits in this study. Most QTLs have had a minor but statistically significant effect, although several major QTLs responsible for more than 8 to 10% of phenotypic variance have been detected for traits such as cold hardiness, biomass, height, and branching.

The detection of QTLs associated to traits of economic importance in Douglas-fir opens up the opportunity to implement MAS. The potential benefit of MAS implementation in Douglas-fir tree-improvement programs has been highlighted in a simulation study by Johnson et al. (2000). In their simulation study, Johnson et al. (2000) found that with large populations (500 trees per family) MAS can increase efficiency of within-family selection considerably, but the result also showed that the combination of small population size (100 trees per family or less) and the action of many moderate-effect QTLs could potentially result in less gain from MAS compared to traditional phenotype-based selection in some situations. More studies will be required before the full potential of MAS can be widely exploited commercially in Douglas-fir.

7.5 SNPs and Trait Association Mapping

Trait association mapping is a powerful alternative to QTL mapping that can identify individual genes and alleles responsible for phenotypic differences in adaptive traits (Neale and Savolainen 2004). Nuclear sequence variation and linkage disequilibrium (LD) were studied in 15 cold-hardiness- and 3 wood-quality-related candidate genes in Douglas-fir by Krutovsky and Neale (2005). The set of candidate genes was selected on the basis of function in other plants and by colocation of ESTs with cold-hardiness-related QTLs in Douglas-fir (Jermstad et al. 2001a, b). Single nucleotide polymorphisms (SNPs) are excellent markers for association mapping of genes controlling complex traits (e.g., Brookes 1999; Rafalski 2002; Carlson et al. 2004). An SNP discovery panel was developed from 24 trees representing six regions in Washington and Oregon, plus parents of the QTL mapping population. SNPs were discovered at a frequency of one per 46 bp on average across both coding and noncoding sequences, contributing to moderately high haplotype and nucleotide diversities. Average nucleotide diversity in Douglas-fir was higher than that reported in humans and soybean, lower than that reported in maize, and similar to that in Drosophila. LD was found to decay relatively steadily, up to 50%, over relatively short segments (within 2,000 bp). The researchers' data confirmed studies in loblolly pine and suggest that conifers have relatively low LD (Brown et al. 2004), making them amenable to trait association by the candidate-gene approach (Neale and Savolainen 2004). This was the first extensive study of nucleotide diversity and LD for candidate genes in Douglas-fir. The SNPs in these and other additional cold-hardiness- and phenology-related candidate genes are currently being used in an association mapping study of ~1300 Douglas-fir trees from Washington and Oregon (Pande et al. 2006). To avoid false associations between phenotypes and genotypes for pooled samples in this association mapping study due to the demographic or population structure, the population differentiation has been carefully estimated using isozyme and theoretically neutral SSR markers. An insignificant subpopulation structure has been found in the Douglas-fir samples based on these markers, which should facilitate the ongoing association mapping (Krutovsky et al. 2006a).

7.6 Comparative Genomics

The completion of the genome sequences in model species and accumulation of numerous EST sequences in many other species provide rich resources for comparative genome analysis. Comparative sequence analysis has been used recently to identify putative single-copy genes and conserved ortholog sets (COS) in three angiosperm species and four conifer species, including Douglas-fir (Krutovsky et al. 2006b). COS genes were identified in the model plant species thale cress (*Arabidopsis thaliana*), rice (*Oryza sativa* ssp. *japonica*), and poplar (black cottonwood, *Populus trichocarpa* (Torr. & Gray ex Brayshaw)) and used to find putative COS in four conifers from the Coniferales order. Using EST sequences, unique transcript sets were assembled in loblolly pine (*Pinus taeda* L.), white spruce (*Picea glauca* (Moench) Voss), Douglas-fir (*Pseudotsuga menziesii* (Mirb.) Franco *var. menziesii*), and sugi (*Cryptomeria japonica* (Thunberg ex Linnaeus f.) D. Don). They were compared with COS genes identified in the three model plant species using comparative sequence analysis. Almost half of the single-copy genes in the herbaceous species (*Arabidopsis* and rice) had additional copies and homologs in poplar and conifers. The identified tentative COS sets have many applications in evolutionary genomics studies, phylogenetic analysis, and comparative mapping (Krutovsky et al. 2006b). Comparative mapping is used to establish the syntenic relationships between genomes of different species. Comparative maps can be used to consolidate genetic maps, to verify QTLs, to identify candidate genes, and in studies of genome evolution (Neale and Krutovsky 2004). Krutovsky et al. (2004) constructed a comparative map between homologous LGs in loblolly pine and Douglas-fir using ESTP and RFLP markers. The comparisons revealed extensive synteny and colinearity of gene order between the two genomes, consistent with the hypothesis of conservative chromosomal evolution among even distantly related species in the Pinaceae family. The study also established a framework for further comparative genomics in the Pinaceae and a working hypothesis that the Pinaceae can be viewed as a single genetic system.

7.7 Molecular Cytogenetics

In conifers ca. 75% of the genome is composed of repetitive DNA. Thus for conifers characterization of repetitive DNA is a significant part of genome analysis. We characterized the organization of 5S rRNA genes in Douglas-fir at both the molecular and chromosome levels using fluorescent in situ hybridization (FISH) (Amarasinghe and Carlson 1998). 5S DNA repeat units containing the coding sequence for 5S rRNA and the nontranscribed spacer (NTS) were cloned using PCR. Sequencing and Southern hybridization revealed repeat units of 888 and 871 bp in length, the latter with a 17-bp deletion in the NTS. The coding region showed high homology with other eukaryotic 5S rRNA genes. A 35-bp region of the NTS immediately upstream of the 5' end of the coding region showed high similarity to the gene in other conifers but not to published 5S rDNA sequences for other plants. Physical mapping of 5S rDNA by FISH using a biotinilated homologous probe revealed a single subtelomeric site on one pair of large metacentric chromosomes. This result would suggest that the organization of major repetitive DNA families may be the same in Douglas-fir as in other higher plants and that Douglas-fir has a true diploid genome. However, as shown in Fig. 2, the number of 18S-5.8S-26S ribosomal RNA sites in Douglas-fir is greater than that observed in angiosperms. While angiosperms typically have nucleolar organizer regions on two to four chromosomes, our preliminary results (V. Amarasinghe and J. E. Carlson unpublished and Fig. 2) suggest that there are at least six and perhaps eight chromosomes in Douglas-fir with sites of major 18S-5.8S-26S ribosomal RNA repeats (also visible as secondary constrictions under certain cytological conditions). This multiplicity of 18S-5.8S-26S ribosomal RNA gene sites has also been observed for several *Pinus* species (Brown and Carlson 1997; Hizume et al 1992, 2002; Doudrick et al. 1995) using the FISH technique. These studies also demonstrate that FISH provides karyotypes that can be used to identify homologous chromosomes among species and that will facilitate comparative genomics among conifers. The FISH results have confirmed that chromosomal differentiation among the Pinaceae is very low, but that regions of the genome do vary noticeably in sequence and organization.

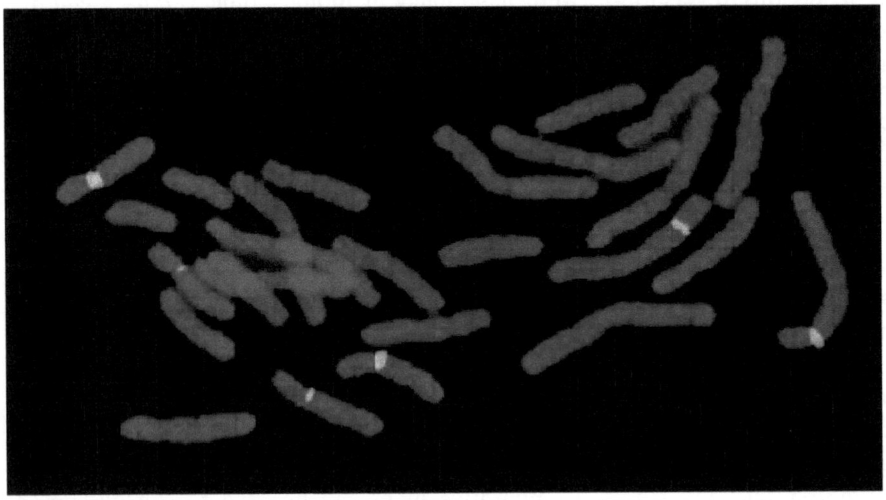

Fig. 2. In situ hybridization of ribosomal RNA genes to a Douglas-fir root tip metaphase chromosome spread. *Yellow bands* are hybridization events for a heterologous 18S-5.8S-26S ribosomal RNA probe and *blue bands* are hybridization events for a 5S RNA probe from Douglas-fir. The chromosomes were stained with propidium iodide, which fluoresces red. Hybridizations were conducted as per Brown and Carlson (1997)

7.8 Future Studies

The development of new technologies such as the 454 system (Margulies et al. 2005) is making DNA sequencing more efficient and affordable. Thus, whole-genome sequences will undoubtedly be available for more plant species in the future. To date, the complete poplar genome sequence and recent establishment of EST databases for Douglas-fir (http://dendrome.ucdavis.edu/dfgp and http://staff.vbi.vt.edu/estap) and other conifers have permitted COS, ESTP, and SNP markers to be used in comparative genomics in the Pinaceae and other forest trees (Neale and Krutovsky 2004; Krutovsky et al. 2004, 2006b). A consortium of scientists working together to develop genomic resources for the species has been founded (e.g., Douglas-fir Genome Project, http://dendrome.ucdavis.edu/dfgp). Current resources include genetic maps, QTL maps, four cDNA libraries (six more planned), ~18,000 ESTs (226,000 additionally planned), SSR markers, SNP markers, and large association mapping populations. Ongoing activities include the development of EST databases, SNP markers in candidate genes, gene-expression microarrays, and association populations. The current EST database is already being used to identify candidate genes for adaptive traits and wood properties (Krutovsky and Neale 2005). Candidate genes will also be identified via gene-expression profiling. Long DNA oligonucleotides (60-mers) have been designed recently for ~11,000 unigenes, and DNA microarrays (NimbleGen) are being synthesized currently for expression analyses to uncover genetic differences among populations and genotypes for traits such as cold tolerance, drought tolerance, disease resistance, and wood properties (Howe et al 2005). As more tree genes and genomes are sequenced and more functional and comparative genomic studies are carried out in Douglas-fir and other conifers, a better understanding of genome organization, structure, and function will emerge for Douglas-fir. Newly developed EST and genome-sequence-based functional markers will be used in population genomic studies to find signatures of natural selection (González-Martínez et al. 2006). Functional genomic studies and subsequent association mapping and comparative genome studies, combined with EST-based marker development and already developed linkage and QTL maps, will ultimately lead to the discovery and characterization of the genes involved in the complex adaptive and economically important traits in Douglas-fir.

References

Adams WT (1983) Application of isozymes in tree breeding. In: Tanksley SD, Orton TJ (eds) Isozymes in plant genetics and breeding. Part A. Elsevier, Amsterdam, pp 381–400

Adams WT, Neale DB, Doerksen AH, Smith DB (1990) Inheritance and linkage of isozyme variants from seed and vegetative bud tissues in coastal Douglas-fir (*Pseudotsuga menziesii* var *menziesii* [Mirb] Franco. Silvae Genet 39:3–4

Agrema H, Carlson JE (1995) DNA markers for selecting high wood density in Douglas-fir. In: Lavereau J (ed) Proc of the 25th Canadian Tree Improvement Association/Western Forest Genetics Association, Hugh John Flemming Forestry Centre, Fredericton, New Brunswick, Canada, p 90

Alexander RR (1985) Major habitat types, community types, and plant communities in the Rocky Mountains. USDA Forest Service, General Technical Report RM-123. Rocky Mountain Forest and Range Experiment Station, Fort Collins, CO

Alexander RR (1988) Forest vegetation on National Forests in the Rocky Mountain and Intermountain regions: habitat types and community types. USDA Forest Service, General

Ali IF, Neale DB, Marshall KA (1991) Chloroplast DNA restriction fragment length polymorphism in *Sequoia sempervirens*. D Don Endl, (*Pseudotsuga menziesii* var *menziesii* [Mirb] Franco, *Calocedrus decurrens* (Torr) and *Pinus taeda* L. Theor Appl Genet 81:83–89

Amarasinghe V, Carlson JE (1998) Physical mapping and characterization of 5S rRNA genes in Douglas-fir. J Hered 89:495–500

Amarasinghe V, Carlson JE (2002) The development of microsatellite DNA markers for genetic analysis in Douglas-fir. Can J For Res 32:1904–1915

Beckman, JS, Soller M (1990) Toward a unified approach to genetic mapping of eukaryotes based on sequence tagged microsatellite sites. Biotechnology 8:930–932

Bennett MD, Smith JB (1991) Nuclear DNA amounts in angiosperms. Phil Trans R Soc Lond Ser B Biol Sci 334:309–345

Botstein D, White RL, Skolnick M, Davis RW (1980) Construction of a genetic linkage map using restriction fragment length polymorphisms. Am J Hum Genet 32:314–331

Brown GR, Carlson JE (1997) Molecular cytogenetics of the genes encoding 18S-5, 8S-26S rRNA and 5S rRNA in two species of spruce (Picea). Theor Appl Genet 95:1–9

Brown GR, Kadel III EE, Bassoni DK, Kiehne KL, Temesgen B, van Buijtenen JP, Sewell MM, Marshal KA, Neale B (2001) Anchored reference loci in loblolly pine (*Pinus taeda* L.) for integrating pine genomics. Genetics 159:799–809

Brown GR, Gill GP, Kuntz RJ, Langley CH, Neale DB (2004) Nucleotide diversity and linkage disequilibrium in loblolly pine. 101:15255–15260

Carlson JE, Tulsieram LK, Glaubitz JC, Luk VW, Kauffeldt C, Rutledge R (1991) Segregation of random amplified DNA markers in F_1 progeny of conifers. Theor Appl Genet 89:758–766

Carlson JE, Agrema HA, Amarsinghe V, Brown GB, Fu Y-B (1994) PCR in tree genome mapping. Forest Biotechnology Workshop, Petawawa National Forestry Institute, Ontario, Canada, 16-17 Oct 1994

Chaparro JX, Werner DJ, O'Malley D, Sederoff RR (1994) Targeted mapping and linkage analysis of morphological, isozyme, and RAPD markers in peach. Theor Appl Genet 87:805–815

Critchfield WB (1984) Impact of the Pleistocene on the genetic structure of North American conifers. In: Lanner R (ed) Proc 8th North American Forest Biology Workshop, Dept of Forest Resources, Utah State University, Logan, UT, pp 70–118

Devey ME, Jermstad KD, Tauer CG, Neale DB (1994) Inheritance of RFLP loci in a loblolly pine three-generation pedigree. Theor Appl Genet 83:238–242

Doudrick RL, Heslop-Harrison JS, Nelson CD, Schmidt T, Nance WL, Schwarzacher T (1995) Karyotype of slash pine (*Pinus elliottii* var. *elliottii*) using patterns of fluorescence *in situ* hybridization and fluorochrome banding. J Hered 86:289–296

El Kassaby YA, Sziklai O, Yeh FC (1982) Linkage relationships among 19 polymorphic allozyme loci in coastal Douglas-fir (*Pseudotsuga menziesii* var '*menziesii*').Can J Genet Cytol 24:101–108

González-Martínez SC, Krutovsky KV, Neale DB (2006) Forest tree population genomics and adaptive evolution. New Phytol 170:227–238

Grattapaglia D, Sederoff R (1994) Genetic linkage maps of *Eucalyptus grandis* and *Eucalyptus urophylla* using a pseudo-testcross: mapping strategy and RAPD markers. Genetics 137:1121–1137

Grattapaglia D, Wilcox P, Chaparro J, O'Malley DM, McCord S, Whetten R, McIntyre L, Sederoff R (1991) A RAPD map of loblolly pine in 60 days. In: International Society for Plant Molecular Biology International Congress, Tucson, AZ

Groover A, Devey M, Fiddler T, Lee J, Megraw R, Mitchell-Olds T, Sherman B, Vujcic S, Williams C, Neale D (1994) Identification of quantitative trait loci influencing wood specific gravity in an outbred pedigree of loblolly pine. Genetics 138:1293–1300

Hamann A, Aitken SN, Yanchuk AD (2004) Cataloguing *in situ* protection of genetic resources for major commercial forest trees in British Columbia. For Ecol Manage 197:295–305

Hanover JW (1992) Applications of terpene analysis in forest genetics. New For 6:159–178

Hizume M, Ishida F, Murata M (1992) Multiple locations of ribosomal RNA genes in chromosomes of pines, *Pinus densiflora* and *P. thunbergii*. Jpn J Genet 67:389–396

Hizume M, Shibata F, Matsusaki Y, Garajova Z (2002) Chromosome identification and comparative karyotypic analyses of four Pinus species. Theor Appl Genet 105:491–497

Howe GT (2006) Bibliography of Douglas-fir ecological genetics, evolutionary genetics, physiological genetics, and tree breeding. Pacific Northwest Research Cooperative Report #24, Oregon State University, Corvallis, OR

Howe DK, Brunner AM, Cherry M, Krutovsky KV, Neale DB, Howe GT (2005) Identifying candidate genes associated with cold hardiness in coastal Douglas-fir using DNA microarrays. In:Proceedings Western Forest Genetics Association 50th Anniversary Meeting: Looking Back – Looking Ahead, 19-21 July 2005, Corvallis, OR. (http://www.fsl.orst.edu/wfga/index_files/2005_Program.pdf)

Howe GT, Jayawickrama KJ, Cherry ML, Johnson GR, Wheeler NC (2006) Breeding Douglas-fir. In: Janick J (ed) Plant breeding reviews. Wiley, New York, 27:245–353

Jermstad KD, Reem AM, Henifin JR, Wheeler NC, Neale DB (1994) Inheritance of restriction fragment length polymorphism and random amplified polymorphic DNAs in Coastal Douglas-fir. Theor Appl Genet 89:758–766

Jermstad KD, Bassomi DL, Wheeler NC, Neale DB (1998) Partial DNA sequencing of Douglas-fir cDNA used for RFLP mapping. Theor Appl Genet 97:771–776

Jermstad KD, Bassomi DL, Jech KS, Wheeler NC, Neale DB (2001a) Mapping of quantitative trait loci controlling adaptive traits in coastal Douglas-fir. I. Timing of vegetative bud flush. Theor Appl Genet 102:1142–1151

Jermstad KD, Bassomi DL, Wheeler NC, Anekonda TS, Aitken SN, Adams WT, Neale DB (2001b) Mapping of quantitative trait loci controlling adaptive traits in coastal Douglas-fir. II. Spring and fall cold-hardiness. Theor Appl Genet 102:1152–1158

Jermstad KD, Bassomi DL, Jech KS, Ritchie GA, Wheeler NC, Neale DB (2003) Mapping of quantitative trait loci controlling adaptive traits in coastal Douglas-fir. III. Quantitative trait loci-by-environment interactions. Genetics 165:1489–1506

Johnson GR, Wheeler NC, Strauss SH (2000) Financial feasibility of marker-aided selection in Douglas-fir. Can J For Res 30:1942–1952

Kinlaw CS, Neale DB (1997) Complex gene families in pine genomes. Trends Plant Sci 2:356–359

Kriebel HB (1993) Molecular structure of forest trees. In: Ahuja MR, Libby WJ (eds) Clonal Forestry. I. Genetics and Biotechnology. Springer, Berlin Heidelberg New York, pp 224–240

Krutovskii KV, Vollmer SS, Sorensen FC, Adams WT, Knapp SJ, Strauss SH (1998) RAPD genome maps of Douglas-fir. J Hered 89:197–205

Krutovsky KV, Neale DB (2005) Nucleotide diversity and linkage disequilibrium in cold-hardiness-wood quality-related candidate genes in Douglas-fir. Genetics 171:2029–2041

Krutovsky KV, Troggio M, Brown GR, Jermstad KD, Neale DB (2004) Comparative mapping in the Pinaceae. Genetics 168:447–461

Krutovsky KV, St. Clair JB, Saich R, Hipkins VD, Neale DB (2006a) Estimation of population structure in the Douglas-fir association mapping study. In: Plant & Animal Genome XIV International Conference on the Status of Plant & Animal Genome Research, Final Program and Abstracts Guide, San Diego, 14-18 Jan 2006, p 34. (http://www.intl-pag.org/14/abstracts/PAG14_W119.html)

Krutovsky KV, Elsik CG, Matvienko M, Kozik A, Neale DB (2006b) Conserved Ortholog Sets in forest trees. Tree Genetics and Genomes (in press)

Litt M, Luty JA (1989) A hypervariable microsatellite revealed by in vitro amplification of a dinucleotide repeat within the cardiac muscle actin gene. Am J Hum Genet 44(3):397-401

Liu Z, Furnier GR (1993) Comparison of allozyme, RFLP, and RAPD markers for revealing genetic variation within and between trembling aspen and bigtooth aspen. Theor Appl Genet 87:97–105

Margulies M, Egholm M, Altman WE, Attiya S, Bader JS, Bemben LA, Berka J, Braverman MS, Chen YJ, Chen ZT, et al. (2005) Genome sequencing in microfabricated high-density picolitre reactors. Nature 437:376–380

Marshall KA, Neale DB (1992) The inheritance of mitochondrial DNA in Douglas-fir (*Pseudotsuga menziesii* var '*menziesii*). Can J For Res 22:73–75

Merkle SA, Adams WT (1987) Patterns of allozyme variation within and among Douglas-fir breeding zones in southwest Oregon. Can J For Res 17:402–407

Morgante M, Olivieri AM (1993) PCR-amplified microsatellites as markers in plant genetics. Plant J 3:175–182

Mukai Y, Suyama Y, Tsumura Y, Kawahara T, Yoshimara H, Kondo T, Tomaru N, Kuramoto N, Murai M (1995) A linkage map for sugi (*Cryptomeria japonica*) based on RFLP, RAPD, and isozyme loci. Theor Appl Genet 90:835–840

Murray B (1998) Nuclear DNA amounts in gymnosperms. Ann Bot 82(Suppl A):3-15

Murray BG, Leitch IJ, Bennett MD (2004) Gymnosperm DNA C-values database (release 3.0, Dec 2004) http://www.rbgkew.org.uk/cval/homepage.html

Nakamura RR, Wheeler NC (1992) Self-fertility variation and paternal success through outcrossing in Douglas-fir. Theor Appl Genet 83:851–854

Neale DB, Krutovsky KV (2004) Comparative genetic mapping in trees: the group of conifers. In: Lörz H, Wenzel G (eds) Biotechnology in Agriculture and Forestry: Molecular Marker Systems. Springer, Berlin Heidelberg New York, pp 267–277

Neale DB, Savolainen O (2004) Association genetics of complex traits in conifers. Trends Plant Sci 9:325–330

Neale DB, Wheeler NC, Adams WT (1984) Inheritance of isozymes in needle tissue of Douglas-fir. Can J Genet Cytol 26:459–468

Neale DB, Wheeler NC, Allard RW (1986) Paternal inheritance of chloroplast DNA in Douglas-fir. Can J For Res 16:1152–1154

Pande B, Krutovsky KV, Jermstad KD, Howe GT, St. Clair JB, Wheeler NC, Neale DB (2006) Association genetics for adaptive traits in Douglas-fir. In: Plant & Animal Genome XIV. The International Conference on the Status of Plant and Animal Genome Research, Final Program and Abstracts Guide, San Diego, 14-18 Jan 2006, p 35 (http://www.intl-pag.org/14/abstracts/PAG14_W120.html)

Piesch RF, Stettler RF (1971) The detection of good selfers for haploid introduction in Douglas-fir. Silvae Genet 20:144–148Ponoy B, Hong YP, Woods J, Jaquish B, Carlson JE (1994) Chloroplast DNA diversity of Douglas-fir in British Columbia. Can J For Res 24:1824–1834

Reiter RS, Williams JGK, Feldmann KA, Rafalski JA, Tingey SV, Scolnik PA (1999) Global and local genome mapping in Arabidopsis thaliana by using recombinant inbred lines and random amplified polymorphic DNAs. Proc Natl Acad Sci USA 89:1477–1481

Sewell MM, Sherman BK, Neale DB (1999) A consensus map for loblolly pine (*Pinus taeda* L.). I. Construction and integration of individual linkage maps from two outbred three-generation pedigrees. Genetics 151:321–330

Slavov GT, Howe GT, Yakovlev I, Edwards KJ, Krutovskii KV, Tuskan GA, Carlson JE, Strauss SH, Adams WT (2004) Highly variable SSR markers in Douglas-fir: Mendelian inheritance and map locations. Theor Appl Genet 108:873–880

Snajberk K, Zavarin E (1976) Mon-and sesqui-terpenoid differentiation of *Pseudotsuga* of United States and Canada. Biochem Syst Ecol 4:159–163

Sorensen FC (1971) "White seedling": a pigment mutation that affects seed dormancy in Douglas-fir. J Hered 62:127–130

Sorensen FC (1973) Frequency of seedlings from natural self-fertilization in coastal Douglas-fir. Silvae Genet 22:20–24

Stam P, van Oojen JW (1995) JOINMAP VERSION 2.0: software for the calculation of genetic linkage maps. CPRO-DLO, Wageningen, The Netherlands

Strauss SH, Tsai C-H (1988) Ribosomal gene number variability in Douglas-fir. J Hered 79:453–458

Tautz D (1989) Hypervariability of simple sequences as a general source for polymorphic DNA markers. Nucleic Acids Res 17:6463–6471

Tsai C-H, Strauss SH (1989) Dispersed repetitive sequences in the chloroplast genome of Douglas-fir (*Pseudotsuga menziesii* var *menziesii* [Mirb] Franco). Curr Genet 16:211–218

Tulsieram LK, Glaubitz JC, Carlson JE (1992) Single tree genetic linkage mapping in conifers using haploid DNA from megagametophytes. BioTechnology 10:686–690 von Rudloff E (1975) Volatile leaf oil analysis in chemosystematic studies of North American conifers. Biochem System Ecol 2:131–167

von Rudloff E, Rehfeldt GE (1980) Chemosystematic studies in the genus Pseudotsuga. IV. Inheritance and geographical variation in the lead oil terpenes of Douglas-fir from the Pacific Northwest Can J Bot 58:546–556

Wheeler NC, Jermstad KD, Krutovsky KV, Aitken SN, Howe GT, Krakowski J, Neale DB (2005) Mapping of quantitative trait loci controlling adaptive traits in coastal Douglas-fir. IV. Cold-hardiness QTL verification and candidate gene mapping. Mol Breed 15(2):145-156

Williams JG, Kubelik AR, Livak KJ, Rafalski JA, Tingey SV (1990) DNA polymorphisms amplified by arbitrary primers are useful as genetic markers. Nucleic Acids Res 18:6531–6535

Yeh FC, O'Malley DM (1980) Enzyme variation in natural population of Douglas-fir (*Pseudotsuga menziesii* var *menziesii* [Mirb] Franco) from British Columbia. 1. Genetic variation patterns in coastal population. Silvae Genet 29:83–92

Zavarin E, Snajberk K (1973) Geographic variability of monoterpenes from cortex of *Pseudotsuga menziesii*. Pure Appl Chem 34:411–434

8 Cryptomeria Japonica

Teiji Kondo[1] and Noritsugu Kuramoto[2]

[1] Forest Tree Breeding Center, 3809-1 Ishi, Juo, Hitachi, Ibaraki 319-1301, Japan
e-mail: kontei@affrc.go.jp
[2] Kyushu Breeding Office, Forest Tree Breeding Center, 2320-5 Suya, Koshi, Kumamoto 861-1102, Japan

8.1 Introduction

8.1.1 Brief History of the Crop

The first author to attempt to divide *Cryptomeria japonica* into distinct lines or varieties was Murai (1947), who recognized two lines, Omote-sugi and Ura-sugi, based on differences in their leaf morphology. Broadly speaking, Omote-sugi grows in the region facing the Pacific Ocean, and Ura-sugi in the region facing the Japanese Sea. Yasue et al. (1987) found similar general geographical differentiation, with some exceptions, based on diterpene hydrocarbon constituents. However, little differentiation among natural populations of *C. japonica* has been detected in analyses of isozyme (Tomaru et al. 1994) and cleaved amplified polymorphic sequence (CAPS) markers (Tsumura and Tomaru 1999). Even the genetic differentiation between populations on the mainland and Yaku Island, located far from the mainland, appears to be very small. Differences between the two lines recognized by Murai were reflected in differences between their 6PGD isozymes in a study by Tomaru et al. (1994), but the distinction was much less clear in a subsequent CAPS marker analysis (Tsumura and Tomaru 1999).

As there is a long history of cultivation of *C. japonica*, many varieties have been developed. Miyajima (1983) classified them into two kinds of varieties. One is a geographical race that has not been improved artificially; the other is a cultivar that has been improved artificially. The first cultivars were selected in the 16th century. Afforestation with the species began in the early 18th century, at which time the main traditionally used cultivars were selected by foresters on Kyushu Island. Since then many cultivars have been developed and most of them have been maintained by propagating cutting.

8.1.2 Botanical Descriptions

C. japonica belongs to the family Taxodiaceae. Although *C. fortunei* was found later as a second species of this genus, no apparent differences from *C. japonica* have been reported. *C. japonica* and *C. fortunei* have almost the same genetic sequences, and *Cryptomeria* forms a clade with *Glyptostrobus* and the *Taxodium* group, based on the sequences of four chloroplast genes, in Taxodiaceae (Kusumi et al. 2000). *C. japonica* grows in warm temperate and cool temperate zones, its natural distribution extending from Yaku island, 30°15' N, to the north end of Honshu island, 40°42' N (Maeda 1983). *C. japonica* prefers habitats with high humidity and rich soil. It is a monoecious plant that bears diclinous flowers in a flowering season extending from March to April. Its needles are lozenge-shaped in cross-section.

The chromosome complement of *C. japonica* consists of 11 pairs of metacentric or submetacentric chromosomes with gradual variations in length (Sax and Sax 1933; Mehra and Khoshoo 1956). Secondary constrictions are found on chromosomes 6 and 10. Chromosome 10 always has secondary constrictions on its short arm, whereas chromosome 6 may occur either with or without secondary constrictions (Toda 1979a, b; Kondo et al. 1985). The heteromorphy of chromosome 6 indicates that heteromorphic chromosomes do not necessarily disrupt meiotic processes (Hizume et al. 1989). These secondary constrictions have been clearly visualized by fluorescent banding with chromomycin A_3 (Kondo and Hizume 1982). The fluorescent bands coincide in position with the secondary constrictions and in number with the maximum number of nucleoli per cell. The fluorescent band of chromosome 10 is larger than that of chromosome 6. The nucleolus formed in chromosome 10 is also larger than that of chromosome 6 (Kondo et al. 1985). Fluorescent signals generated by in situ hy-

Genome Mapping and Molecular Breeding in Plants, Volume 7
Forest Trees
C. Kole (Ed.)

bridization with the wheat 18S-5.8S-26S rDNA genes appear at the same position as the fluorescent bands generated with chromomycin A_3 (Hizume et al. 1998). Furthermore *Arabidopsis*-type telomere sequence repeats (TTTAGGG)n can hybridize to both ends of each chromosome (Hizume et al. 2000). Compared with the chromosomes of *Pinus*, those of *C. japonica* seem to have a simpler, less rearranged morphology.

The genome size of *C. japonica* has been determined by flow cytometry of isolated nuclei stained with propidium iodide using *Hordeum vulgare* nuclei as an internal standard (Hizume et al. 2001). The resulting value (mean for five plants) was 22.09 pg/2C, corresponding to ca. 10 Gbp (1×10^{10} bp) according to conversion factors published by Mukai (1998), equivalent to 100 times more than the *Arabidopsis* genome.

8.1.3 Economic Importance

C. japonica is one of the most important timber species in Japan, favored for its straight bole and rapid growth, and has been planted on 4.53 million ha, corresponding to 45% of the artificial forest in Japan. In the Kanto region, which covers central parts of Honshu island including Tokyo, the average height, DBH (diameter at breast height), and volume per hectare of artificial stands are typically 18.8 m, 24.9 cm, and 428.7 m^3 at 40 years of age, and 21.5 m, 29.2 cm, and 515.7 m^3 at 50 years of age, respectively (Ohtomo 1983).

8.1.4 Breeding Objectives

Systematic breeding of *C. japonica* began in the late 1950s. A Tree Breeding Station network has been established under the Forestry Agency of the Ministry of Agriculture and Forestry, covering the whole of Japan. The main initial breeding objective was to improve its growth. Later, resistance to Sugi bark borer and snow damage was added, and recently the breeding objectives have become diverse. As the wood of *C. japonica* is softer than that of imported timber, improvement of wood strength is an urgent goal to promote domestic forestry. Reductions in pollen production, increased rates of CO_2 fixation, and shade tolerance have also been added to the objectives, to address issues associated with pollinosis, global warming, and multistoried forests, respectively.

8.1.5 Classical Mapping Efforts

Linkage was first evaluated in relation to morphological characters when Ohba et al. (1974) investigated relationships among the dominant *twisted-leaf* gene and two recessive genes (an albino gene and a gene responsible for green coloration in winter) but found no linkage among them. However, Kuromaru et al. (1983) found linkage between two loci encoding peroxidase isozymes with a recombination value of 0.167 in repulsion phase. In addition, Kuramoto et al. (1996) found a linkage relationship between a dwarf gene and a gene associated with leaf whitening in summer, with a recombination value of 0.315 in coupling phase. Furthermore, Kuramoto et al. (1997) found several linkage relationships among five isozyme loci and the dwarf gene. Because of the presence of lethal genes, the segregation ratio was distorted in some progeny families, especially selfed families. Although morphological traits and isozymes were studied, only their linkage relationships were examined; no linkage maps were constructed because of the limited number of markers.

8.1.6 Classical Breeding Achievements

As *C. japonica* is the most important tree species in Japan, ca. 3,500 plus tree clones of the species, more than for any other major species bred in Japan, have been selected for mass selection breeding. As mentioned above, the main initial breeding objective was to improve its growth. Compared to local varieties, we have obtained a 15% increment in volume. Problems caused by the sugi bark borer, *Semanotus japonicus*, whose larvae feed on bark and xylem, have also been addressed. To help prevent damage by this insect, an inoculation test has been established and a resistant variety has already been released. A further problem is that crooked trees are often found in regions with heavy snowfall, caused by the pressure exerted by snow sliding down slopes. Two clones, which grow straight even in heavy snow regions, have already been developed after field trials in a region with heavy snowfall.

8.2 Construction of Genetic Maps

8.2.1 Brief History of Mapping Efforts

Since it is the leading species in Japanese forestry and more genetic information has been gathered since the early linkage studies, Japanese tree geneticists and breeders have been eager to construct a linkage map of *C. japonica*. This has been a major challenge for the scientists involved, who have struggled to keep up with advances in breeding major food crop species, such as wheat and rice, and model species (especially *Arabidopsis thaliana*). Mukai et al. (1995) constructed the first linkage map, mainly based on RFLP markers. Although this map provided limited coverage, it was well constructed. CAPS markers were added to it by Iwata et al. (2001), and it was finally integrated, in a consensus map, by Tani et al. (2003). In addition, linkage maps based on dominant RAPD and AFLP markers were constructed by Kuramoto et al. (2000) and Nikaido et al. (2000), respectively. The linkage map by Kuramoto et al. (2000) was designed to facilitate QTL analysis of wood strength. The distances covered by these maps were longer than those based on RFLP or CAPS markers.

8.2.2 First-Generation Maps

The first linkage map was constructed by Mukai et al. (1995) using an F_2 progeny from a cross between two cutting cultivars, Kumotooshi and Okinoyama. A total of 91 (77 RFLP, 12 RAPD, and 1 isozyme) markers were distributed among 13 linkage groups (LGs), covering 887.3 cM (Table 1). The average interval between adjacent markers was 12.3 cM. Thirty-five markers distributed in six clusters showed distorted segregation, presumably due to the presence of (an) embryonic lethal gene(s) (Ohba 1979) or other deleterious genes, the effects of which would appear mainly in selfed progeny. LG 5 was assigned to chromosome 10 using trisomics (Suyama et al. 1996). Iwata et al. (2001) developed CAPS markers and applied them to the same population (Mukai et al 1995). In the revised linkage map, a total of 167 markers (46 CAPS, 101 RFLP, 17 RAPD, and 2 isozyme markers and 1 dwarf gene) were distributed among 15 LGs, covering 1,109.1 cM, with an average interval between adjacent markers of 8.7 cM. As CAPS markers were added to the map, 30 markers consisting of RFLP, RAPD, and isozyme markers without confirmed map positions in the previous map were mapped to precise positions. Thus, the linkage map was extended to 1,109.1 cM and its density was increased. The cited authors developed 217 CAPS markers, and the polymorphisms of the CAPS markers were found to be associated more strongly with intron than with exon regions. The CAPS markers proved to be very useful in the integration of different maps.

Kuramoto et al. (2000) constructed a linkage map of RAPD markers using an F_1 progeny between two cutting cultivars, Boka and Iwao, for QTL analysis of wood strength. As a pseudotestcross mapping strategy (Grattapaglia and Sederoff 1994) was adopted, linkage maps of both parents were obtained. In the linkage map of Iwao (Fig. 1), 119 RAPD markers were distributed among 21 LGs, covering 1,756.4 cM, with an average interval between adjacent markers of 14.8 cM. In the linkage map of Boka (Fig. 2) 84 RAPD markers were distributed among 14 LGs, covering 1,111.9 cM, with an average interval between adjacent markers of 13.2 cM. The genome length was initially estimated as ca. 2,800 cM using the moment estimator method according to Hulbert et al. (1988). All these maps were constructed using the MAPMAKER/EXP 3.0 computer program (Lander et al. 1987; Lincoln et al. 1992a). Nikaido et al. (2000) constructed longer linkage maps based on AFLP markers and a small number of CAPS markers using an F_1 progeny from reciprocal crosses between two cutting cultivars, Kumotooshi and Haara, and (again) a pseudotestcross mapping strategy. In the linkage map of Kumotooshi a total of 132 (123 AFLP and 9 CAPS) markers were distributed among 23 LGs, covering 1,992.3 cM, with an average interval between adjacent markers of 17.9 cM. In the linkage map of Haara a total of 91 markers, consisting of 83 AFLP and 8 CAPS markers, were distributed among 19 LGs, covering 1,266.1 cM, with an average interval between adjacent markers of 16.0 cM. Segregation distortion of AFLP markers was found in half of them, attributed to the presence of lethal genes and fragment-complexes originating from different loci.

8.2.3 Second-Generation Maps

Tani et al. (2003) integrated linkage maps of two unrelated F_2 populations using JoinMap 3.0 software (Van

Table 1. Linkage maps of Cryptomeria japonica

Authors	Population*	Markers	Linkage groups	Total map distance (coverage %)	Marker interval	Genome length
Mukai et al. (1995)	F_2 (Kumotooshi × Okinoyama)	91 markers (77 RFLP, 12 RAPD, 1 isozyme, 1 dwarf gene)	13 groups and 6 pairs	887.3 cM	12.3 cM**	Not estmated
Iwata et al. (2001)	Same population as Mukai et al. (1995) F_2 (Kumotooshi × Okinoyama)	167 markers (46 CAPS, 101 RFLP, 17 RAPD, 2 isozyme, 1 dwarf gene)	15 groups	1,109.1 cM	8.7 cM	Not estimated
Nikaido et al. (2000)	F_1 (Haara, Kumotooshi) reciprocal crosses	Kumotooshi: 132 markers (123 AFLP, 9 CAPS) Haara: 91 markers (83 AFLP, 8 CAPS)	Kumotooshi: 23 groups Haara: 19 groups	Kumotooshi: 1,992.3 cM (80%) Haara: 1,266.1 cM (50%)	Kumotooshi: 17.9 cM Haara: 16.0 cM	Length of 2,500 cM was used
Kuramoto et al. (2000)	F_1 (Boka × Iwao)	Iwao: 119 RAPD markers Boka: 84 RAPD markers	Iwao: 21 groups Boka: 14 groups	Iwao: 1,756.4 cM (62%) Boka: 1,111.9 cM (45%)	Iwao: 14.8 cM Boka: 13.2 cM	2,800 cM
Tani et al. (2003)	Two populations • Same population as Mukai et al. (1995) F_2 (Kumotooshi × Okinoyama) • F_2 from sibcross of F_1 (Yabukuguri × Iwao)	438 markers (172 CAPS, 200 RFLP, 37 microsatellite, 5 SNP, 22 RAPD, 1 isozyme, 1 dwarf gene)	11 large groups and 1 small group	1,372.2 cM	3.0 cM	• F_2 (Kumotooshi × Okinoyama): 1,395.5 cM • F_2 from sibcross of F_1 (Yabukuguri × Iwao): 1,810.1 cM, 2,168.5 cM

* All parents were cutting cultivars such as Kumotooshi, Okinoyama, Haara, Iwao, Boka
** Marker interval was calculated by authors

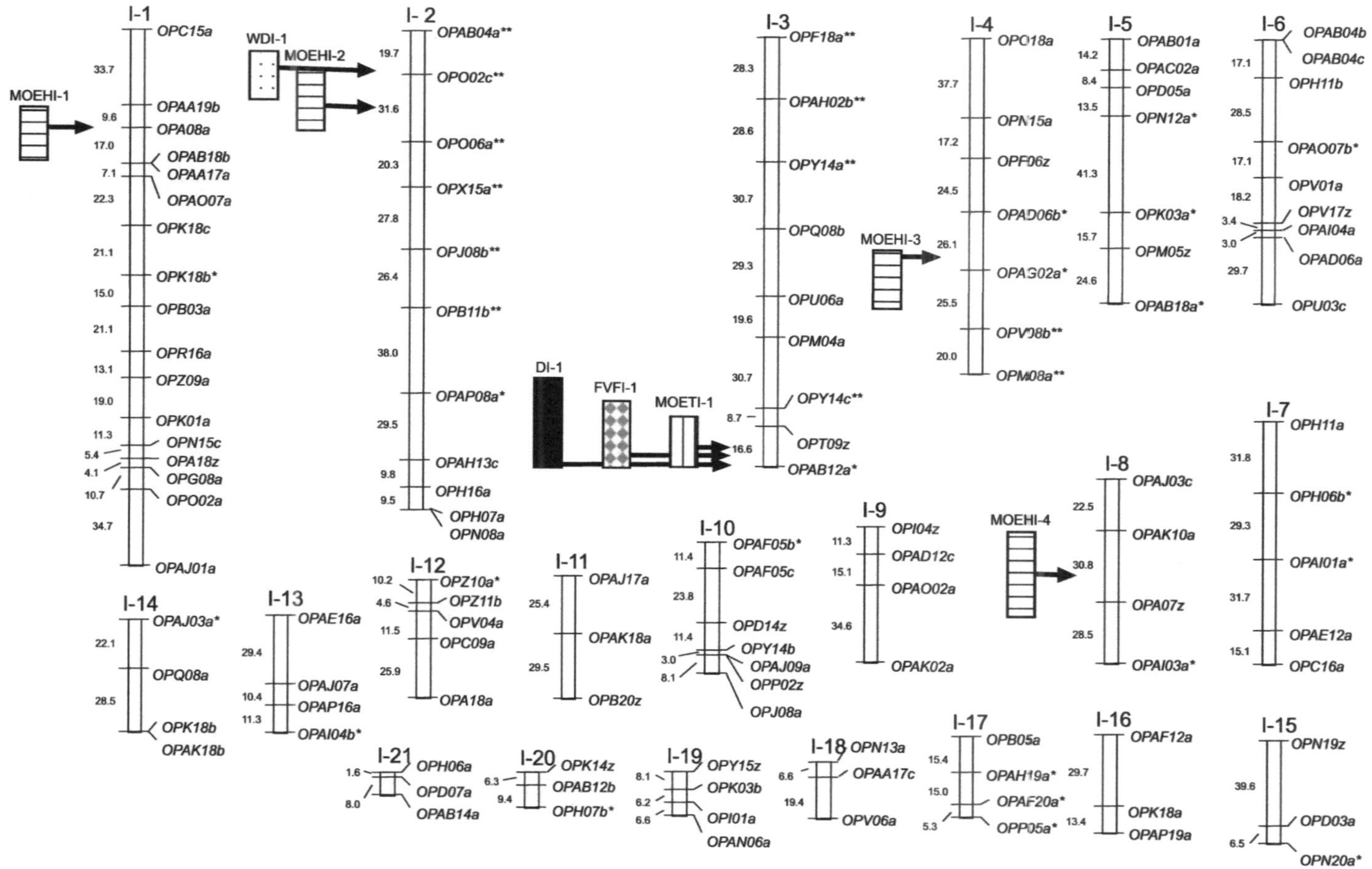

Fig. 1. Linkage map and estimated locations of QTLs in Iwao-sugi, using RAPD markers. A total of 177 linked markers were distributed among 25 LGs. Four LGs that consisted of only two markers were not included in this figure. The markers are listed on the *right* and map distances in centiMorgan are shown on the *left*. One *asterisk* (*) and two *asterisks* (**) designate markers with distorted segregation ($0.01 < P < 0.05$ and $P < 0.01$, respectively). One hundred nineteen RAPD markers with confirmed map positions were assigned to 21 LGs, covering 1,756.4 cM. *Bars* to the *left* of the LGs correspond to 2.0 LOD support intervals for the QTL locations. *Arrows* extending from bars indicate the most likely QTL positions estimated using MAPMAKER/QTL analysis (Table 1). MOEH, modulus of elasticity measured by the hanging method; MOET, modulus of elasticity measured by the tapping method; FVF, fundamental vibration frequency; WD, wood density in green condition; D, diameter (Kuramoto et al 2000, courtesy Can J For Res)

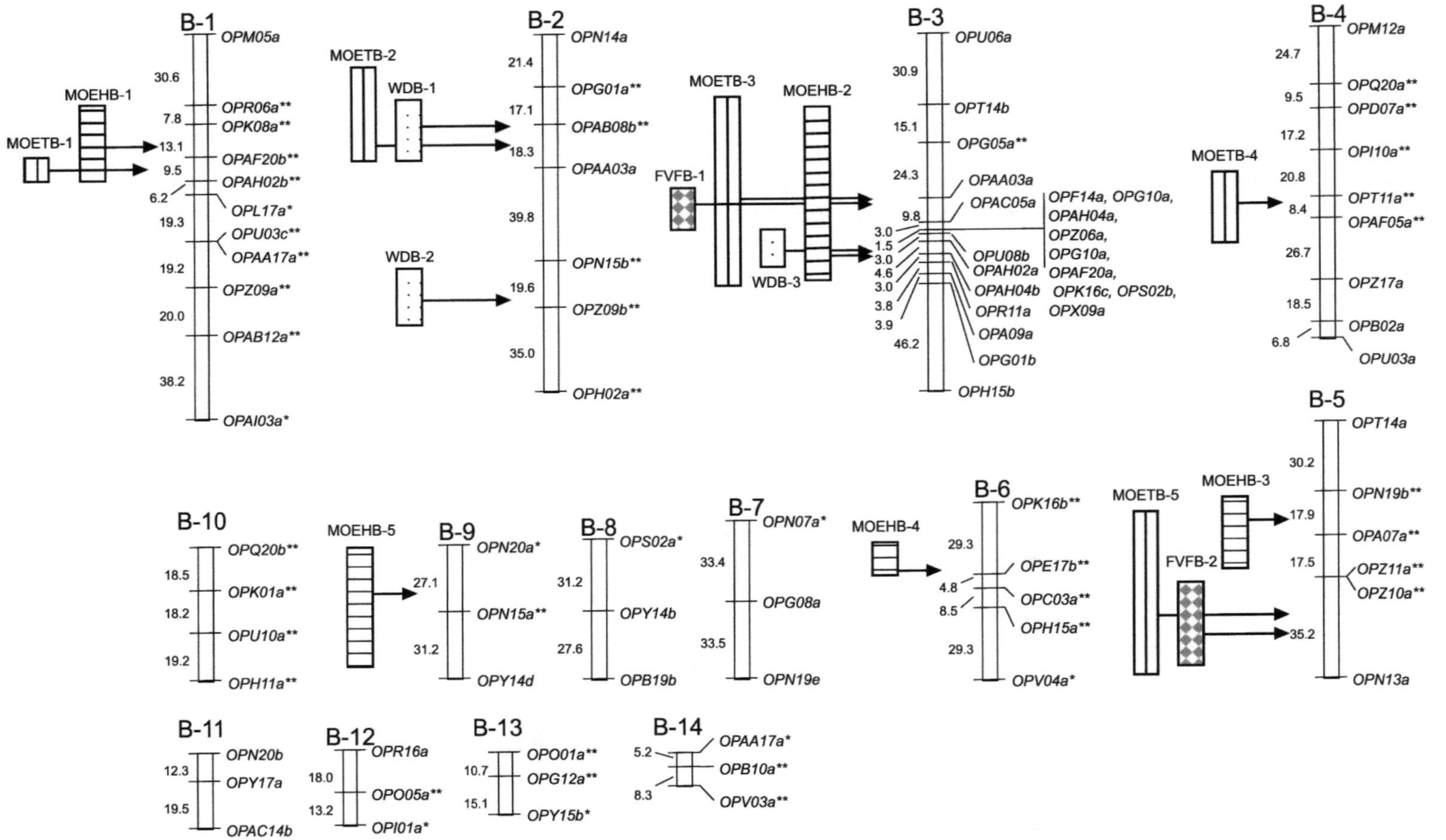

Fig. 2. Linkage map and estimated locations of QTLs in Boka-sugi, using RAPD markers. A total of 117 linked markers were distributed among 21 LGs. Seven LGs that consisted of only two markers were not included in this figure. The markers are listed on the *right* and map distances in centiMorgan are shown on the *left*. One *asterisk* (*) and two *asterisks* (**) designate markers with distorted segregation ($0.01 < P < 0.05$ and $P < 0.01$, respectively). Ninety-eight RAPD markers with confirmed map positions were assigned to 14 LGs, covering 1,111.9 cM. *Bars* to the *left* of the LGs correspond to 2.0 LOD support intervals for the QTL locations. *Arrows* extending from *bars* indicate the most likely QTL positions estimated using MAPMAKER/QTL analysis (Table 2). MOEH, modulus of elasticity measured by the hanging method; MOET, modulus of elasticity measured by the tapping method; FVF, fundamental vibration frequency; WD, wood density in green condition (Kuramoto et al 2000, courtesy Can J For Res)

Table 2. Gene mapping in Cryptomeria japonica

Reference	Trait	Population	Marker type	Gene symbol	Linkage group	Flanking marker	Distance
Mukai et al. (1995)	Dwarf	F_2 progeny from a cross between Kumotooshi and Okinoyama	RAPD, RFLP	MT-d	2	RAPD marker, K08b RFLP marker, CD0461R	8.1 cM 33.5 cM
Tani et al. (2003)	Dwarf	F_2 progeny from a cross between Kumotooshi and Okinoyama	CAPS, RFLP, SNP, microsatellite	MT dwarf	KO7	RAPD marker, K08b RFLP marker, CD0461R	6.8 cM 13.5 cM
		Two unrelated F_2 populations, F_2 from a cross between Kumotooshi and Okinoyama, and F_2 population from a sibcross of F_1 (Yabukuguri × Iwao) (integrated map)	CAPS, RFLP, SNP, microsatellite	MT dwarf	YI5&KO7	RFLP marker, CD0195R Microsatellite marker, CJG0083M	1.7 cM 4.7 cM
Goto et al. (2003)	Allergen, Cry j 1	F_1 progeny between Boka and Iwao	RAPD	CRYJ1-352	I-18	RAPD marker, OPN13a	0 cM
Tani et al. (2003)	Allergen, Cry j 1	F_2 progeny from a cross between Kumotooshi and Okinoyama	CAPS, RFLP, SNP, microsatellite	Cry j 1C	KO7	Microsatellite marker, CJS0002M RFLP marker, CC1778R	2.7 cM 2.6 cM
		Two unrelated F_2 populations, F_2 from a cross between Kumotooshi and Okinoyama, and F_2 population from a sibcross of F_1 (Yabukuguri×Iwao) (integrated map)	CAPS, RFLP, SNP, microsatellite	Cry j 1C	YI5&KO7	CAPS marker, CC2989C Microsatellite marker, CJS002M	0.5 cM 2.1 cM
Tani et al. (2003)	Allergen, Cry j 2	F_2 progeny from a cross between Kumotooshi and Okinoyama	CAPS, RFLP, SNP, microsatellite	Cry j 2C	YI2	RFLP markers, CD0440R RFLP markers, CC1488R1	3.7 cM 4.2 cM
		Two unrelated F_2 populations, F_2 from a cross between Kumotooshi and Okinoyama, and F_2 population from a sibcross of F_1 (Yabukuguri × Iwao) (integrated map)	CAPS, RFLP, SNP, microsatellite	Cry j 2C	YI2&KO11,12	CAPS marker, CC2657C RFLP marker, CD0344R	14.7 cM 10.8 cM

Ooijen and Voorrips 2001). One F_2 population originated from the cross between Kumotooshi and Okinoyama used by Mukai et al. (1995) and Iwata et al. (2001). The other originated from a sibcross of an F_1 progeny (Yabukuguri × Iwao). In the integrated map a total of 438 markers (172 CAPS, 200 RFLP, 37 microsatellite, 5 SNP, 22 RAPD markers, 1 isozyme marker, and 1 dwarf gene) were distributed in 11 large LGs and 1 small group, covering 1,372.2 cM in total. The average interval between adjacent markers was 3.0 cM. Although large numbers of markers were mapped (including microsatellites), resulting in a high-density map, the total distance covered by it was thought to be less than half of the total genome length, and it was still shorter than maps constructed with RAPD markers (Kuramoto et al. 2000) and AFLP markers (Nikaido et al. 2000). Tani et al. (2003) speculated that their map did not cover regions where genes were sparsely distributed and that it would be helpful to add extensive microsatellite markers or random genetic markers, such as AFLP and RAPD markers, to fill in the less dense regions (Tani et al. 2003). As the libraries were derived from limited sources, 3-day imbibed embryos and inner-bark tissues, it should be noted that this map may not have covered all of the gene-dense regions. Further integration of the maps by Kuramoto et al. (2000) and Nikaido et al. (2000) is likely to extend the coverage toward saturation. A saturated linkage map is essential for accurate QTL analysis and effective marker-assisted selection (MAS). Over 200 CAPS markers have already been developed and should facilitate the integration of different maps (Iwata et al. 2001). Thirty-seven CAPS markers have been used as bridges to integrate the two maps thus far, but this is too few (fewer than 3 bridging CAPS markers per LG for 9 out of the 12 LGs) for full integration. Thus, the CAPS markers (thus far developed, at least) were not as powerful as hoped, and the development of other kinds of markers for integration is awaited to overcome current limitations.

8.3 Gene Mapping

In the first linkage map by Mukai et al. (1995) a dwarf gene, *MT-d*, was mapped using the F_2 progeny from a cross between two cutting cultivars, Kumotooshi and Okinoyama. *MT-d* was flanked by an RAPD marker, K08b, and an RFLP marker, CD0461R, at distances of 8.1 cM and 33.5 cM, respectively. These distances were shortened to 6.8 cM and 13.5 cM, respectively, in the revised map by adding CAPS, RFLP, SNP, and microsatellite markers (Tani et al. 2003). Furthermore, *MT-d* was flanked by an RFLP marker, CD0195R, and a microsatellite marker, CJG0083M, at distances of 1.7 cM and 4.7 cM, respectively, in the integrated map. Although the distance between *MT-d* gene and CD0195R was the smallest at 1.7 cM compared with other markers, in the previous map it was 34.6 cM.

Cry j 1 and Cry j 2 are major allergens involved in *C. japonica* pollinosis. Goto et al. (2003) mapped the gene encoding Cry j 1 using a CAPS marker derived from isoform information acquired using F_1 progeny from a cross between two cutting cultivars, Boka and Iwao, which had already been used in QTL analysis of wood strength (Kuramoto et al. 2000). The CAPS marker cosegregated with (and thus was located in the same position, or very close to) an RAPD marker, OPN13a. Tani et al. (2003) also mapped the genes encoding Cry j 1 and Cry j 2 in their integrated map, but did not describe how they were mapped. According to Tani et al. (2003) Cry j 1 was flanked by a microsatellite marker, CJS0002M, and an RFLP marker, CC1778R, at distances of 2.7 cM and 2.6 cM, respectively. Cry j 2 was flanked by two RFLP markers, CD0440R and CC1488R1, at distances of 3.7 cM and 4.2 cM, respectively. The locations of these genes in the integrated map are shown in Table 2.

8.4 Detection of Quantitative Trait Loci

Improvement of the wood strength of *C. japonica* is an urgent objective. Kuramoto et al. (2000) analyzed QTLs associated with wood strength using a linkage map derived from RAPD markers in the F_1 progeny of a cross between Boka and Iwao. Effective QTLs were associated with several traits related to wood strength, such as the modulus of elasticity (MOE; an indicator of wood strength), wood density, fundamental vibration frequency, and stem diameter using MAPMAKER/QTL 1.1 software (Paterson et al. 1988; Lincoln et al. 1992b). Five and ten QTLs for MOE were detected in the linkage maps of Iwao and Boka, respectively (Figs. 1 and 2). Since these QTLs explained about 45% of the total phenotypic variance, they were thought to be sufficiently effective for use in breeding programs. MOE was evaluated by two methods, which gave similar values but were based on different principles. In this analysis some QTLs detected by one method did not overlap with those detected by the other one. It was postulated that the differences in the mapping positions of QTLs related to MOE might reflect differences in the bases of these methods. Furthermore, QTLs associated with wood density and fundamental vibration frequency overlapped with some of the QTLs associated with MOE. Therefore, it was concluded that the QTL analysis revealed important components of wood strength, and relationships among them, via their positions on the linkage maps.

Yoshimaru et al. (1998) analyzed QTLs associated with the juvenile growth, flower bearing, and rooting ability of *C. japonica* using the linkage map constructed by Mukai et al. (1995) on the basis of RFLP, RAPD, and isozyme loci using the F_2 progeny of a cross between Kumotooshi and Okinoyama. Very effective QTLs were detected for all the traits, using the MAPMAKER/QTL 1.1 program as a free model of QTL effects (Paterson et al. 1988; Lincoln et al. 1992b). Since major QTLs for early growth were detected near the dwarf gene in LG 2, the authors considered that these QTLs are pleiotropic effects of the dwarf gene. In LG 2 effective QTLs for male and female flower bearing were also detected. Another QTL associated with female flower bearing was also detected near the dwarf gene in LG 2. This QTL was thought to have some genetic relationship with QTLs for early growth. A less effective QTL for female flower bearing was detected in LG 5.

8.5 Advanced Works

The cDNA libraries have been derived from the inner bark and strobili (male and female) of *C. japonica*. From the inner-bark library, 1,583 out of 2,231 clones were assigned to putative functions (Ujino-Ihara et al. 2000). The remaining 648 clones did not show significant homology to any known sequences. Sequences representing genes concerned with cell wall formation and stress responses were abundant. From the cDNA clones 67 sequence tagged site (STS) markers were developed (Ujino-Ihara et al. 2002). In addition, 1,210 expressed sequence tags (ESTs) representing 1,173 transcripts were obtained from male and female strobili (Ujino-Ihara et al. 2003), 807 of which were assigned to putative functions, including those of genes expressed in developing flower tissues of other plant species, such as *CONSTANS* and the genes encoding MADS-domain proteins.

8.6 Future Scope of Works

As the genome size of *C. japonica* is huge as compared to *Arabidopsis*, it will be laborious and costly to undertake comprehensive standard genome analyses, such as making a complete physical map and reading all the sequences. Therefore, it may be more efficient to adopt two other strategies. One is to increase the density of the linkage map and to map useful genes on it using conventional test crosses. Another is to isolate functional genes from the DNA library according to information obtained from intensively examined plants such as *Arabidopsis*, which would require the establishment of a DNA library that covers the whole genome.

References

Goto Y, Kondo T, Kuramoto N, Ide T, Yamamoto K, Inaoka K, Yasueda H (2003) Mapping the gene encoding Cry j 1: a major *Cryptomeria japonica* pollen allergen. Silvae Genet 52:97–99

Grattapaglia D, Sederoff RR (1994) Genetic linkage maps of *Eucalyptus grandis* and *Eucalyptus urophylla* using a pseudo-testcross strategy and RAPD markers. Genetics 137:1121–1137

Hizume M, Kondo T, Miyake T (1989) Meiosis in pollen mother cells of 15 *Cryptomeria japonica* clones with special emphasis on the behavior of the 6th chromosome pair heteromorphic to chromomycin A_3 band. Jpn J Genet 64:287–294

Hizume M, Shibata F, Kondo T (1998) Fluorescence *in situ* hybridization of ribosomal RNA gene in *Cryptomeria japonica*, Taxodiaceae. Chrom Sci 2:99–102

Hizume M, Shibata F, Matsusaki Y, Kondo T (2000) Chromosome localization of telomere sequence repeats in five gymnosperm species. Chrom Sci 4:39–42

Hizume M, Kondo T, Shibata F, Ishizuka R (2001) Flow cytometric determination of genome size in the Taxodiaceae, Cupressaceae *sensu stricto* and Sciadopityaceae. Cytologia 66:307–311

Hulbert S, Ilott T, Legg EJ, Lincoln S, Lander E, Michelmore R (1988) Genetic analysis of fungus, *Bremia lactucae*, using restriction length polymorphism. Genetics 120:947–958

Iwata H, Ujino-Ihara T, Yoshimura K, Nagasaka K, Mukai Y, Tsumura Y (2001) Cleaved amplified polymorphic sequence markers in sugi, *Cryptomeria japonica* D. Don, and their locations on a linkage map. Theor Appl Genet 103:881–895

Kondo T, Hizume M (1982) Banding for the chromosomes of *Cryptomeria japonica* D. Don. J Jpn For Soc 64:356–358

Kondo T, Hizume M, Kubota R (1985) Variation of fluorescent chromosome bands of *Cryptomeria japonica*. J Jpn For Soc 67:184–189

Kuramoto N, Tomaru N, Murai M, Ohba K (1996) Genetic analysis of four isozyme loci and a linkage relationship between dwarf gene (*dw*) and leaf whiting gene in summer (*swl*) of sugi (*Cryptomeria japonica*) (in Japanese). Breed Sci 46:379–384

Kuramoto N, Tomaru N, Murai M, Ohba K (1997) Linkage analysis of isozyme and dwarf loci, and detection of lethal genes in sugi (*Cryptomeria japonica* D. Don) (in Japanese). Breed Sci 47:259–266

Kuramoto N, Kondo T, Fujisawa Y, Nakata R, Hayashi E, Goto Y (2000) Detection of quantitative trait loci for wood strength in *Cryptomeria japonica*. Can J For Res 30:1525–1533

Kuromaru M, Kawasaki H, Ohba K (1983) Genetic analysis of the peroxidase isozyme of seedlings derived from a twisted-leaf sugi (*Cryptomeria japonica* D. Don) (in Japanese). J Jpn For Soc 65:253–257

Kusumi J, Tsumura Y, Yoshimaru H, Tachida H (2000) Phylogenetic relationships in Taxodiaceae and Cupressaceae *sensu stricto* based on *matK* gene, *chlL* gene, *trnL-trnF* IGS regions, and *trnL* intron sequences. Am J Bot 87:1480–1488

Lander ES, Green P, Abrahamson J, Barlow A, Daly M, Lincoln S, Newberg L (1987) MAPMAKER: an interactive computer package for constructing primary genetic linkage maps of experimental and natural populations. Genomics 1:174–181

Lincoln S, Daly M, Lander E (1992a) Constructing genetic maps with MAPMAKER/EXP 3.0. Whitehead Institute Technical Report, 3rd edn. Cambridge, MA

Lincoln S, Daly M, Lander E (1992b) Mapping genes controlling quantitative traits with MAPMAKER/QTL 1.1. Whitehead Institute Technical Report, 2nd edn. Whitehead Institute, Cambridge, MA

Maeda T (1983) Natural distribution (in Japanese). In: Sakaguchi K (ed), Suginosubete, Zenkoku Ringyo Kairyo Fukyu Kyokai, Tokyo, Japan, pp 8–27

Mehra PN, Khoshoo TN (1956) Cytology of conifers. J Genet 54:165–185

Miyajima H (1983) Variety (in Japanese). In: Sakaguchi K (ed) Suginosubete, Zenkoku-rinngyou-kairyou-fukyu-kyoukai, Tokyo, Japan, pp 126-140

Mukai Y (1998) Measuring the genome size of trees (in Japanese). Shinrin-kagaku 24:36

Mukai Y, Suyama Y, Tsumura Y, Kawahara T, Yoshimaru H, Kondo T, Tomaru N, Kuramoto N, Murai M (1995) A linkage map for sugi (*Cryptomeria japonica*) based on RFLP, RAPD, and isozyme loci. Theor Appl Genet 90:835–840

Murai S (1947) Major forestry tree species in the Tohoku region and their varietal problems. In: Kokudo Saiken Zourin Gijutsu Kouensyu, Aomori-rinyukai, pp 131–151 (in Japanese)

Nikaido A M, Ujino T, Iwata H, Yoshimura K, Yoshimaru H, Suyama Y, Murai M, Nagasaka K, Tsumura Y (2000) AFLP and CAPS linkage maps of *Cryptomeria japonica*. Theor Appl Genet 100:825–831

Ohba K (1979) Detection of embryonic lethal genes in sugi, *Cryptomeria japonica* D. Don (in Japanese). In: Transactions of the 90th Meeting of the Japanese Forestry Society, pp 257–258

Ohba K, Maeda T, Fukuhara N (1974) Inheritance of twisted-leaf sugi, *Cryptomeria japonica* D. Don and linkage between the twisted-leaf gene and two recessive genes, albino and green (Midori sugi) (in Japanese). J Jpn For Soc 56:276–281

Ohtomo E (1983) Growth characteristics of sugi stand - species comparison (in Japanese). In: Sakaguchi K (ed) Suginosubete, Zenkoku Ringyo Kairyo Fukyu Kyokai, Tokyo, Japan, pp 486–501

Paterson A, Lander E, Hewitt J, Peterson S, Lincoln S, Tanksley S (1988) Resolution of quantitative trait into Mendelian factors using a complete RFLP linkage map. Nature 335:721–726

Sax K, Sax HJ (1933) Chromosome number and morphology in the conifers. J Arnold Arbor 14:356–375

Suyama Y, Mukai Y, Kondo T (1996) Assignment of RFLP linkage groups to their respective chromosomes in aneuploid of sugi (*Cryptomeria japonica*). Theor Appl Genet 92:292–296

Tani N, Takahashi T, Iwata H, Mukai Y, Ujino-Ihara T, Matsumoto A, Yoshiura K, Yoshimaru H, Murai M, Nagasaka K, Tsumura Y (2003) A consensus linkage map for sugi (*Cryptomeria japonica*) from two pedigrees, based on microsatellites and expressed sequence tags. Genetics 165:1551–1568

Toda Y (1979a) The karyotype of *Cryptomeria japonica* D. Don (in Japanese). Trans 90th Mtg Kyushu Branch, Jpn For Soc:261-262

Toda Y (1979b) A karyotype of *Cryptomeria japonica* D. Don. IV (in Japanese). La Kromosomo II-14:404–407

Tomaru N, Tsumura Y, Ohba K (1994) Genetic variation and population differentiation in natural populations of *Cryptomeria japonica*. Plant Sp Biol 9:191–199

Tsumura Y, Tomaru N (1999) Genetic diversity of *Cryptomeria japonica* using co-dominant DNA markers based on sequence-tagged sites. Theor Appl Genet 98:396–404

Ujino-Ihara T, Yoshimura K, Ugawa Y, Yoshimaru H, Nagasaka K, Tsumura Y (2000) Expression analysis of ESTs derived from the inner bark of *Cryptomeria japonica*. Plant Mol Biol 43:451–457

Ujino-Ihara T, Matsumuto A, Iwata H, Yoshimura K, Tsumura Y (2002) Single-strand conformation polymorphism of sequence-tagged site markers based on partial sequences of cDNA clones in *Cryptomeria japonica*. Genes Genet Syst 77:251–257

Ujino-Ihara T, Taguchi Y, Yoshimura K, Tsumura Y (2003) Analysis of expressed sequence tags derived from developing seed and pollen cones of *Cryptomeria japonica*. Plant Biol 5:600–607

Van Ooijen JW, Voorrips RE (2001) JoinMap Version 3.0, software for the calculation of genetic linkage maps. Plant Research International, Wageningen, The Netherlands

Yasue M, Ogiyama K, Suto S, Tsukahara H, Miyahara F, Ohba K (1987) Geographical differentiation of natural Cryptomeria stands analyzed by diterpene hydrocarbon constitutions of individual trees. J Jpn For Soc 69:152–156

Yoshimaru H, Ohba K, Tsurumi K, Tomaru N, Murai M, Mukai Y, Suyama Y, Tsumura Y, Kawahara T, Sakamaki Y (1998) Detection of quantitative trait loci for juvenile growth, flower bearing and rooting ability based on a linkage map of sugi (*Cryptomeria japonica* D. Don). Theor Appl Genet 97:45–50

Subject Index

MIX
Papier aus verantwortungsvollen Quellen
Paper from responsible sources
FSC® C105338

If you have any concerns about our products,
you can contact us on
ProductSafety@springernature.com

In case Publisher is established outside the EU,
the EU authorized representative is:
Springer Nature Customer Service Center GmbH
Europaplatz 3, 69115 Heidelberg, Germany

Printed by Libri Plureos GmbH
in Hamburg, Germany